Springer Series in Synergetics

Editor: Hermann Haken

Synergetics, an interdisciplinary field of research, is concerned with the cooperation of individual parts of a system that produces macroscopic spatial, temporal or functional structures. It deals with deterministic as well as stochastic processes.

Volumes 1–39 are listed on the back inside cover

Hajime Takayama (Ed.)

Cooperative Dynamics in Complex Physical Systems

Proceedings of the
Second Yukawa International Symposium,
Kyoto, Japan, August 24–27, 1988

With 181 Figures

Springer-Verlag Berlin Heidelberg New York
London Paris Tokyo

Professor Hajime Takayama

Research Institute for Fundamental Physics, (Yukawa Hall),
Kyoto University, Kyoto 606, Japan

Series Editor:

Professor Dr. Dr. h. c. Hermann Haken

Institut für Theoretische Physik der Universität Stuttgart, Pfaffenwaldring 57/IV,
D-7000 Stuttgart 80, Fed. Rep. of Germany and
Center for Complex Systems, Florida Atlantic University,
Boca Raton, FL 33431, USA

ISBN-13:978-3-642-74556-0   e-ISBN-13:978-3-642-74554-6
DOI: 10.1007/978-3-642-74554-6

© Springer-Verlag Berlin Heidelberg 1989
Softcover reprint of the hardcover 1st edition 1989

2154/3150-543210 – Printed on acid-free paper

Preface

The second Yukawa International Seminar (YKIS'88), devoted to cooperative dynamics in complex physical systems, was held from August 24 to 27, 1988, at Kansai Seminar House in Kyoto. The meeting was aimed at clarifying various cooperative aspects of complex physical systems that appear in different scientific disciplines, e.g., frustrated and random systems, polymers, spin glasses, glasses, neural networks, chemical and biological systems, and fluids. The various ideas developed so far were critically examined, thus preparing the way for possible breakthroughs in the field.

During the past decade complex physical systems, in which local, topological constraints such as competition between microscopic interactions (frustration) cause novel macroscopic cooperative phenomena, have become a subject of current interest in statistical and solid-state physics. A typical example is a spin glass, in which frustrations are distributed randomly. The consequences of its study are varied: the free energy landscape having a hierarchical structure, glassy dynamics, and so on. These spin-glass ideas have influenced considerably not only the field of related random systems, but also other areas of science, such as optimization and the modeling of neural networks.

Different cooperative aspects of interest are phenomena arising from dynamical constraints, which in turn are due to nonlinearity of the dynamics of the systems. Besides the well-examined chaotic behavior appearing even in systems of low dimensions, the nonlinear dynamics inherent to systems with many degrees of freedom has attracted much attention recently. It is expected to yield a tremendous variety of novel cooperative phenomena, such as turbulence in fluids, various pattern formations, the traditional 1/f noise, and so on.

In the present seminar, 27 invited papers and 60 contributed papers on these subjects were presented. We believe that the excellent lectures of the invited speakers, as well as the well-prepared poster presentation of each contributor and lively discussions among many participants, led the meeting to a successful conclusion.

This volume contains most of the invited and contributed papers presented at YKIS'88. We are very grateful to all authors for their efforts in preparing such excellent manuscripts. Reflecting the activity of our country in the field, so many contributed papers were submitted that we were forced to severely limit the length of the manuscripts. We believe, nevertheless, that they will aid the readers, as much as the invited papers will do, in understanding current developments in the field.

The YKIS'88 was organized by the Research Institute for Fundamental Physics, Kyoto University. The meeting was supported by the Yukawa Foundation, the Japan Society for the Promotion of Science, the Nishina Memorial Foundation, the Kajima Foundation, the Murata Science Foundation, and the Inoue Foundation for Science. The organizing committee gratefully acknowledges their generous financial support. Finally, thanks are due to Miss K. Horino for her invaluable assistance.

Kyoto, Japan   *H. Takayama*
October 1988

Opening Address

Ladies and gentlemen, it is my great pleasure to welcome all the participants of the second Yukawa International Seminar, this time on cooperative dynamics in complex physical systems. I am grateful to all of you for your participation in this seminar, in particular to those who have come to Kyoto all the way from abroad, to make this meeting a fruitful one.

Last year a new series of seminars started under the title "Yukawa International Seminars", which is a continuation, in spirit, of the old series of Kyoto Summer Institutes. This series is organized by the Research Institute for Fundamental Physics founded in honor of the late Professor Hideki Yukawa, and the subject of the seminar is selected each year from among the fields covered by the activities of this institute. The title of the first seminar, held last year, was "Mesons and Quarks in Nuclei", which has a direct bearing on Yukawa's work, and the subject for this year was selected from solid-state physics, to make this series known to a wider circle of physicists.

Last but not least, I would like to thank the many sponsors of this seminar, namely, the Yukawa Foundation, the Japan Society for the Promotion of Science, the Nishina Memorial Foundation, the Kajima Foundation, the Murata Science Foundation, and the Inoue Foundation for Science.

I apologize for having failed to be here in person to convey my message since I am attending another symposium held simultaneously in Nagoya. However, I wish you a pleasant stay in Kyoto.

Thank you.

Kazuhiko Nishijima,
Director
Research Institute for Fundamental Physics,
Kyoto University

1. M. Kawashima
2. H. Ito
3. T. Uezu
4. K. Nemoto
5. S.C. Müller
6. R.D. Young
7. W. Goldburg
8. P. Bak
9. P.B. Littlewood
10. M.A. Moore
11. J.H. Oh
12. J. Souletie
13. M. Ocio
14. I. Morgenstern
15. P. Manneville
16. H. Kitatani
17. H. Sakaguchi
18. M. Mézard

19. Y. Natsume
20. A.J. Bray
21. M. Katori
22. K. Horino*
23. Y. Fujinaka*
24. Hiraku Nishimori
25. S. Chikazawa
26. T. Aoyagi*
27. H. Aruga
28. S. Nasuno
29. K. Yakubo
30. T. Goto
31. K. Honda
32. M. Makita
33. M. Matsushita
34. P. Goldbart
35. R. Tao
36. X. Hu

37. R.G. Palmer
38. T. Konishi
39. S. Murayama
40. T. Miyashita
41. K. Iio
42. F. Matsubara
43. H. Malchow
44. T. Shirakura
45. K. Koyama
46. T. Suzuki
47. Y. Sawada
48. H. Hayakawa
49. K. Sekimoto
50. N. Ito
51. Y. Ozeki
52. H. Takano
53. H. Daido
54. S. Kabashima

---

* Secretariat

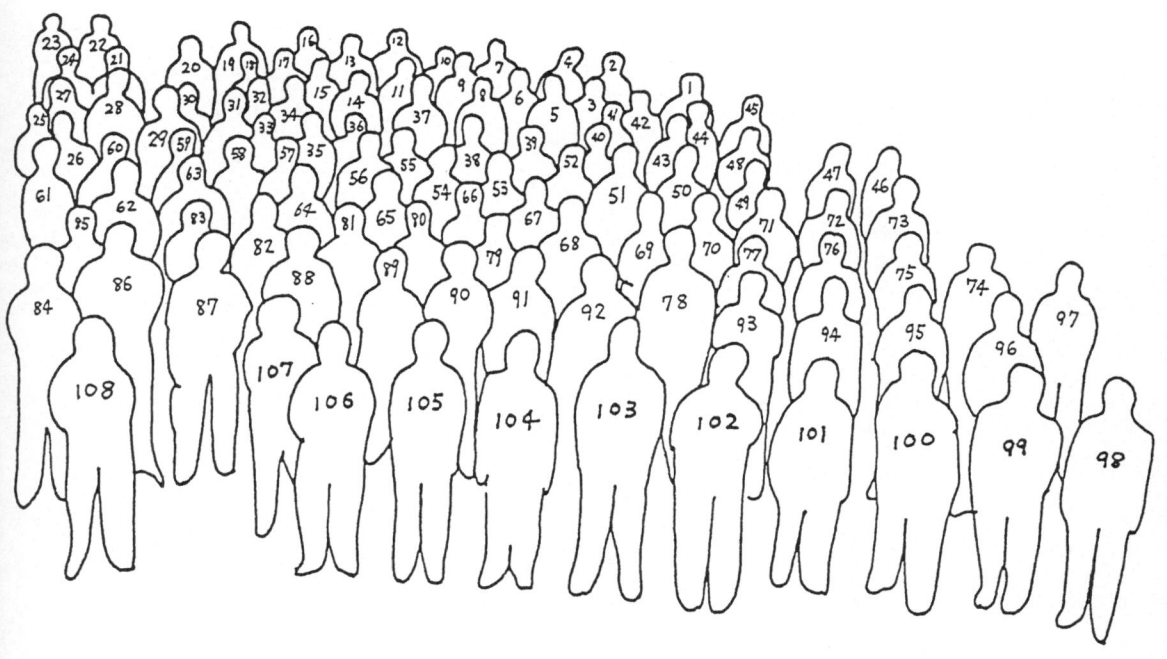

55. Kazuko Kawasaki	73. H. Ishii	91. Y. Iba
56. M. Schreckenberg	74. Hidetoshi Nishimori	92. A. Ito
57. Y. Aizawa	75. T. Chikyu	93. H. Yoshizawa
58. K. Mitsubo	76. S. Miyashita	94. K. Katsumata
59. T. Kamai	77. Y. Kasai	95. H. Kimura
60. M. Hagiwara	78. S. Takesue	96. S. Ozawa
61. T. Iwamoto*	79. Y. Kimura	97. F. Tanaka
62. H.E. Stanley	80. Kyozi Kawasaki	98. T. Kawasaki
63. R.L. Orbach	81. Y. Kuramoto	99. C. Kawabata
64. T. Tsuzuki	82. S. Katsura	100. H. Takayama
65. K. Toko	83. E. Knobloch	101. T. Haseda
66. H. Mitani	84. M. Mekata	102. R. Kubo
67. Y. Harada	85. A. Onuki	103. B. Hess
68. S. Maegawa	86. E.F. Wassermann	104. M. Suzuki
69. Y.Y. Suzuki	87. Y. Ajiro	105. S. Amari
70. Y. Taguchi	88. A.P. Malozemoff	106. Y. Miyako
71. T. Ikegami	89. I. Ono	107. T. Sasada
72. S. Fujiki	90. S. Shinomoto	108. M.A. Virasoro

---

* Secretariat

Contents

Part II   Spin Glasses and Related Random Systems

Simulation of Spin Dynamics for FMR in Compounds with Competing
Anisotropies

Part III   **Optimization Problems and Neural Network Models**

Part IV   **Nonlinear Dynamics in Fluids, Chemical and Biological
                Systems, etc.**

Part I

New Type of Phase Transitions
in Frustrated Systems, Polymers, etc.

Statistical Physics of Domain Walls and Grain Boundaries in Ordering Kinetics

Kyozi Kawasaki[1], *T. Nagai*[2], *and K. Nakashima*[1]

[1]Department of Physics, Faculty of Science, Kyushu University 33,
 Fukuoka 812, Japan
[2]Department of General Education, Kyushu Kyoritsu University,
 Kitakyushu 807, Japan

1. INTRODUCTION

It is now well recognized that topological defects like domain walls, vortex lines etc. play crucial roles at late stages of ordering kinetics [1]. However, treatments of generally interconnected and interacting topological defects present formidable problems of cooperative dynamics in complex physical systems. In this talk we take up two concrete examples studied recently in our research group.

2. QUASI-ONE-DIMENSIONAL SYSTEMS

Here we are dealing with strongly anisotropic systems typified by layered magnets where domain walls are almost perpendicular to some fixed axis. Since our work in this area has already been published and space is limited, we must refer the reader to our publications for details [2]. Here we just remark that the problem is characterized by (1) crucial roles played by normally hidden short range attractive forces between adjacent domain walls, (2) instability of domain walls against long wavelength distortions, and (3) existence of dimensional cross-over.

3. GRAIN GROWTH

3.1 Introductory Remarks

A class of problems closely related to the problem of random domain walls is concerned with time evolution of random cellular patterns which can be seen widely in nature [3]. Examples are soap froths, biological cell patterns, fracture patterns in lava, etc. Here we take up an example from metallurgy, namely, the grain growth in polycrystalline materials where the main driving force for change is the tendency to minimize the free energy stored in the grain boundaries. This problem has been recognized since early days as an outstanding example where chaos must compromise with order due to topological constraints such as space filling. Since excellent review articles exist [3,4], no further comments on the recent developments will be given. It is clear now, however, that despite efforts over many years the fundamental questions still remain unanswered. We do not yet know definitely whether there exists just one or several asymptotical universal statistical behaviors at long times (i.e. attractors) or how they do (or do not) depend on initial conditions and system parameters. Thus it should be quite useful if models can be constructed which retain some basic features of realistic grain growth dynamics but are still simple enough so as to permit large scale computer simulations. Here we describe such models, named the vertex models of 2 and 3 dimensions, and some computer simulation results in 2 dimensions.

Springer Series in Synergetics, Vol. 43   **Cooperative Dynamics in Complex Physical Systems**
Editor: H. Takayama   © Springer-Verlag Berlin, Heidelberg 1989

3.2 Two-Dimensional Vertex Models

Our starting point is very simple. If we look at examples of cellular patterns in nature which have been evolving for long times, cell boundaries appear to be almost flat (straight in 2 diminsions). Thus our basic assumption is to neglect curvature of grain boundaries entirely. The 2-dimensional cellular pattern is then completely specified by locations of all the vertices and the way they are connected by grain boundaries (here to be referred to as edges). This is not a new idea but the dynamical law for the model has not been properly formulated [5].

The dynamical law for change is obtained if we assume that dynamics is completely dissipative and hence the equation of motion can be expressed as the balance of dissipative and static forces. Such forces can be readily found if the Rayleigh dissipation function and the free energy function expressed in terms of the coordinates and velocities of the vertices are written down. The latter is proportional to the total length of grain boundaries whereas the former is obtained by noting that the dissipation is associated with the rates of displacements of grain boundary elements in the normal directions. This work is partly described in [6] and the full details will be published elsewhere [7]. The resulting dynamical law is expressed as the dynamically coupled set of equations of motion of all the vertices with friction tensors which reflect generally anisotropic environments of individual vertices. Simple dynamical models can be generated from this set of equations by introducing additional approximations. If we smooth out anisotropies of the friction tensors, our Model II below is obtained where the friction tensor reduces to a scalar friction constant which still depends on the environment of the vertex under consideration. If a further mean-field-like approximation is introduced for the friction constant, we obtain our Model I below. Here we give the equations of motion of these models for vertices located at $\vec{r}_i(t)$ in suitable dimensionless units:

Model I $\qquad \frac{1}{2} r_B(t) \frac{d}{dt} \vec{r}_i(t) = -\partial V(\{\vec{r}(t)\}) / \partial \vec{r}_i(t)$ (1)

Model II $\qquad \eta_i(\{\vec{r}(t)\}) \frac{d}{dt} \vec{r}_i(t) = -\partial V(\{\vec{r}(t)\}) / \partial \vec{r}_i(t)$ (2)

where $V(\{\vec{r}\})$ is the total length of grain boundary which depends on $\{\vec{r}\} = (\vec{r}_1, \vec{r}_2, \ldots)$, $r_B(t)$ is the average edge length and η_i is the

friction constant for the vertex i given by

$$\eta_i(\{\vec{r}\}) \equiv \frac{1}{6} \sum_j^{(i)} | \vec{r}_j - \vec{r}_i |,$$ (3)

the sum being over the vertices j connected to i by single edges. Here only topologically stable vertices with three edges are considered.

The above equations of motion must be supplemented with the collision rule when a pair of vertices come together. Two types of processes are possible:
(i) reconnection process (neighbor switching process) which occurs when the edge which disappears upon collision does not border on triangular cells,
(ii) annihilation of a triangular cell when two of its vertices collide.

Here we briefly describe some results of the computer simulations. The initial state is chosen to be a Voronoi cell pattern with 24,000 cell centers randomly distributed in the 1×1.2 rectangular domain. The run was repeated 20 times.

Some insights into the validity of replacing actual curved edges by straight bonds between vertices are gained by considering highly symmetric configurations with a regular polygon at the center whose edges can be curved and whose vertices are attached to straight edges which extend to infinity like rays. For these particular configurations vertices move in radial directions at the same speed, which can be computed either analytically or numerically for various evolution models of cellular structure like the curvature-driven grain growth model (G), the soap froth model (S) and the present vertex model (V) [7]. The outward velocities of vertices $v_K(n)$ with K= G,S,V, and n the number of vertices are shown in Table 1.

Table 1. Velocities of vertices in symmetric configurations
          [arbitrary units]

n	$v_G(n)$	$v_S(n)$	$v_V(n)$
5	-0.246	-0.241	-0.268
6	0	0	0
7	0.170	0.173	0.163
∞	$\pi/3$	$2/3^{1/2}$	1

We see that the vertex model approximation tends to enhance shrinking rates and suppress growing rates of cells. These tendencies will cancel somewhat for random cell patterns.

First we show in Fig.1 the average area of a cell as a function of time. The expected linear growth law [4] is confirmed.

Next we show in Fig.2 the normalized size distribution functions of cells, which are averages over 20 runs and at different times excluding short initial transient periods.

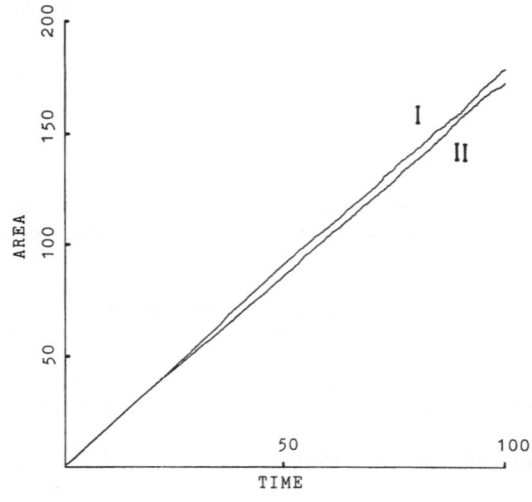

Fig.1 The average areas of single cells as functions of time divided by their initial values for Model I and Model II

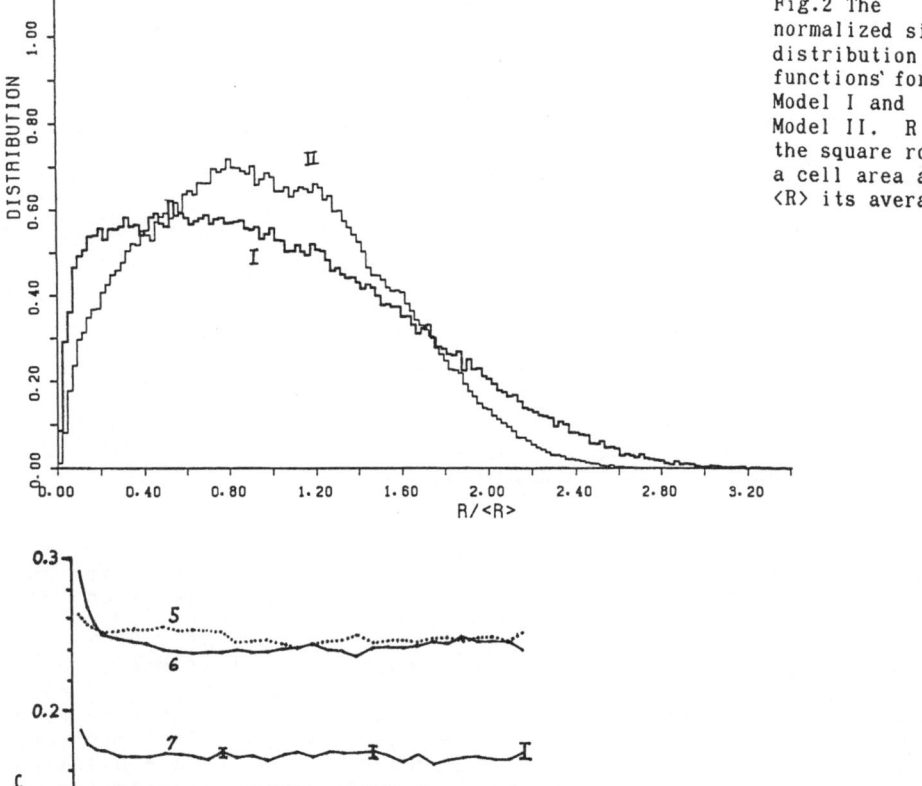

Fig.2 The normalized size distribution functions' for Model I and Model II. R is the square root of a cell area and ⟨R⟩ its average.

Fig.3 Time variations of cells having fixed numbers of edges which are indicated by the numbers attached to the lines for Model II. Typical sizes of error bars are shown by vertical line segments

The difference between the two models can be readily understood as arising from dependence of the friction forces on local environments of vertices which, in turn, affects their velocities. This effect is included only in Model II.

We now examine the shape distribution which tells us the fractions of cells with certain numbers of edges. Fig.3 shows time evolution of the fractions of cells with certain numbers of edges. The peculiar feature disclosed for the first time here is a small amplitude oscillation superimposed on seemingly steady state behavior. We made 10 additional runs and the same behavior was seen. Although oscillatory behaviors in some kinetic growth models are reported [8], and arguments are put forward for intermittent time evolution [9], we do not understand the origin of this behavior. More computer simulations under different conditions (initial states, system sizes, etc.) are needed here. Putting aside this peculiarity

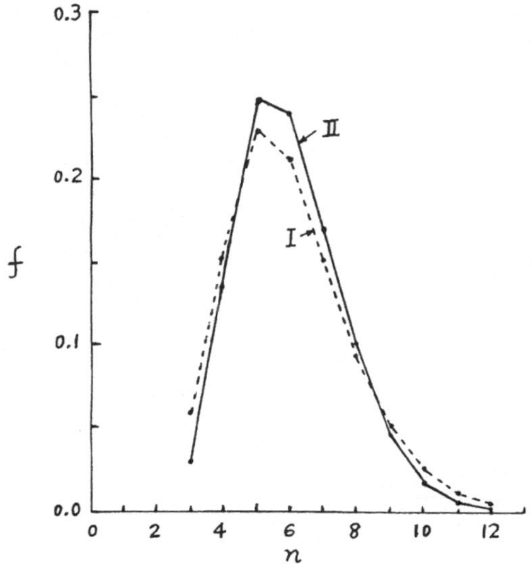

Fig.4 The normalized frequencies f of the number of cell edges n for Model I and Model II

temporarily we show in Fig.4 the shape distribution function averaged over the 20 runs and over the time excluding initial short transient periods.

We note the close similarity of this result to that of Monte Carlo simulations of the 64-state Potts model [10] and also the significant difference from Beenakker's recent result [11] which neglects recombination processes. In fact we can directly find from our simulation data the frequencies of recombination and annihilation processes, whose ratio turned out to approach a finite value somewhere between 1 and 1.4 at long times.

3.3 Three-Dimensional Vertex Models

When one attempts to extend the 2-dimensional vertex model to 3 dimensions one encounters the obstacle that a face cannot be made flat in general for fixed positions of vertices and straight edges connecting them. A way out would be to replace a face by a set of connected triangular planes, which, unfortunately, is not unique. However, if we focus on the motion of a single vertex, this ambiguity does not bother us, as will be shown below. Here we are dropping from the outset the dynamical coupling among different vertices.

Consider now a vertex i at $\vec{r}_i$ where two or more faces meet (6 faces for a topologically stable vertex). Consider also a portion of one of these faces surrounded by the vertex i and two other vertices j and k which are connected to i by two edges. We replace this portion by the flat triangular face whose area and unit normal are denoted as A and $\vec{n}$, respectively. When the vertex i moves with the velocity $\vec{v}_i$ there is an energy dissipation associated with the motion of the triangle with the rate 2R. R can be readily evaluated and becomes in some dimensionless units

$$R = \frac{A}{12} (\vec{n} \cdot \vec{v}_i)^2 .$$
(4)

There is also the free energy change which is equal to the area change of the triangle in some dimensionless units, which is

6

associated with a small displacement $\delta \vec{r}_i$ of the vertex i, and is given by

$$\frac{1}{2} \delta \vec{r}_i \cdot \vec{n} \times (\vec{r}_k - \vec{r}_j)$$ (5)

where $\vec{n} \times (\vec{r}_k - \vec{r}_j)$ was so chosen as to be directed toward $\vec{r}_i$ from the edge (j,k). Using expressions like (4) and (5), the vertex equation of motion can be obtained as a balance of frictional and static forces acting on the vertex i. The result is

$$(\frac{1}{6} \sum_\alpha^{(i)} A_\alpha \vec{n}_\alpha \vec{n}_\alpha) \cdot \vec{v}_i = - \frac{1}{2} \sum_\alpha^{(i)} \vec{n}_\alpha \times \vec{r}_\alpha$$ (6)

where the sum is over all the triangular portions of faces touching the vertex i which are specified by α, and r_α is the edge of the triangle α facing the vertex i whose direction was chosen as described above. This equation can be simplified by averaging over all the directions of n_α on the left hand side, which yields

$$(\frac{1}{18} \sum_\alpha^{(i)} A_\alpha) \vec{v}_i = - \frac{1}{2} \sum_\alpha^{(i)} \vec{n}_\alpha \times \vec{r}_\alpha .$$ (7)

This equation corresponds to Model II, (2), in 2 dimensions. The equation corresponding to Model I, (1), also follows if we replace the friction constant in (7) by $\bar{A}/18$ where $\bar{A}$ is the system average of the sum of areas of triangles touching on each vertex.

The vertex equation of motion like (6) or (7) must be supplemented with the collision rule of a pair of vertices which come together. Again there are two kinds of processes: (i) reconnection process and (ii) annihilation of a tetrahedron, which are depicted in Fig.9 (b) and Fig.9 (a) of [12], respectively. During such processes unusual types of structures can appear temporarily. For instance during a process of type (i) a configuration shown in Fig.5(a) appears where dots and lines represent vertices and edges, respectively, and the double line is an unusual type of edge where 4 cells meet. (Normally, 3 cells meet on an edge). Likewise during a process of type (ii) the state shown in Fig.5(b) appears where the dashed line is not a genuine edge but is merely the intersecting line of two planar sections of a face and similarly the encircled dots are not genuine vertices but are knicks of edges. These pseudo-vertices can be assumed to obey the same equation of motion (6) or (7) by treating these pseudo-edges or unusual edges as ordinary edges. Alternatively one may be able to simplify the models so as to dispense with these unusual types of processes.

Before concluding this section we must comment on dissipations and free energies associated with edges and vertices themselves, which we have neglected entirely. Our rationale for this is that we are interested primarily in late stages where domain sizes are so large that contributions from faces overwhelm those from edges and vertices.

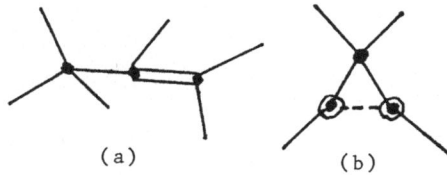

(a)   (b)

Fig.5 Unusual structures that appear during reconnection process (a) and annihilation of a tetrahedron (b)

4. CONCLUDING REMARKS

In this paper we discussed two examples which illustrate enormous advantages offered by reductions of complex dynamical systems to simpler ones. The original dynamical systems before reduction were the kinetic Ising model or time-dependent Ginzburg-Landau type model for the case of section 2 and the kinetic Potts model with infinite degeneracy or the curvature-driven interfacial system connected at vertices for the case of section 3.

We have greatly benefited from enlightening discussions with Professor D. Weaire on grain growth.

References

1. K. Kawasaki: In Proceedings of the 1986 Summer School on Statistical Mechanics, ed. by C.-K. Hu (Institute of Physics, Academia Sinica,Tapei 1987)
   J. D. Gunton: In Fluctuations and Stochastic Phenomena in Condensed Matter,
   ed. by L. Garrido (Springer, Heidelberg 1987)
2. K. Kawasaki, A. Ogawa and T. Nagai: Physica B 149, 97 (1988) and earlier references quoted therein
3. D. Weaire and N. Rivier: Contemporary Phys. 25, 59 (1984)
4. H. V. Atkinson, Acta Metall. 36, 469 (1988)
5. R. L. Fullman, In Metal Interfaces (American Society for Metals, Cleveland 1952)
   A. Soares, A. C. Ferro and M. A. Fortes, Scr. Metall. 19 1491 1985)
6. K. Kawasaki: In Ordering and Organization in Ionic Solutions, ed. by N. Ise and I. Sogami (World Scientific, Singapore 1988)
   T. Nagai, K. Kawasaki and K. Nakamura, J.Phys. Soc. Jpn. 57, 2221 (1988)
7. K. Kawasaki, T. Nagai and K. Nakashima, to be published
8. R. Savit: In Competing Interactions and Microstructures: Statics and Dynamics, ed. by R. LeSar, A. Bishop and R. Heffner, (Springer,Heidelberg 1988)
9. H. Furukawa, Phys. Rev. A29, 2160 (1984) and A30, 1052 (1984)
   C. W. J. Beenakker Phys. Rev. Lett. 57, 2454 (1986)
10. P. S. Sahni, D. J. Srolovitz, G. S. Grest, M. P. Anderson and S. A. Safran,
    Phys. Rev. B28 2705 (1983)
11. C. W. J. Beenakker, Physica 147A, 256 (1987) and to be published [Phys. Rev.A]
12, M. A. Fortes and A. C. Ferro, Acta Metall. 33, 1697 (1985)

Super-Effective-Field CAM Theory of Dynamical Complexity

M. Suzuki

Department of Physics, Faculty of Science, University of Tokyo,
Hongo, Bunkyo-ku, Tokyo 113, Japan

The main purpose of the present paper is to review the basic ideas of
the super-effective-field theory of critical phenomena[1-3] and the
coherent-anomaly method[4-19], and to explain how to apply these new
methods to dynamical complexity.

1. First we discuss here   what the dynamical complexity is. This
   is typically represented here by spin glasses, neural systems,
   turbulence, topological order(chiral order), critical dynamics or
   self-organized critical phenomena and other exotic phase
   transitions such as the resonating-valence-bond (RVB) problem and
   the high-T_c superconductivity.

   How do we study these exotic and complex systems? The answer of
   the present author is to apply the super-effective-field CAM
   theory to these dynamical complex systems.

2. The basic idea of the coherent-anomaly method (CAM) is to extract
   non-classical criticality from mean-field critical coefficients
   obtained by systematic mean-field approximations.

   More explicitly, we discuss here the dynamical susceptibility
   of ferromagnets. It takes the form

$$\chi(T,\omega) = \frac{\chi_{cl}(T,\omega)}{1-\mathcal{F}(T,\omega)} \tag{1}$$

in dynamical mean-field approximations[4,5,18,20]. Here, $\chi_{cl}(T,\omega)$
denotes the dynamical susceptibility of the relevant cluster used
for the cluster mean-field approximation, and $\mathcal{F}(T,\omega)$ the dynamical
feedback term[1,4,18]. Near the critical point T_c, Eq.(1) may be
reduced to the following form:

$$\chi(T,\omega) \simeq \frac{\bar{\chi}(T_c)}{\varepsilon+i\omega\bar{\tau}(T_c)} \tag{2}$$

in the low-frequency limit, where

$$\varepsilon = (T-T_c)/T_c \ , \ \ \bar{\chi}(T_c) = \chi_{cl}(T_c,0) \ \ \text{and} \ \ \mathcal{F}(T_c,0) = 1 \ . \tag{3}$$

According to the general CAM theory [4], we have

$$\bar{\chi}(T_c) \to \infty \ \text{and} \ \bar{\tau}(T_c) \to \infty, \ \text{as} \ T_c \to T_c^* \tag{4}$$

when the criticality of the relevant system is non-classical,
namely when

$$\chi(T) \sim \frac{1}{(T-T_c*)^\gamma} \ \ ; \ \ \gamma > 1 \ , \tag{5}$$

and

$$\tau \sim \frac{1}{(T-T_c^*)^{\Delta}} \quad ; \quad \Delta > 1 \; . \tag{6}$$

Then, it is quite natural to assume that

$$\bar{\chi}(T_c) \sim \frac{1}{(T_c-T_c^*)^{\psi_\chi}} \text{ and } \bar{\tau}(T_c) \sim \frac{1}{(T_c-T_c^*)^{\psi_\tau}} \tag{7}$$

when $\hat{\delta}(T_c) \equiv (T_c-T_c^*)/T_c^* \ll 1$.

The general CAM theory [4,5] yields the following coherent-anomaly relations

$$\gamma = 1 + \psi_\chi \text{ and } \Delta = 1 + \psi_\tau \; . \tag{8}$$

Thus, the fractional parts of the critical exponents γ and Δ come from intrinsic fluctuation.

As is seen in the above explanation, the existence of the feedback term $\mathcal{F}(T,\omega)$ is essential in the CAM, because it includes some systematic effect of the infinite degree of freedom. In fact, Eq.(1) is expanded as[5]

$$\chi(T,\omega) = \chi_{cl} + \chi_{cl}\mathcal{F} + \chi_{cl}\mathcal{F}^2 + \cdots \cdot \tag{9}$$

This infinite series expresses the response of the corresponding "cluster-cactus tree", whose lattice structure resembles a real cactus. Of course, such a real cactus is finite, but the connectivity is quite similar to that of our system. Thus, the existence of this feedback term yields the transition point T_c even for a finite cluster. In other words, Eq.(1) can be interpreted[21] as a fixed point of the following iteration:

$$\chi_{n+1}(T,\omega) = \chi_{cl}(T,\omega) + \mathcal{F}(T,\omega)\chi_n(T,\omega) \; . \tag{10}$$

Namely,

$$\chi(T,\omega) = \lim_{n\to\infty} \chi_n(T,\omega) \; . \tag{11}$$

Here, $\chi_n(T,\omega)$ denotes the dynamical susceptibility of a finite cactus tree with "size" n in the unit of the elementary cluster size.

3. Then, one might ask whether it is possible to construct mean-field approximations even in exotic phase transitions such as spin glasses and chiral orders. This question has been essentially solved by the present author[1-3]. That is, if we introduce the concept of a super-effective-field in a general cluster, then we may construct an extended effective-field theory to describe even novel phase transitions in complex systems.

The main idea of the super-effective-field theory[1-3] is to go beyond the decoupling of the relevant Hamiltonian as in the ordinary mean-field theory, and to introduce ad hoc an effective field Λ_s conjugate to a possible order parameter Q_j. Then the super-effective Hamiltonian is given by

$$\tilde{\mathcal{H}} = \mathcal{H}_{cl} - \Lambda_s \sum_{j \in \partial\Omega} Q_j \; , \tag{12}$$

10

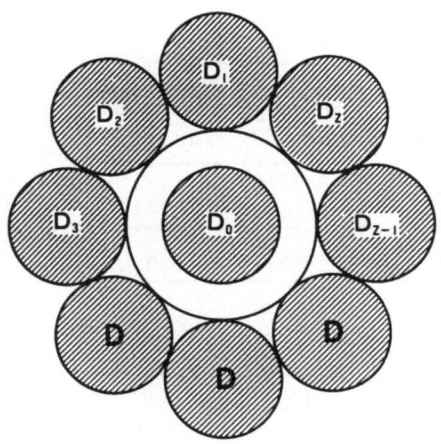

Fig.1 A general super-effective cluster [1].

where $\mathcal{H}_{cl}$ denotes the cluster Hamiltonian and $\partial\Omega$ the boundary cells of the cluster shown in Fig.1.

The self-consistency condition is given by

$$< Q_0 > = \varepsilon_{01} < Q_1 > \; ,$$
(13)

where ε_{01} is the modular factor to express the symmetry of the order parameter[1].

The above scheme seems to be a straightforward extension of the ordinary effective-field theory, but it is quite general and very powerful as will be discussed later.

4. Thus, if we combine the CAM with the super-effective-field theory, then we may have a unified theory of phase transitions in the following sense :

a) A unified method to study both classical type(i.e.,mean-field type) and non-classical type(i.e.,fractional) critical phenomena.

b) To study both ordinary and exotic phase transitions.

c) A unification of the ordinary decoupling mean-field approximation and perturbational study of critical phenomena. This aspect is easily realized through the power-series and continued-fraction CAM theories[12,13]. Mean-field type approximations can be constructed[12,13] even from the perturbational expansions of response functions, if we apply the CAM idea to the inverse functions or continued fractions of physical quantities.

d) To be able to study both classical and quantum systems.

e) A very effective combination of Kubo's linear response theory[22] and Fisher's concept of finite-size scaling[23,24].

The above situations will be summarized schematically in Fig.2.

5. Here we explain the super-effective-field CAM theory of spin glasses[1,11,25].

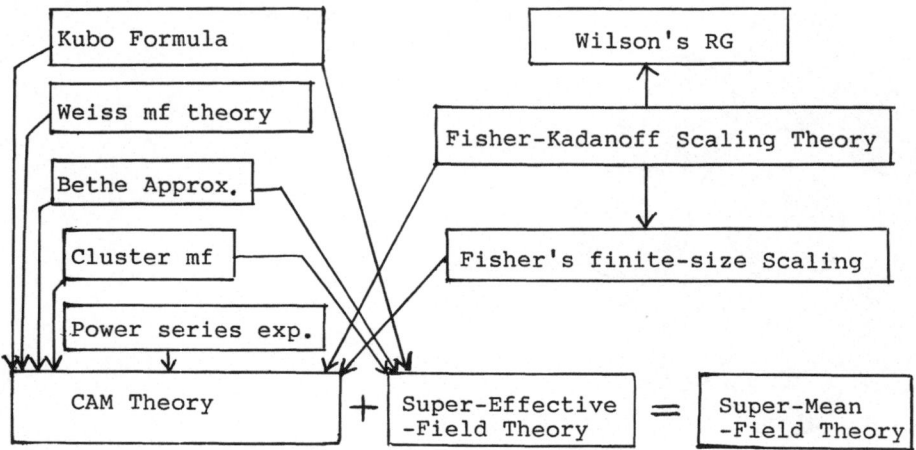

Fig.2 Schematic Relations of Theories on Fluctuations

(i) First we review the super-effective-field real replica theory of spin glasses in equilibrium[1,11,25]. We introduce[26-30] the two replicated systems $\mathcal{H}^{(1)}$ and $\mathcal{H}^{(2)}$ and also the super-effective field term $\Lambda_s \Sigma S_j S_j'$ between the two replica systems only on the boundary $\partial \Omega$ of the two replicated clusters. Namely our effective cluster Hamiltonian is

$$\tilde{\mathcal{H}} = \mathcal{H}_{cl}^{(1)} + \mathcal{H}_{cl}^{(2)} - \Lambda_s \sum_{j \in \partial \Omega} S_j S_j' \; . \tag{14}$$

The order parameter is defined[26] by

$$Q = \sum_j Q_j \text{ and } Q_j = S_j S_j' \; . \tag{15}$$

Now we have the following self-consistency equation

$$\langle\langle Q_0 \rangle\rangle_J = \langle\langle Q_j \rangle\rangle_J \; , \tag{16}$$

where $\langle \cdots \rangle_J$ denotes the average over the random distribution of the coupling constant J_{ij}. Then, Eq.(16) is rewritten as

$$\langle\langle S_0 S_0' \rangle\rangle_J = \langle\langle S_j S_j' \rangle\rangle_J \; . \tag{17}$$

Thus, the equation to determine the spin glass transition point T_{sg} is given by[1,25]

$$\sum_{k \in \partial \Omega} \langle\langle S_0 S_k \rangle^2\rangle_J = \sum_{k \in \partial \Omega} \langle\langle S_j S_k \rangle^2\rangle_J \; . \tag{18}$$

This is our fundamental equation[1,25] for the cluster shown in Fig.1 in which a single super-effective-field is applied. When there are several super-effective-fields, the above equation (18) to determine T_{sg} becomes a matrix equation[31].

In order to combine the above super-effective theory with the CAM, it is convenient to study the nonlinear susceptibility χ_2 [27] defined by

$$m = \chi_0 H + \chi_2 H^3 + \cdots \; , \tag{19}$$

where m and H denote the magnetization and external field, respectively. It is negatively proportional[27] to the Edwards-Anderson susceptibility, namely the response function of the order parameter Q with respect to the super-effective-field Λ_s ; $\chi_2 = -\beta\chi_{sg}$.

It is quite easy to calculate χ_{sg} of the ±J Ising spin glass in our formulation. The result is

$$\chi_{sg} = (\beta\mu_B^2)^2 \frac{1+<<S_0S_j>^2>_J}{1-(z-1)<<S_0S_j>^2>_J} , \qquad (20)$$

where z denotes the number of nearest neighbors. As

$$<<S_0S_j>^2>_J = \tanh^2 K \qquad (21)$$

in the ±J Ising spin glass, we obtain the critical point $K_{sg} = J/k_BT_{sg}$ in the form

$$\tanh^2 K_{sg} = \frac{1}{z-1} . \qquad (22)$$

This agrees with that obtained by Katsura et al.in a different context[32]. If we take the limit $z \to \infty$, then we obtain the mean-field result or the SK result[33]

$$k_BT_{sg} = \sqrt{z} \, J . \qquad (23)$$

Thus, we obtain easily the coherent anomaly of the nonlinear susceptibility[1,11,31]

$$\begin{cases} \overline{\chi}_{sg}^{(mf)} = 0.08333 & , \; k_BT_{sg}^{(mf)}/J = 2.4495 \\[2ex] \overline{\chi}_{sg}^{(Bethe)} = 0.16140 & , \; k_BT_{sg}^{(Bethe)}/J = 2.0781 \\[2ex] \overline{\chi}_{sg}^{(square)} = 0.18714 & , \; k_BT_{sg}^{(square)}/J = 2.0330 \\[2ex] \text{and} \; \overline{\chi}_{sg}^{(cubic)} = 0.21685 & , \; k_BT_{sg}^{(cubic)}/J = 1.9857 . \end{cases} \qquad (24)$$

These yield[1,11]

$$k_BT_{sg}^* \simeq 1.2J \; \text{and} \; \gamma_s \simeq 2.89 , \qquad (25)$$

where γ_s is the critical exponent of susceptibility defined in

$$\chi_{sg} \sim \frac{1}{(T-T_{sg}^*)^{\gamma_s}} . \qquad (26)$$

This agrees with the results obtained in different methods[34-36]. However, calculations of larger clusters will be necessary to obtain a more definite conclusion. Such a study is now in progress.

It will be also quite interesting to apply the present formulation to the two-dimensional ±J Ising model, in order to investigate whether there exists a phase transition in two dimensions or not. For this purpose, we have made the plot log $\overline{\chi}_{sg}(T_{sg})$-log T_{sg} using the following critical coefficient data[31]

$$
\begin{cases}
\overline{\chi}_{sg}^{(Bethe)} = 0.3802 \quad , \quad k_B T_{sg}^{(Bethe)}/J = 1.5187 \\[2ex]
\overline{\chi}_{sg}^{(square)} = 0.4892 \quad , \quad k_B T_{sg}^{(square)}/J = 1.4554 \\[2ex]
\overline{\chi}_{sg}^{(cross)} = 0.7797 \quad , \quad k_B T_{sg}^{(cross)}/J = 1.3438 \; .
\end{cases}
\tag{27}
$$

We find that the above log-log plot makes nearly a straight line, namely

$$
\log \overline{\chi}_{sg}(T_{sg}) = \psi_s \log T_{sg} + \text{constant}, \tag{28}
$$

with $\psi_s \simeq 5.8$. This yields

$$
\chi_{sg}(T) \sim T^{-\gamma_s} \tag{29}
$$

with $\gamma_s = 1 + \psi_s \simeq 7$ in two dimensions. Thus, we arrive at the conclusion that there exists no spin-glass transition in two dimensions, as is expected.

By the way, a similar log-log plot for the three-dimensional $\pm J$ Ising spin glass is found to deviate remarkably from a straight line, which corresponds to the existence of a phase transition as was shown in Eq.(26).

It is also quite easy to study the critical behavior of spin glass in higher dimensions. In fact, the CAM gives the result[31] that γ_s decreases as the dimensionality d increases in the above three different kinds of mean-field approximations. Much larger clusters have to be studied in order to investigate whether there exists the upper dimensionality d_u above which the critical behavior of spin glass is classical. It will be studied in the near future.

(ii) Next we give here a new formulation of the dynamical super- effective-field CAM theory of spin glasses.

For simplicity we consider here the case $T > T_{sg}$. Then, we may use the following uniform order parameter:

$$
Q = \sum_{j=1}^{N} S_j S_j' \; . \tag{30}
$$

The relaxation time of this order parameter is defined by

$$
\tau_{sg} = \int_0^\infty \frac{<<Q(t)Q>>_J}{<<Q^2>>_J} \, dt = \int_0^\infty \frac{\sum_j <<S_0(t)S_j>^2>_J}{\sum_j <<S_0 S_j>^2>_J} \; . \tag{31}
$$

Here we introduce the dynamical real replica

$$
\mathcal{H}(t) = \mathcal{H}_\ell^{(1)} + \mathcal{H}_\ell^{(2)} - \Lambda_s(t) \sum_j^{\partial\Omega} S_j S_j' - \Lambda(t) \sum_k^\Omega S_k S_k' \; . \tag{32}
$$

When $\Lambda(t) = \Lambda e^{i\omega t}$, we obtain $Q(t) = \chi_Q(\omega)\Lambda e^{i\omega t}$. Here the dynamical susceptibility $\chi_Q(\omega)$ is calculated in the form

$$\chi_Q(\omega) \simeq \frac{\bar{\chi}_{sg}(T_{sg})}{\varepsilon + i\omega\, \bar{\tau}_{sg}(T_{sg})} \tag{33}$$

near T_{sg} for small ω . Then we may have

$$\bar{\tau}_{sg}(T_{sg}) \simeq \frac{C}{(T_{sg}-T_{sg}^*)^{\psi_{sg}}} \tag{34}$$

near the critical region $(T_{sg}-T_{sg}^*)/T_{sg}^* \ll 1$. The coherent-anomaly exponent ψ_{sg} can be estimated from several cluster-mean-field approximations corresponding to the above static case. Thus, the critical slowing down exponent Δ_{sg} defined in

$$\tau_{sg} \sim (T-T_{sg}^*)^{-\Delta_{sg}} \tag{35}$$

can be obtained through the coherent-anomaly relation

$$\Delta_{sg} = z\nu = 1 + \psi_{sg} . \tag{36}$$

The calculation of Δ_{sg} is now in progress, using the stochastic Ising model[37,38]

6. There are many applications to other complex systems. It is possible to formulate dynamical super-effective-field CAM theories of topological orders and other exotic phases. These will be reported in the near future.

7. The summary and discussion will be given here. The basic ideas of the CAM and super-effective-field theory(SEFT) have been explained with an emphasis on critical dynamics. A scheme of the dynamic super-effect-field CAM theory of spin glasses has been formulated.

Our future projects concerning applications of the above theory to dynamical complexity are the following :

a) To estimate the dynamical exponent of spin glasses, $\Delta_{sg} = z\nu$.

b) To study the relaxation of topological orders such as chiral order.

c) To construct a super-effective-field CAM theory of the high-T_c superconductivity.

d) To estimate the critical exponents of self-organized critical phenomena[37,38].

e) To apply our new theory to other dynamical complexity.

References

1. M.Suzuki, J.Phys.Soc. Jpn. 57(1988) 2310.
2. M.Suzuki, J.Phys.Soc. Jpn. 57(1988) 683.
3. M.Suzuki, the Proceeding of the 19th Yamada Conference on ordering and Organizatin in Ionic Solutions, held at Kyoto, Nov.9-12, 1987, ed Ise,N. (World Sci. Pub.).
4. M.Suzuki, J.Phys.Soc. Jpn. 55(1986) 4205, to be referred to as I. See also M.Suzuki, Phys.Lett. 116A(1986) 375, and Quantum Field Theory (Proc.Int.Symp.Positano, Salerno, Italy, June 5-7, 1985)ed.
   Mancini,F. (North-Holland, Amsterdam, 1986).

5. M.Suzuki, M.Katori, and X.Hu, J.Phys.Soc. Jpn. 56(1987) 3092.
6. M.Katori, and M.Suzuki, J.Phys.Soc. Jpn. 56(1987) 3113.
   See also M.Suzuki,and M.Katori, J.Phys.Soc. Jpn. 55(1986) 1.
7. X.Hu, M.Katori, and M.Suzuki, J.Phys.Soc. Jpn. 56(1987) 3865.
8. X.Hu, and M.Suzuki, J.Phys.Soc. Jpn. 57(1988) 791.
9. M.Katori, and M.Suzuki, J.Phys.Soc. Jpn. 57(1988) 807.
10. M.Suzuki, Prog.Theor.Phys. suppl. 87(1986) 1.
11. M.Suzuki, Phys.Lett.   127A(1988) 410.
12. M.Suzuki, J.Phys.Soc. Jpn. 56(1987) 4221.
13. M.Suzuki, J.Phys.Soc. Jpn. 57(1988) 1.
14. N.Ito, and M.Suzuki, Int.J.Modern Phys.B. vol.2(1988) 1.
15. T.Oguchi, and H.Kitatani, J.Phys.Soc. Jpn. (1988).
16. M.Takayasu, and H.Takayasu, Phys.Lett. 128A(1988) 45.
17. X.Hu, and M.Suzuki, Physica 150A(1988) 310.
18. M.Katori, and M.Suzuki, J.Phys.Soc. Jpn. 57(1988) No.11.
19. J.L.Monroe, Phys.Lett.A(1988).
20. M.Suzuki and R.Kubo, J.Phys.Soc. Jpn. 24(1968) 51.
21. M.Suzuki, J.Stat.Phys. (1988) Oct. or Nov.
22. R.Kubo, J.Phys.Soc. Jpn. 12(1957) 570.
23. M.E.Fisher and M.N.Barber, Phys.Rev.Lett. 28(1972) 1516.
24. M.Suzuki, Prog.Theor.Phys. 58(1977) 1142.
25. M.Suzuki, J. de Physique, Dec.1988(Invited paper of ICM,
    July,1988   in Paris).
26. S.F.Edwards and P.W.Anderson, J.Phys.F; Metal Phys. 5(1975) 965.
27. M.Suzuki, Prog.Theor.Phys. 58(1977) 1151.
28. C.De Dominicis and T.Garel, J. de Phys. 40(1979) L575.
29. A.Blandin, M.Gabay and T.Garel, J.Phys. C : Solid State Phys.
    13(1980) 403.
30. R.H.Swendsen and J.-S.Wang, 1986.
31. M.Suzuki and N.Hatano, J.Phys.Soc. Jpn.
32. S.Katsura, Prog. Theor. Phys. 55(1976) 1049. S.Fujiki, and
    S.Katsura, Prog.Theor. Phys. 65(1981) 1130.
33. D.Sherrington, and S.Kirkpatrick, Phys.Rev.Lett. 35(1975)
    1792.
34. A.T.Ogielski and I.Morgenstern, Phys.Rev.Lett. 54(1985) 928.
35. R.N.Bhatt and A.P.Young, Phys.Rev.Lett. 54(1985) 924.
36. R.R.P.Singh and S.Chakravarty, Phys.Rev.Lett. 57(1986) 245.
37. R.J.Glauber, J.Math. Phys. 4(1963) 294.
38. M.Suzuki and R Kubo, J.Phys.Soc. Jpn. 24(1968) 51.
39. P.Bak, C.Tang and K.Wiesenfeld, Phys.Rev.Lett. 59(1987) 381,
    Phys.Rev.38A (1988) 364.
40. C.Tang and P.Bak, J.Stat. Phys. 51(1988) 797.

Some Recent Experiments in Turbulently Stirred Fluids

W. Goldburg and P. Tong*

Department of Physics and Astronomy, University of Pittsburgh,
Pittsburgh, PA 15260, USA

I. Introduction

It is well known that strong fluctuations in composition appear in the vicinity of the critical point of binary liquid mixtures and simple fluids.[1] The equilibrium properties of such systems exhibit scaling that is now understood. Similarly, much progress has been made in our understanding of nucleation and spinodal decomposition when the system is quenched below its critical point.[2] In this paper we are concerned with a different type of experiment, in which the near-critical system is in a steady state. This state is produced by continuous, turbulent stirring.

One might anticipate that turbulent shear will strongly modify both the wave number spectrum and the lifetime of composition fluctuations. In the simple case where the shear is uniform (for example, $\gamma = \frac{\partial v_x}{\partial y} = \text{const}$) and the system is in the one-phase region, the shear will stretch, and thereby suppress, the composition variations c_k in the mixture. Here c_k is a Fourier component of the local deviation in composition $c(\mathbf{r})$ from its equilibrium value. This smoothing effect of the shear decreases the intensity I(k) of light scattered by the system.[3,4] The angular distribution of I(k) is proportional to $< |c_k|^2 >$, and the scattering vector k has an amplitude $k = [4\pi n/\lambda]sin(\theta/2)$. Here n is the refractive index of the mixture, λ is the wavelength of the incident laser light, and θ is the scattering angle. In addition, the distortion of the fluctuations produces an anisotropic angular distribution of the scattered light. It also changes the critical exponents, and even produces a slight decrease in the critical temperature.[4]

In the more complicated case of turbulent stirring,[5,6,7] we might expect the phase transition to be broadened, since the shear is random in magnitude as well as direction.

As the mixture is cooled through the critical point, two competing effects come into play. Thermodynamics favor the development of two phases of well-defined composition; opposing this phase separation, local shear stretches and dissolves incipient droplets or domains.

In this paper we describe light scattering experiments which show the effect of turbulence on the amplitude of the fluctuations and on their lifetime. We find that

*Now at Exxon Research and Engineering Company, Route 22 East, Annandale, New Jersey 08801, USA.

strong shear broadens the transition and produces a striking decrease in the lifetime of composition fluctuations.

From the theory of critical phenomena,[1] the thermal equilibrium correlation length ξ diverges at the critical point as

$$\xi = \xi_0(T/T_c - 1)^{-\bar{\nu}} \tag{1}$$

with $\bar{\nu} = 0.625$. For simple fluids and mixtures, ξ_o is one or two Angstroms, making ξ of the order of $10\ nm$ when $|T - T_c| = 1K$. The relaxation rate, Γ_ξ, of composition fluctuations of size ξ has the form

$$\Gamma_\xi = \frac{k_B T}{6\pi\eta\xi^3}, \tag{2}$$

where k_B is Boltzmann's constant, η is the viscosity of the mixture and T is the absolute temperature.

The effect of shear depends upon its strength. When the shear rate γ is small compared to Γ_ξ (weak shear), a fluctuation in composition decays before it has had a chance to "feel" the shear, so that its effect is small. In the opposite limit, $\gamma \gg \Gamma_\xi$, the velocity field cannot be treated as a weak perturbation on the composition field, $c(\mathbf{r}, t)$.

In the next section we will discuss the effect of stirring on the wave number spectrum for a binary mixture. The effect of turbulence on the temporal fluctuations in composition is reported in Sec. III, and the work is summarized in Sec. IV.

II. The Effect of Stirring on the Wave Number Spectrum

We have studied the effect of turbulent shear on binary mixtures contained in sample cells several centimeters in diameter and about $8\ cm$ high. The turbulence is produced by a magnetic stirring bar sealed within the sample.[7] In such small containers the turbulence generated by stirring is far from being homogeneous and isotropic. Nevertheless we can make progress in interpreting the results by invoking the Kolmogorov theory of fully developed turbulence, which implicitly assumes the above conditions to be satisfied over some length scales R. It is therefore appropriate to summarize the Kolmogorov theory briefly at this point.

In the Kolmogorov view of turbulence,[8] the kinetic energy is continuously transferred from large scales down to smaller scales, until dissipation occurs. In our case stirring the fluid generates eddies of size R_0, which is of the order of the container size or perhaps the radius of the stirrer. The continuous stirring causes the large eddies to break up into smaller and smaller ones, of size $R_0/2$, $R_0/2^2$, etc., until finally a smallest size R_d is reached. The cut-off R_d is defined such that the viscous damping time, R_d^2/ν, is equal to the eddy lifetime, or "turnover time", $\tau(R_d) = R_d/u(R_d)$. Here $u(R)$ is the characteristic velocity associated with eddies of size R, and ν is the kinematic viscosity of the fluid. In the inertial range $R_d < R < R_0$, the cascade proceeds without loss of energy into heat.

Let ϵ be the rate (in ergs/gm/sec) at which the stirring supplies energy to create the turbulent cascade. From a dimensional argument ϵ should be of order $u(R_0)^2/\tau(R_0) = u(R_0)^3/R_0$. Since it is assumed that the turbulent energy is conserved in the cascade, the relationship between ϵ and $u(R)$ holds for all R in the inertial range. Thus one has

$$u(R) \sim (\epsilon R)^{\frac{1}{3}}. \tag{3}$$

For future reference we note that the turbulent shear rate γ (R) associated with eddies of size R is, from dimensional considerations, u(R)/R. The strength of the turbulence is often expressed in terms of the Reynolds number, $Re = R_0 u(R_0)/\nu$, instead of ϵ. The above arguments then give the maximum turbulent shear rate $\gamma(R_d)=\nu Re^{3/2}/R_0^2$, and $R_d = R_0 Re^{-3/4}$. In our experiments the stirring frequency f ranged from 4 to 14 Hz, which corresponds to $Re = 2\pi f R_0^2/\nu \sim 10^4$. Because the turbulence was very far from ideal,[7] a more realistic value of Re, for the purpose of estimating $\gamma(R_d)$ and R_d, might better be taken as roughly 1000. This corresponds to $R_d = 50 \mu$, and $\gamma(R_d) = \gamma_{max} = 10^4 sec^{-1}$.

To observe an effect of turbulence on the angular distribution of the scattered intensity $I(k)$, it was necessary to make measurements at temperatures quite close to the critical temperature T_c of our critical mixture. The sample was isobutyric acid + water (IBW) for the experiment now to be discussed. The sample cell was a sealed cylindrical container filled with a critical mixture and containing a stirring bar to produce the turbulence. The cell was several cm. in diameter and 8 cm. in height. The stirring bar, which was 2 cm. in length, rotated about the axis of the cell and was driven by an external magnet which rotated the bar at a frequency in the range $3Hz \leq f \leq 14Hz$. The incident light beam, from a He-Ne laser, traversed the cell roughly 1 cm. above the stirring bar, where the turbulence was maximal. In some cases the cell contained ribs to increase the degree of turbulent mixing. The scattered light was recorded in a plane perpendicular to the axis of the sample cell. Further experimental details can be found in Ref. 7.

Figure 1 is a log-log plot of the measured $I(k)$ vs. the scattering vector, k, for various stirring frequencies, including $f = 0$ (thermal equilibrium). The measurements were made at $T = T_c + 15\ mK$, where $T_c = 26.4°C$. The upper curve shows $I(k)$ in the absence of stirring, which fits an Ornstein-Zernike form $I(k) \sim (\xi^{-2} + k^2)^{-1}$.

The curves below it, which correspond to the three indicated stirring frequencies, show that increasing Re decreases $I(k)$ at small k only. This finding is consistent with the notion that turbulence cannot strongly affect those composition fluctuations which are too small in spatial extent or decay too fast. The stirring only suppresses the long-lived, small wave number fluctuations whose relaxation rate, Γ_k is comparable to or smaller than the maximum shear rate $\gamma(R_d)$. Near T_c, $\Gamma_k = (k_B T/6\pi\nu)k^3$. Let the value of k at which the suppression of $I(k)$ occurs be k^*. From Fig. 1, k^* is about $8 \times 10^5 cm^{-1}$ at $f = 14.2\ Hz$. This is, within 50%, the value of k^* obtained from equating Γ_k with $\gamma(R_d)$, if one calculates $\gamma(R_d)$ at $Re = 10^4$.

So far we have concentrated attention on the influence of turbulent shear on a one-phase system. Now we consider the additional effect of random shear on

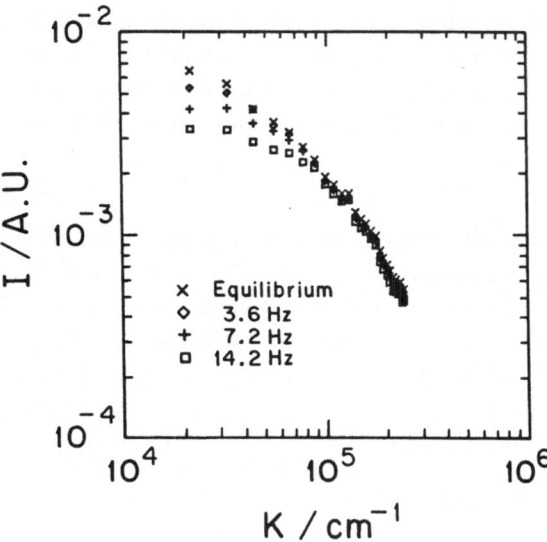

Figure 1: Scattered intensity I(k) in IBW for various stirring frequencies at 15 mK above T_c. The statistical errors are too small to be shown.

interfaces between phases. As discussed in section I, domains of opposing phases will be stretched and distorted by the shear. Were the system in equilibrium, it would seek to reduce its free energy by minimizing the interfacial area between the separated phases. In the presence of gravity there will be a flat interface between the two bulk phases. With stirring, domains of opposing composition will form highly contorted and stretched interfaces. The system may, in fact, choose to exist in a single homogeneous phase, even though the temperature is well into the two phase region (of the unstirred system). Such a configuration presumably reduces the rate of entropy production in this steady-state.

In one set of measurements on IBW, $I(k)$ was measured at various temperatures, above and below T_c at a fixed stirring rate, $f = 14.2\,Hz$.[7] In the temperature range $-25.4\ mK \leq T - T_c \leq +9.8\ mK$, the measured $I(k)$ was of the same functional form. There is no evidence, visual or otherwise, that two phases were present in this temperature range. Rather, the scattered intensity increased smoothly and monotonically for every k-value, as the temperature was lowered. At even lower temperatures, the system became so cloudy, that the scattering measurements no longer provided useful information about the composition fluctuations, $< |c_k|^2 >$.

Special attention should be given to the fact that in this temperature range, $I(k)$ has a simple scaling form $I(k) = A(Re,T)S[k/B(Re,T)]$. The measured scaling function $S[x]$ describes the system both above and below T_c in the temperature range mentioned above. At present there is no way to derive this scaling result. It is notable, however, that our data fit very well to a scaling function derived by Onuki.[5] In his analysis it is assumed that the system is in the two-phase region, where the osmotic compressibility is negative. He predicts, as indeed we observed, that the sharpness of the transition will be destroyed by the turbulent velocity field. However, his method has no justification for $T > T_c$, where we also observe the self-similar form of S[x].

III. The Effect of Stirring on Temporal Fluctuations

In addition to its influence on the wave number spectrum, the turbulent shear also affects the temporal fluctuations in composition. This effect can be probed by measuring the autocorrelation function, $g(t) = < I(t')I(t'+t) > / < I(t') >^2$, where $I(t)$ is the scattered light intensity. The brackets designate an average over time t'. In the absence of stirring, the spontaneous fluctuations in composition give rise to an exponentially decaying contribution to $g(t)$ with the decay rate being $2Dk^2$, where $D = k_BT/6\pi\eta\xi$ is the diffusivity for fluctuations of size ξ when $k\xi << 1$. When the mixture is weakly stirred, the composition fluctuations are carried along by the turbulent velocity, which Doppler shifts the light. A difference in Doppler shifts from many pairs of scatterers in the beam causes rapid decay of intensity fluctuations.

It has been shown[9] that when the turbulent velocity fluctuations dominate the decay of the homodyne autocorrelation function, $g(t)$ can be written as [9] $g(t) = 1 + G(t)$, where

$$G(t) = \int_0^L dR \ h(R) \int_{-\infty}^{\infty} dV \ P(V, R) \ cos(kVt) \ . \tag{4}$$

Here $\mathbf{V}$ (R) is the velocity difference of two points in the fluid that are separated by a distance R, and V is the projection of $\mathbf{V}$ (R) on the direction $\vec{k}$. The inner integral is an ensemble average of the phase factor, $cos(kVt)$ (due to frequency beating), over the velocity distribution function $P(V, R)$ for a fixed R. The outer integral sums over contributions from eddies of all sizes in a quasi-one-dimensional scattering volume of length L. The R-average is weighted by the function $h(R) = (2/L)(1 - R/L)$, which is the number fraction of pairs of scatterers separated by a distance R within the slit width L. Clearly there is little contribution to $g(t)$ from pairs of scatterers separated by a distance R which is almost as large as L. The small-scale statistics of the velocity fluctuations is characterized by the velocity distribution function $P(V, R)$. In Eq. (4) we have neglected a diffusive contribution to the decay of G(t).

The present authors have used this homodyne correlation spectroscopy technique. to study turbulent pipe flow and grid flow.[9] It was found that when the Reynolds number is large enough, the measured correlation function, $G(t)$, obeys a scaling form $G(t) = G(ku(L)t)$, where $u(L)$ is given in Eq.(3). Therefore the characteristic decay time of $G(t)$ is of the order of $(ku(L))^{-1}$.

We describe here a variant of the above flow experiment, in which the fluid (aqueous suspension of polystyrene spheres of 0.05 μ in diameter) is contained in a closed cell, and the turbulence is produced by a magnetic stirring bar sealed within the sample. The seed particles in the fluid, which are the source of the scattered light, merely sample the local velocity. In a cell of identical shape, the same stirring experiment was repeated, using a critical mixture of 2,6-lutidine and water (LW). In this cell there were no seed particles. The source of the scattering was the concentration fluctuations, as in sec. II. The goal of this experiment is to compare G(t) in the critical mixture and the solution of polystyrene spheres, under conditions of identical stirring frequency f, scattering vector $\vec{k}$, and slit width L. Since the viscosity of LW is very close to that of water, P(V,R) is almost the same for the two systems, though the source of the intensity fluctuations is entirely different.

A standard light scattering apparatus and a digital correlator were used to measure the correlation function $g(t)$. Figure 2 is a plot of the measured $G(t)$ for both flow systems. The closed circles show the effect of turbulently stirring the aqueous suspension of polystyrene spheres. The stirring frequency is 8 Hz, the scattering angle is 90^0, and the slit width L = 1 mm. The decay time of $G(t)$ is roughly 10 μ sec, and is independent of the temperature. This decay time is an order of magnitude faster than the diffusive decay time in the absence of stirring, which justifies our neglecting this contribution to G(t) in Eq. (4).

The crosses in Fig. 2 show the effect of stirring on a critical LW mixture at the same stirring frequency, slit width and scattering angle. This mixture has an inverted coexistence curve, so that phase separation is produced by heating the sample above $T_c(= 33^0C)$. The temperature of the sample was 170 mK below T_c, where the mixture is in one phase.

The data in Fig. 2. illustrate a feature shared by all our measurements: sufficiently far from T_c, the functional form and decay time of $G(t)$ is the same for

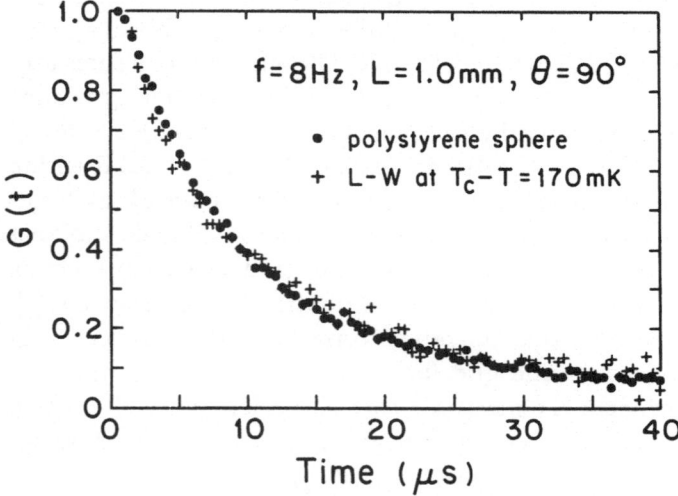

Figure 2: Intensity correlation function G(t) for stirred solutions of polystyrene spheres of diameter 0.05 μ (solid circles). The crosses show G(t) for a critical mixture of 2, 6-lutidine and water when the system is in one phase at $T_c - T =$ 170 mK. For both systems the stirring frequency, scattering angle, and slit width are the same, and G(t) has been normalized so that $G(0) = 1$.

the mixture as for simple fluid (water seeded with small particles), when all the controllable parameters are identical. This result is expected in the weak shear regime,[10] where the composition fluctuations can be assumed to be independent of the turbulent velocity fluctuations. This decoupling of the composition and turbulent velocity fluctuations allows one to separate the thermal average and the average over the velocity distribution. Therefore the correlation function $g(t)$ can be written as a product of two decay functions: one is from velocity fluctuations as shown in Eq. (4), and the other is from composition fluctuations. (In the case of aqueous suspension, the contribution from composition fluctuations is replaced by a diffusive contribution from seed particles).

There is reason to expect[10] that far from T_c, where the shear is weak, the measured $g(t)$ for a stirred binary mixture should have the same character as that associated with turbulent moving seed particles in water, if the velocity field is the same for the two systems. If this is so, G(t) in the mixture will be independent of temperature. This is indeed what we observe. When the temperature of the binary mixture approaches to T_c, the lifetime of the composition fluctuations increases (critical slowing down), and therefore the shear effect becomes strong. In this strong-shear regime, one might expect a strong coupling between the velocity and composition variables. Indeed, we found that when $|T-T_c| \leq 30\,mK$ there appears a very short relaxation time in the measured correlation function $g(t)$. This short decay time for the IBW sample at $T-T_c = 10\,mK$, is about six times smaller than the decay time associated with the turbulent velocity fluctuations. We have not yet understood the origin of the short time decay in $g(t)$. We can only say that this speeding up of the fluctuations near the critical point in a stirred mixture, arises from a strong interaction between the turbulent velocity field and the composition fluctuations.

IV. Summary

We have described light scattering experiments which demonstrate that turbulent stirring of a critical mixture can strongly affect its fluctuation spectrum near the critical point. As expected, the angular distribution measurements show that stirring mainly suppresses those fluctuations of low wave number, since they live long enough to be significantly distorted by the shear. We have also shown that turbulent stirring affects temporal fluctuations in composition. Far from the critical temperature, the turbulent velocity fluctuations dominate the decay of the correlation function $g(t)$, and the stirred binary mixture behaves similarly to the stirred water that has been seeded with particles to scatter light. On the other hand, when the shear rate becomes sufficiently large (a condition that can be achieved by approaching T_c or by increasing the Reynolds number Re), there appears a very fast decay in the measured $g(t)$. It is surely signaling a strong coupling between the turbulent velocity field and the composition fluctuations.

Acknowledgments

This work would not have come about without continuing interaction with A. Onuki. We have also benefited from many exchanges with J. V. Maher, C. K. Chan and J. Stavans. This research was supported by the National Science Foundation under Grant No. DMR 8611666.

References

1. H. E. Stanley, "Introduction to Phase Transitions and Critical Phenomena" (Oxford University Press, 1971); S. K. Ma, "Modern Theory of Critical Phenomena" (W. A. Benjamin, Reading, MA, 1976).

2. J. D. Gunton, M. San Miguel, and P. A. Sahni in "Phase Transitions and Critical Phenomena", Vol. 8, edited by C. Domb and J. L. Lebowitz (Academic, London, 1983).

3. A. Onuki and K. Kawasaki, Ann. Phys. 121, 456 (1979); K. Kawasaki and A. Onuki, in "Light Scattering Near Phase Transitions", edited by H. Z. Cummins and A. P. Levanyuk (North Holland, Amsterdam, 1983), p. 455.

4. D. Beysens, M. Gbadamassi, and L. Boyer, Phys. Rev. Lett. 43, 1253 (1981).

5. A. Onuki, Phys. Lett. 101A, 286 (1984).

6. D. J. Pine, N. Easwar, J. V. Maher and W. I. Goldburg, Phys. Rev. A29, 308 (1984).

7. C. K. Chan, W. I. Goldburg and J. V. Maher, Phys. Rev. A35, 1756 (1987).

8. A. N. Kolmogorov, C. R. Acad. Sci. USSR 30, 301 (1941).

9. P. Tong, W. I. Goldburg, C. K. Chan, and A. Sirivat, Phys. Rev. A37, 2125 (1988).

10. A. Onuki (private communication).

A Numerical Study of Defect Dynamics in a Three-Dimensional Complex Field

Hiraku Nishimori and T. Nukii

Department of Applied Physics, Tokyo Institute of Technology,
Meguro-ku, Tokyo 152, Japan

1. Introduction

There has been remarkable progress in the research of the dynamics of a non-conserved quenched field by the advance of an efficient theoretical method, that is, the u-field theory /1/. TOYOKI and HONDA extended the theory to a complex field /2/. They analytically obtained the dynamics of vortices in a quenched field. However there remain many important problems: 1) formation processes of vortices from a high temperature disordered phase, 2) elementary process of vortex dynamics, such as reconnection of two vortex lines, 3) influence of an external field.

We perform the numerical calculation of a 3-dimensional complex field. We particularly study the dynamics of vortices and find more detailed processes than those analytically obtained by the previous theories.

2. Model

We adopt the following spatially and temporally discretized equation:

$$\Psi_{i+1,n} = (1-m\varepsilon)f\{\Psi_{i,n}\} + \varepsilon \sum_{n'} f\{\Psi_{i,n'}\} \; ; \; f\{\Psi\} = \Psi + (1-|\Psi|^2)\Psi \; .$$

Here Ψ is a order parameter, the suffix i is the time, n is a lattice point, n' are the nearest neighbors of n, m is the number of nearest neighbors, ε denotes the quench depth and $f(\Psi)$ determines the local dynamics of an individual site under rotationally symmetric (ϕ^4 type) potential. The above equation is not the directly discretized version of the physics of the TDGL eq. /3/. This type of discretized model is called Coupled-Map-Lattice (CML) model which is proposed by KANEKO et al. and by OONO et al. /3,4/.

3. Observation

Firstly, we study the evolution of vortices by starting with initial conditions in which we prepare certain configulations of vortex lines, and we find some elementary processes i.e. reconnection, smoothing and shrinking as shown in Fig. 1. Particularly in the shrinking process of a vortex loop, the radius of the loop decreases with time as $\sqrt{c-t}$. This is consistent with the phenomenological theory $\dot{r}=\Gamma k$ where r is the position of the line element of vortex, k is the curvature at r and Γ is the kinetic coefficient.

Secondly, we make a numverical simulation of the evolution of a quenched system, whose size is 20×20×20. As an initial condition, the system is prepared in a high temperature disordered phase, namely, a complex random number whose amplitude is sufficiently smaller than 1 (the final value) is assigned at each site as an order parameter. The time evolution of vortices is shown in Fig. 2. We can see the time region in which the total vortex line length in the system decreases with time as $t^{-0.75} \sim t^{-0.80}$. This time region is seen in the larger system (64×64×64 and 124×124×124).

Springer Series in Synergetics, Vol. 43   **Cooperative Dynamics in Complex Physical Systems**
Editor: H. Takayama   © Springer-Verlag Berlin, Heidelberg 1989

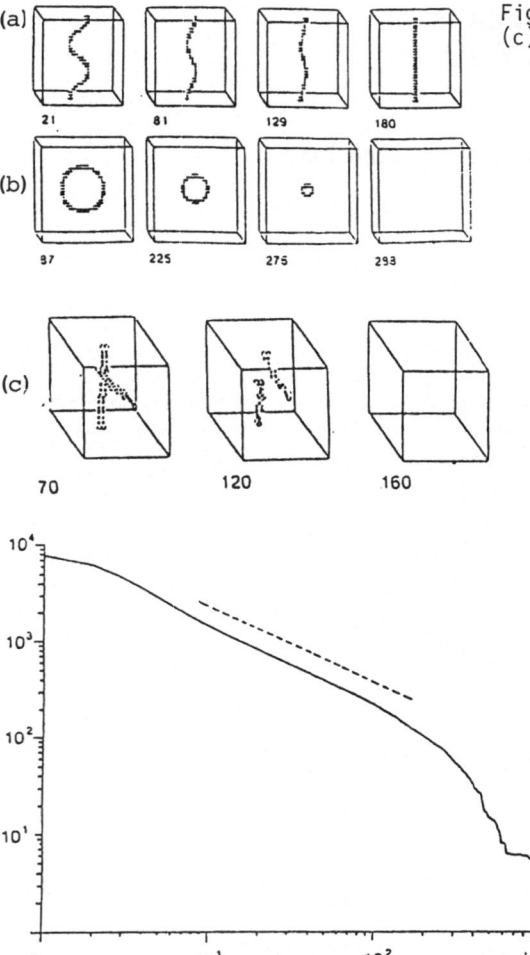

Fig. 1. (a) smoothing, (b) shrinking, (c) reconnection

(a)

21   81   129   180

(b)

97   225   275   293

(c)

70   120   160

Fig. 2. Time step vs total vortex length. The dashed line has a slope of -0.75.

Next, we extend the above CML model to the system including an external field as follows

$$\Psi_{i+1,n} = (1-m\epsilon)f\{\Psi_{i,n}\} + \epsilon\sum_{n'}\exp(i\vec{r}_{n'}\cdot\vec{A})f\{\Psi_{i,n'}\} \; ; \; f\{\Psi\} = \Psi + (1-|\Psi|^2)\Psi$$

where $\vec{r}_{n'}$ are the positions of nearest neighbor site n' and $\vec{A}$ is the external field. In the system described by the above equation, the force acting on the line element is not only the tension but also the Magnus force perpendicular to $\vec{A}$ and to the tangent of the vortex line. The vortex loop is enlarged or shrunk depending on its angle to $\vec{A}$. The external field is related to the relative velocity between normal phase and superfluid phase in the two fluid model of ^{4}He. So, we may expect the occurrence of phase transition to a turbulent state which is not yet clearly observed in our numerical study.

4. Discussion

We studied the dynamics of vortices in a 3-dimensional quenched complex field. By using a computationally efficient method, various behaviors of vortices are observed. The power law of the decreasing process for the total vortex length is observed. This power law deviates from the theoretical prediction based on dimensional analysis /2,5/. Two reasons are considered; 1) frequent occurrence of reconnection between vortices, 2) spatially intermittent distribution of vortices at the stage of vortex formation from the disordered phase. As for the decreasing process of interface area in a scalar field, the initial fractal dimension of interface is one of the decisive elements for the exponent of the power law /6/.

Acknowledgement

The authors thank Prof.K.KITAHARA, Prof.Y.θONO and Dr.K.KANEKO. Part of the numerical calculations were performed on HITAC S-820 at the Institute for Molecular Science.

References

1. T.Ohta, D.Jasnow and K.Kawasaki: Phys. Rev. Lett. 49(1982)1223
2. H.Toyoki and K.Honda: Prog. Theor. Phys. 78(1987)237
3. Y.Oono and S.Puri: Phys. Rev. Lett. 58(1987)836
4. K.Kaneko: Phys. Lett. A26(1987)25
5. K.Kawasaki: Phys. Rev. A31(1985)3880
6. H.Toyoki and K.Honda: Phys. Lett. 111(1985)367

Monte Carlo-Molecular Dynamics Simulation in Two-Dimensional Spin Systems

C. Kawabata[1], M. Takeuchi[1], T. Nakanishi[1], and A.R. Bishop[2]

[1]Okayama University Computer Center, Okayama 700, Japan
[2]Center for Non-Linear Studies, Los Alamos National Laboratory,
Los Alamos, NM 87545, USA

In a previous report, using a combined Monte Carlo-molecular dynamics technique, we have investigated the dynamic structure factor $S(q,w)$ on a square lattice for the isotropic Heisenberg and planar ferromagnetic Hamiltonian with nearest neighbor interaction. We have obtained dynamical signatures of the Kosterlitz-Thouless transition in the XY model case [1, 2].   More recently, we have also discussed the vortex signatures in dynamic structure factors for two-dimensional easy-plane ferromagnets [3].
.Here we focus on studying the ferromagnetic Heisenberg model with next nearest neighbor antiferromagnetic interaction,

$$H = - J_1 \sum_{i \neq j} (S_i^x S_j^x + S_i^y S_j^y + \lambda S_i^z S_j^z)$$

$$- J_2 \sum_{i \neq k} (S_i^x S_k^x + S_i^y S_k^y + \lambda S_i^z S_k^z) \qquad (1)$$

where (i,j) label nearest neighbor site and (i,k) label second nearest neighbor site on a two-dimensional square lattice, and J_1 and J_2 are nearest neighbor and next nearest neighbor exchange coupling constants.   The XY and isotropic Heisenberg limits correspond to $\lambda = 0$ and 1, respectively.
For a 50x50 square lattice, we have typically used 10^4x5 Monte Carlo steps per spin in order to equilibrate random initial spin

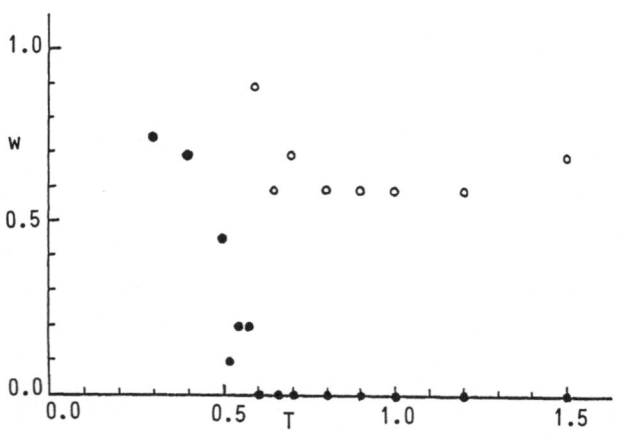

Fig. 1. Spin wave frequency w versus temperature T

Springer Series in Synergetics, Vol. 43   **Cooperative Dynamics in Complex Physical Systems**
Editor: H. Takayama   © Springer-Verlag Berlin, Heidelberg 1989

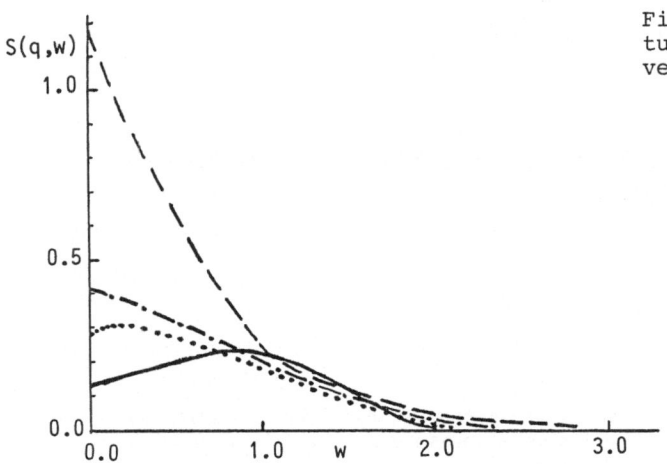

Fig. 2. Dynamic struc-
ture factors S(q,w)
versus frequency w

configurations; molecular dynamics with 4th order Runge-Kutta, time
step 0.05 and total time 512x0.05, and averaging over 5 ensembles.
   Our results in the present case are summarized in Figs.1 and 2.
In Fig.1, spin wave frequency w (●) at the wave vector q(1,1)
versus temperature T with J_1 = -1 (ferro) and J_2 = 1 (antiferro)
demonstrates strong evidence for a rapid spin-wave softening to
w ≅ 0 as T = 0.55 ± 0.05 (T_s), contrasted with results for the
isotropic Heisenberg model with nearest neighbor ferromagnetic
interaction [1] .   It is shown that a central peak half-width
(O) appears above T_s .   Fig.2 shows a representative dynamic
structure factor S(q,w) for T = 0.4 (——), 0.55 (·····),
0.6 (—·—·—), 0.8 (----) .
A detailed report of these computer studies will be given elsewhere.

References

1. C.Kawabata,M.Takeuchi and A.R.Bishop,
   Journal of Magnetism and Magnetic Materials.54-57(1986)871,
   and references therein.
2. C.Kawabata,M.Takeuchi and A.R.Bishop,
   Journal of Statistical Physics,Vol.43,Nos.5/6(1986)869.
3. F.G.Mertens,A.R.Bishop,G.M.Wysin and C.Kawabata,
   Physical Review Letters.59(1987)117 and to be published in
   Physical Review B and references therein.

Can Rod-like Molecules Form a Smectic Phase Without Attractive Forces?

H. Kimura

Department of Applied Physics, Faculty of Engineering, Nagoya University, Chikusaku, Nagoya 464-01, Japan

1. Introduction

It has become well established that systems of long rod-like molecules can exhibit the oriented nematic liquid crystal phase due to the hard core repulsion even without intermolecular attraction.

Theoretical evidence for the possibility of the nematic to smectic A phase transition (onset of a static one-dimensional density wave along the direction of molecular alignment) in the same systems has been presented previously by us [1]. Recently, a computer simulation on a system of hard parallel spherocylinders [2] confirms quite well the essence of our mean field theory on cylinders and rectangular parallelipipeds.

Here, we present a somewhat more simplified theory than the previous one and show that the hard core repulsion favours the smectic A phase in systems of cylinders and spherocylinders with large length to width ratio, while this is not the case in systems of ellipsoids of revolution.

2. Stability of Smectic A Phase in Hard Rod Systems

We consider a fluid composed of N hard rod molecules in volume V. In order to focus on the formation of density wave, we assume always the molecules align completely parallel to one another, then we express the molecular distribution function as

$$\rho(z) = V^{-1} \{ 1 + 2 \sigma \cos(2 \pi z/d) \}, \tag{1}$$

which corresponds to the density wave along the z-axis. The amplitude σ is the order parameter of the smectic phase, so the state $\sigma = 0$ is the nematic phase. In a mean field approximation, we investigate whether such a density wave can be stabilized due to the effect of hard core repulsion or not. The wave length d is calculated from the condition of free energy minimum.

Molecules of three different shapes : cylinder, spherocylinder and ellipsoid of revolution, with the length to width ratio L/D are studied.

(1) Cylinders. When the packing fraction c = (N/V)x(molecular volume) exceeds a certain value c(AN), the fluid shows the second order nematic to smectic A phase transition. This prediction and calculated values c(AN) = 0.57 and d = 1.40L agree fairly well with experiments. The experimental values are c(AN) = 0.35 and d = 1.27L [2].

(2) Spherocylinders (cylinders capped at each end by a hemisphere of the same diameter) also have the stable smectic A phase for c > c(AN), but in this case c(AN) depends on the ratio L/D as shown in Fig.1.

Springer Series in Synergetics, Vol. 43   **Cooperative Dynamics in Complex Physical Systems**
Editor: H. Takayama   © Springer-Verlag Berlin, Heidelberg 1989

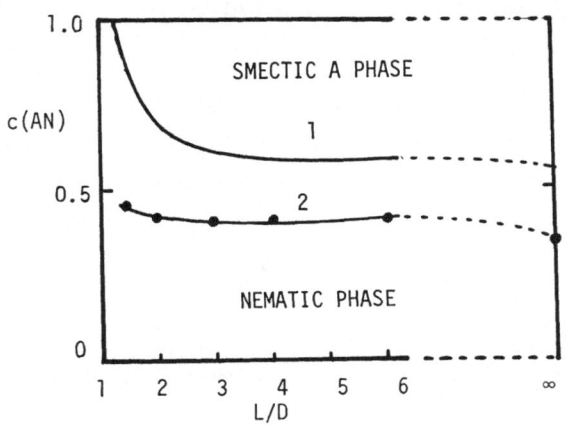

Fig.1. Phase diagram of parallel spherocylinders. 1 ——— present result, 2 —•— simulation[2].

(3) Ellipsoids cannot form stable density wave order irrespective of the ratio L/D, because the calculated c(AN) exceeds the physical limit 1. This explains the reason why the smectic state could not be found on the phase diagram of this system obtained by a simulation [3].

3. Discussion

If we consider the orientational degrees of freedom, we should find a certain lower limit of L/D (∼ 5) for the existence of c(AN) in cylinders and spherocylinders [1]. Therefore, attractive force as discussed by MCMILLAN [4] may be needed to stabilize the smectic phase in real liquid crystals, because many of the molecules have values of L/Ds smaller than 5.

It is shown that the hard core repulsion does not favour the smectic C phase (a density wave in which the molecules orient obliquely with respect to the z-axis) for all systems studied here. Hence, in order to have a stable smectic C phase, it is necessary to introduce certain attractive interactions different from those considered by MCMILLAN which cannot determine the type A or C of the smectic phase.

References

1. M.Hosino, H.Nakano, H.Kimura: J.Phys.Soc.Japan 46,1709 (1979); ibid 47,740 (1981); ibid 51,741 (1982)
2. A.Stroobants, H.N.W.Lekkerkerker, D.Frenkel: Phys.Rev. A36,2929 (1987)
3. D.Frenkel, B.M.Mulder, J.P.McTague: Phys.Rev.Lett. 52,287 (1984)
4. W.L.McMillan: Phys.Rev.A4,1238 (1971)

Martensitic Transformation: Nonlinear Order Parameter

T. Suzuki and S. Kojima

Institute of Applied Physics, University of Tsukuba, Tsukuba 305, Japan

The second order structural phase transformation observed in a perovskite type crystal as $SrTiO_3$ or $BaTiO_3$ has been successfully described in terms of an order parameter which corresponds to a certain specific normal mode of lattice vibrations. However, in the martensitic transformation, which is defined as the first order structural transformation observed in metals and alloys, the presence of an order parameter has not yet been established. In the second order structural phase transformation of $SrTiO_3$, freezing of the Γ_{25} phonon at R point has been established by neutron inelastic scattering experiment[1]. In the martensitic transformation, although some tendency towards softening of phonons in some area of the k-space has been observed[2], such complete freezing as observed in $SrTiO_3$ has never been observed.

If all atoms in a crystal were arranged in perfect periodicity, thermal vibration of atoms in the crystal should be described as superposition of normal mode lattice vibrations. Any crystal which is in thermal equilibrium at a finite temperature contains a specific concentration of vacancies. A vacancy migrate within a crystal. Atomic displacements involved in the migration of vacancy is entirely different from the lattice vibration described by normal modes in the following respects[3]: the displacement is highly localized around the vacancy and the magnitude of the displacement is not infinitesimal but comparable to lattice spacing. Hence, any crystal which is in thermal equilibrium at finite temperature should involve two entirely different kinds of atomic displacements: one associated with the normal mode lattice vibration and the other associated with the migration of vacancies.

The present authors wish to propose that while the order parameter pertinent to the second order structural transformation corresponds to a specific normal mode lattice vibration, the nonlinear order parameter pertinent to the martensitic transformation corresponds to the finite and localized atomic displacement associated with the migration of vacancies.

The criterion for the stability of a crystal against infinitesimal deformation is determined by quadratic expansion of the elastic deformation energy in terms of atomic displacement coordinates. This in turn can be expressed as the sum over normal coordinates with different wave vector k's, $\Sigma \omega_k^2 Q_k Q_K^*$. Hence, the criterion is expressed in terms of the phonon dispersion relationship between ω and k. If ω is positive for all values of k, the crystal is positive against any infinitesimal deformation. Because the amplitude of a normal mode is inversely proportional to the number of atoms in the crystal, it is always infinitesimal compared to the lattice spacing a. The crystal is not called into changing its structure by the atomic displacements associated with the normal mode lattice vibrations.

The criterion for the stability of a crystal against a finite and localized deformation is entirely different from that against infinitesimal deformation

Springer Series in Synergetics, Vol. 43   **Cooperative Dynamics in Complex Physical Systems**
Editor: H. Takayama   © Springer-Verlag Berlin, Heidelberg 1989

discussed above. A simple example of a system, which is stable for infinitesimal deformation but unstable against finite and localized deformation, is provided by a line of dominoes placed at proper equal spacings.

A more realistic model for a crystal is provided by an extension of the Fermi-Pasta-Ulam one dimensional nonlinear lattice[4]. In this model, a crystal is represented as a stacking of atomic plane and the stability of the model is studied against the transverse displacement u_i of i-th atomic plane. A nonlinear interatomic potential between neighboring atomic planes

$$V_2 = (1/2)(u_i - u_{i+1})^2 + (P_4/4)(u_i - u_{i+1})^4 + (P_6/6)(u_i - u_{i+1})^6 \qquad (1)$$

is introduced. For the choice of the nonlinear parameters $P_4 = -(n+1)$ and $P_6 = n(n > 3)$, the potential has one shallow minimum at $u_i - u_{i+1} = 0$ and a pair of deeper minima at $u_i - u_{i+1} = \pm 1$. This model is stable against any infinitesimal deformation around $u_i - u_{i+1} = 0$ and has a standard phonon dispersion relationship

$$\omega^2 = 4 \sin^2(ka/2). \qquad (2)$$

However, it is shown by a molecular dynamical simulation that once one of the atomic plane is displaced over the maximum of the potential V_2, a group of atomic planes slides into the configuration corresponding to $u_i - u_{i+1} = +1$ or -1. Besides this configuration, another configuration corresponding to 9R stacking sequence is shown to be generated by introducing additional three body interaction potential defined by

$$V_3 = q_3/2(u_{i+3} - 2u_i + u_{i-3})^2 , \qquad (3)$$

with a positive constant q_3 for three body interaction. The introduction of the additional interaction potential modifies the phonon dispersion relationship as

$$\omega^2 = 4 \sin^2(ka/2) + 16 q_3 \sin^4(3ka/2) . \qquad (4)$$

This dispersion relationship has a hump at $k = \pi/3a$ and a dip at $k = 2\pi/3a$. The position of the dip in the dispersion relationship coincides with what is expected from the soft mode picture of the 9R structure. This indicates that the presence of the dip in the phonon dispersion relationship does not necessarily imply the presence of the soft mode.

The authors would like to thank Dr.J.W.Cahn, Professor T.Sakudo, Professor Y.Yamada and Professor Manfred Wuttig for stimulating discussions and comments. The present work is supported by a Grant-in-Aid for Scientific Research from the Ministry of Education (B-62460185).

References

1. G.Shirane and Y.Yamada: Phys. Rev. 177, 16 (1969).
2. M.Mori, Y.Yamada and G.Shirane: Solid State Comm. 17, 127 (1975).
3. C.P.Flynn and A.M.Stoneham: Phys. Rev. B1, 3966 (1970).
4. T.Suzuki: Met. Trans. A 12, 709 (1981).

Successive Magnetic Ordering in a Triangular Lattice Antiferromagnet CsNiCl₃

S. Maegawa[1], *T. Goto*[1], *and Y. Ajiro*[2]

[1]Department of Physics, College of Liberal Arts, Kyoto University, Kyoto 606, Japan
[2]Department of Chemistry, Faculty of Science,Kyoto University, Kyoto 606, Japan

The hexagonal compound CsNiCl₃ is one material typical of triangular lattice antiferromagnets of Heisenberg-type with small single-ion anisotropy (D/k_B= $-0.033 \sim -0.13$K). The Ni²⁺ ions along the c-axis have the antiferromagnetic intrachain interaction (J/k_B=16.6K) and the chains form the triangular lattice in the c-plane with a weak antiferromagnetic interchain interaction (J'/k_B=0.28K). As early as in 1972, CLARK and MOULTON [1] observed two magnetic phase transitions at T_{N1}=4.8K and T_{N2}=4.4K from NMR experiment and proposed an interesting model for these successive transitions. In their model, the spin structure in the low temperature phase below T_{N2} is that the spins on the triangular lattice in the c-plane form a 120° spin structure in the ac-plane. In the intermediate phase between T_{N1} and T_{N2}, only the $S_{\parallel}$ component parallel to the c-axis is ordered, leaving the $S_{\perp}$ component in the c-plane to be disordered (component disorder model), as has been supported by subsequent neutron diffraction measurements [2]. However, a recent ESR experiment has shown that the model should be improved to include a dynamical character. In order to get more understanding, we have performed a detailed NMR experiment of ¹³³Cs in the single crystals by a coherent pulsed-NMR method [3]. The most interesting result is that the 120° spin structure remains to have a freedom of rotation even below T_{N2}.

The temperature and angular dependences of the resonance fields at an operating frequency of 4.0MHz were examined. The experimental results are summarized as follows. i) Paramagnetic Phase ($T > T_{N1}$) : The NMR spectra for both $H_\theta \parallel$ c-axis and $H_\theta \perp$ c-axis have a single peak at the same resonance field as that for a free ¹³³Cs. ii) Intermediate Phase ($T_{N2} < T < T_{N1}$) : The spectrum for $H_\theta \parallel$ c-axis remains with a single peak, while the spectrum for $H_\theta \perp$ c-axis splits into two peaks, whose resonance fields are independent of the direction of the external magnetic field in the c-plane (isotropic). iii) Low Temperature Phase I ($T_D <\_ T < T_{N2}$) : The spectra for both $H_\theta \perp$ c-axis and $H_\theta \parallel$ c-axis have two peaks. The resonance fields for $H_\theta \perp$ c-axis remain isotropic. iv) Low Temperature Phase II ($T < T_D$) : For $H_\theta \perp$ c-axis, new peaks appear between the original two peaks in the spectra around a temperature designated by T_D and grow bigger as temperature is lowered. The spectrum consists of maximum six peaks and finally these resonance fields depend on the direction of the external field, as shown in Fig.1.

The observed angular dependence of the resonance fields (Fig.1) in the low temperature phase II can be interpreted by calculating the dipolar fields H_d at Cs nuclear sites from the Ni²⁺ spins which form a 120° spin structure proposed previously. However though, below T_{N2} this model cannot explain the isotropic angular dependence in the low temperature phase I. Next we consider the result in the intermediate phase. Following the component disorder model proposed for this phase, the ordered and fixed values of $S_i{}''$ components give, at Cs nuclear sites, the internal fields with constant values and with fixed directions in the c-plane, which results in an angular dependence of the resonance fields for $H_\theta \perp$ c-axis. The present isotropic result, therefore, can not be explained also in this phase by the simple component disorder model.

A new interpretation is required for the intermediate phase and the low temperature phase I. According to MIYASHITA [4], the energy difference between

Springer Series in Synergetics, Vol. 43   **Cooperative Dynamics in Complex Physical Systems**
Editor: H. Takayama   © Springer-Verlag Berlin, Heidelberg 1989

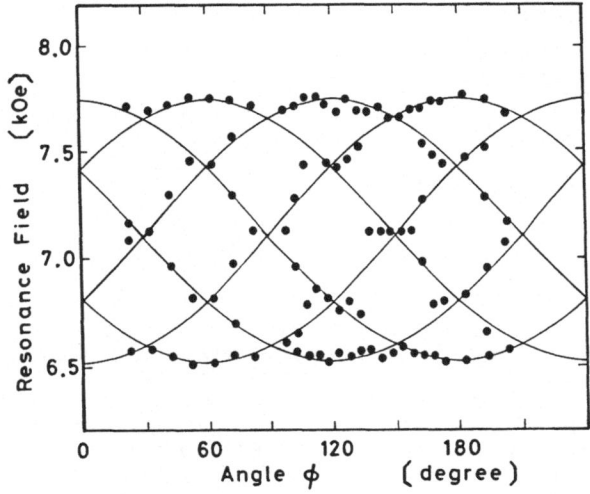

Fig. 1
The angular dependence of the resonance fields in the c-plane at 0.49K in the low temperature phase II at 4.0MHz. Solid curves show the values calculated on the 120° spin structure with $\theta_1 = 0$.

the 120° spin structure in the ac-plane with $\theta_1 = 0$ and that with $\theta_1 = \pi/2$, where θ_1 is a tilting angle of the first sublattice spin, is estimated as $-\Delta E/J' = d^3/(1-d^2)$, where $d = D/J'$. If $d \ll 1$, the 120° structure may have a freedom of a continuous rotation (quasi-degeneracy) in the ac-plane. In the intermediate phase, besides the quasi-degenerate situation, there exists another freedom of rotation around the c-axis. Consequently the components of spins are described as $S_i^\perp = 0$ and $S_i^{//} = \cos(\overleftarrow{k} \cdot \overleftarrow{r}_i + \theta_1)$ where $\overleftarrow{k} = (\pm 4\pi/3, 0, 2\pi)$. The slow and continuous variation of θ_1 in time and/or space owing to the quasi-degeneracy would give the isotropic angular dependence of the resonance fields. The free rotation around the c-axis ceases at T_{N2}, but the quasi-degeneracy survives down to T_D, which corresponds to the low temperature phase I. The appearance of the new peaks for $H_0 \perp$ c-axis below about T_D suggests that the 120° structure with $\theta_1 = 0$ becomes dominant gradually in the low temperature phase II. The characteristic time constant τ_C associated with the variation of θ_1 is estimated to be $\tau_C \gg 1/\gamma H_d \sim 10^{-6}$ sec. This slow motion has not been detected by neutron experiments whose associated time scale is of the order of 10^{-10} sec [2]. From our present work, however, it is not possible to distinguish whether the variation of θ_1 comes from the spacial distribution of θ_1 or the time variation of θ_1.

Another interesting result is that T_{N1}, T_{N2} and T_D are sensitively influenced by a small amount of Co^{2+} impurity. For instance, $T_{N1}=4.8K$, $T_{N2}=4.4K$ and $T_D \simeq 1K$ for pure $CsNiCl_3$, while for $CsNiCl_3$ containing 0.6% of Co^{2+} $T_{N1}=5.30K$, $T_{N2}=3.87K$ and $T_D \simeq 3K$. This fact implies the quasi-degeneracy is easily destroyed by this impurity.

In conclusion, the successive magnetic ordering in $CsNiCl_3$ can be understood by the successive freezings of the free rotations of the 120° spin structure. Below T_{N1} the rotation around the a-axis is frozen but the 120° structure has quasi-degeneracy in the ac-plane due to the small anisotropy D and also it rotates freely around the c-axis. The rotation around the c-axis freezes below T_{N2}, however, the freedom in the ac-plane survives down to far below T_{N2}. The latter freezing is not accompanied by the transition. These successive freezings are found to be much influenced by Co^{2+} impurity.

References

1. R.H.Clark and W.G.Moulton: Phys.Rev. B5, 788 (1972)
2. H.Kadowaki, K.Ubukoshi and K.Hirakawa: J.Phys.Soc.Jpn. 56, 751 (1987)
3. S.Maegawa, T.Goto and Y.Ajiro: J.Phys.Soc.Jpn. 57, 1402 (1988)
4. S.Miyashita: J.Phys.Soc.Jpn. 55, 3605 (1986)

Evidence for Z_2-Vortex Excitations in the Quasi-Two-Dimensional Triangular Lattice Antiferromagnets HCrO$_2$ and LiCrO$_2$

Y. Ajiro and H. Kikuchi

Faculty of Science, Kyoto University, Kyoto 606, Japan

The title compounds are the family of the AMO$_2$-type ($R\bar{3}m$) crystal which has a pronounced layer structure and are promising candidates for the two-dimensional Heisenberg triangular lattice antiferromagnet (2DH TAL-AF). One of the characteristic features of this frustrated system with continuous spin symmetry is that it undergoes a new type of phase transition at $T_{KM}=0.66JS^2$ driven by the pairing of the topological point defects known as Z_2 vorticies [1]. In the real substances accessible to the experimental study, three-dimensionally stacked TAL-AF exhibits a phase transition which belongs to a new universality class characterized by novel critical exponents as has been demonstrated recently [2,3]. It is still challenging, however, to give experimental evidence for the existence of such a unique vortex in ideal 2DH TAL-AF from the experiments on the real substances. Here we show [4], for the first time, that the electron paramagnetic resonance (EPR) experiments can be used to extract valuable information about Z_2 vortex under some reasonable assumptions and provide experimental evidence for the existence of Z_2 vortex excitations in the highly 2D ordered temperature region of the quasi-2D substances.

EPR measurements were carried out on powder samples at X-band (9.05 GHz) between 4.2 K and 300 K, using a standard reflection type spectrometer. The first derivative of the absorption curve was detected by modulating the magnetic field at 100 kHz. The observed EPR linewidth increases with decreasing temperatures and shows an anomaly at $T_N=20\pm2$ K for HCrO$_2$ and 65 ± 2 K for LiCrO$_2$, respectively, as shown in Fig. 1.

A noticeable fact is that the observed T_N's from the EPR measurements are only about 10 % higher than the theoretical value of T_{KM} estimated from the known values of the relevant exchange constants. The fact suggests that the present substances behave as 2DH TAL-AF over a wide range of temperature down to very close to T_N, although the real transition must be caused inevitably by some non-ideal effects such as interlayer interaction or Ising anisotropy. In this situation, one can expect that the concept of the Z_2 vortex is still meaningful as the basic mechanism of the ordering process in the quasi-2DH TAL-AF over the wide range of temperature in which 2D short-range ordering is highly developed.

Another noticeable fact is that the temperature dependence of the EPR linewidth can not be understood in terms of widely-accepted critical broadening. The most logical approach to interpret the present result is to consider a broadening mechanism based on the 2D model. When the temperature is above T_{KM}, unbounded free vorticies will be thermally excited and the spin state at a given site may fluctuate through the passage of vortex. The spin-flipping rate is given by the probability of collision with the vortex which freely moves in the 2D lattice. In the most simple case, one can assume that the characteristic decay rate of the time-dependent spin-correlation is proportional to both the vortex density n_v and the averaged vortex velocity $\langle V_v \rangle$, and the number density is given by

$$n_v \propto \exp(-E_v/k_B T),$$

Springer Series in Synergetics, Vol. 43 **Cooperative Dynamics in Complex Physical Systems**
Editor: H. Takayama   © Springer-Verlag Berlin, Heidelberg 1989

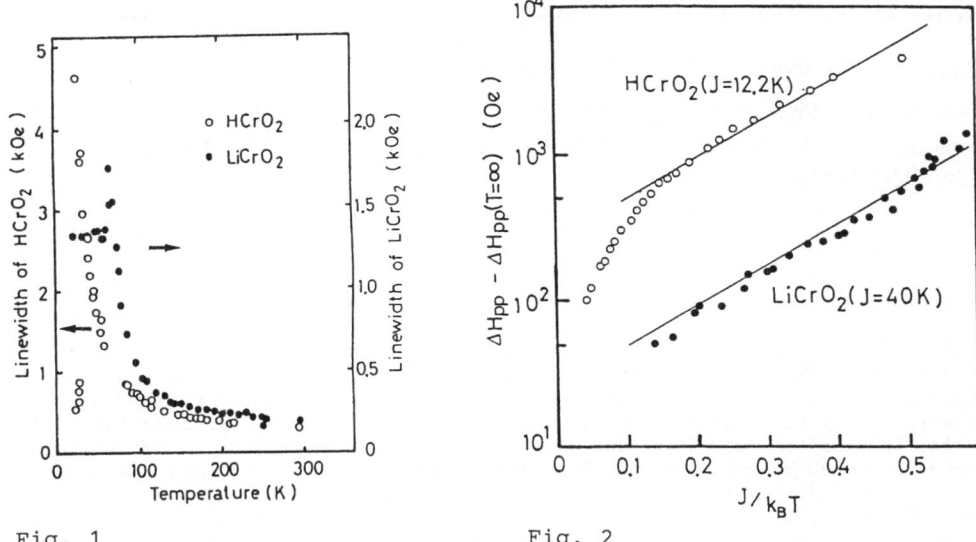

Fig. 1

Fig. 2

Fig. 1 Temperature dependence of the linewidth in HCrO$_2$ and LiCrO$_2$

Fig. 2 Linewidth as a function of the scaled inverse temperature, J/kT

where E_v is the activation energy of the free vortex. We finally obtained the following form of the linewidth:

$$\Delta H_{pp} = A' \exp (E_v / k_B T).$$

Experimental data fit quite well the expected simple exponential law as shown in Fig. 2. More importantly, the activation energy for the free Z_2 vortex takes a universal value, $E_v = 2.7 \pm 0.3$ JS$^2 \simeq 4kT_{KM}$, which is in reasonable agreement with the theoretical prediction.

References

1. H. Kawamura and S. Miyashita: J. Phys. Soc. Jpn. 53, 4138 (1984)
2. H. Kadowaki, S.M.Shapiro, T. Inami and Y. Ajiro: J. Phys. Soc. Jpn. 57, 2640 (1988)
3. Y. Ajiro, T. Nakashima, Y. Unno, H. Kadowaki, M. Mekata and N. Achiwa: J. Phys. Soc. Jpn. 57, 2648 (1988)
4. Y. Ajiro et al.: J. Phys. Soc. Jpn. 57, 2268 (1988)

Soliton Dynamics in Impure Ising-like Antiferromagnetic Chains

T. Goto[1], T. Kohmoto[1], S. Maegawa[1], and M. Mekata[2]

[1]Department of Physics, College of Liberal Arts, Kyoto University, Kyoto 606, Japan
[2]Department of Applied Physics, Faculty of Engineering, Fukui University, Fukui 910, Japan

Solitons in one-dimensional magnetic systems have recently been receiving considerable interest. The soliton behaves like a free particle when the mean free path λ is much longer than the inverse of the soliton density n_s ("coherent model"). On the contrary, if the condition $n_s\lambda \ll 1$ holds due to scatterings from impurities or defects, the soliton moves diffusively ("incoherent model"). Experimetal evidence for such behavior was first presented with regard to the sine-Gordon soliton in TMMC doped with Cu^{2+} or Cd^{2+} by the ^{14}N nuclear relaxation experiment by BOUCHER et al.[1]. The purpose of the present work is to study from the nuclear magnetic relaxation the behavior of the soliton (propagating domain boundary) in impure Ising-like antiferromagnetic chain of S=1/2 in the hexagonal compound $CsCoCl_3$. This compound exhibits successive phase transitions due to the frustration effect at $T_{N1} \fallingdotseq 22K$ and $T_{N2} \fallingdotseq 9K$. Short range order is well developed above T_{N1}, and, in the intermediate phase ($T_{N2} < T < T_{N1}$), one-third of the magnetic chains still remain disordered. The presence of the coherent soliton in $CsCoCl_3$ for $T > T_{N2}$ has been shown by the experiments of ESR [2] and neutron scattering [3].

We have measured, using a coherent pulsed-NMR method, the relaxation time T_1 of ^{133}Cs in the single crystals doped with Mg^{2+} or Fe^{2+}. Figure 1 shows the temperature dependences of T_1^{-1} for $T > T_{N1}$ for $CsCo_{1-x}Mg_xCl_3$ (x=0.02 and 0.005). The data of T_1 for the pure crystal [4] are presented for the comparison. No NMR signal was detected in the temperature range between ~7K and T_{N1}, because of the short transverse relaxation time T_2. (Note that T_{N1} decreases in the presence of non-magnetic impurity [5].) The relaxation rate exhibits much more remarkable temperature dependence as compared with that for the pure compound for 20K<T<30K, and becomes almost constant in the higher temperatures. Figure 2 shows the data for $CsCo_{1-x}Fe_xCl_3$ (x=0.03). The behavior of T_1 for $T > T_{N2}$ is very different qualitatively from that of Mg^{2+}-doped compound.

According to the soliton model, the time correlation of local order decays as a result of flippings due to the random passings of the solitons. Then, the auto-correlation function of each electron spin is expressed, in the long-time approximation, as $f(t)=\exp[-2N(t)]$, where $N(t)$ is the average number of spin flips occuring during the time t. The nuclear relaxation rate T_1^{-1} is obtained by taking the Fourier-transform of $f(t)$ with respect to the Larmor frequency ω_N. In the coherent model, we have $N(t)=n_sV_ot$ and $f(t)=\exp(-\Gamma t)$; $\Gamma=2n_sV_o$, where $n_s=\exp(-|J|/kT)$, and $V_o \approx (4/\pi)\varepsilon|J|/\hbar$ (ε is exchange anisotropy) is the average velocity of the soliton. Then we obtain, in the high temperature limit,

$$T_1^{-1} \sim \Gamma^{-1} \sim V_o^{-1}\exp(|J|/kT). \qquad (n_s\lambda \gg 1) \qquad (1)$$

The dashed line in Fig.1 represents the best fit of (1) with $|J|/k=75K$ to the data for x=0. The gap at T_{N1} corresponds to the change in the number of the disordered chains. The good agreement between the theory and the experiment proves the validity of the coherent soliton model in the pure $CsCoCl_3$. Apparently, however, (1) cannot explain our results for the Mg^{2+}-doped compound.

Now we shall consider the effect of the impurity. If we assume that the motion of the soliton becomes diffusive due to the presence of the impurity, and apply 1D random walk approximation [1], we find $N(t)=n_s\sqrt{Dt}$ where D is the diffusion coefficient given by $D=V_o\lambda$. This results in an $\exp(-\sqrt{t})$-type decay as $f(t)=\exp(-\sqrt{\Delta t})$; $\Delta=4Dn_s^2$, and we obtain

Springer Series in Synergetics, Vol. 43   **Cooperative Dynamics in Complex Physical Systems**
Editor: H. Takayama   © Springer-Verlag Berlin, Heidelberg 1989

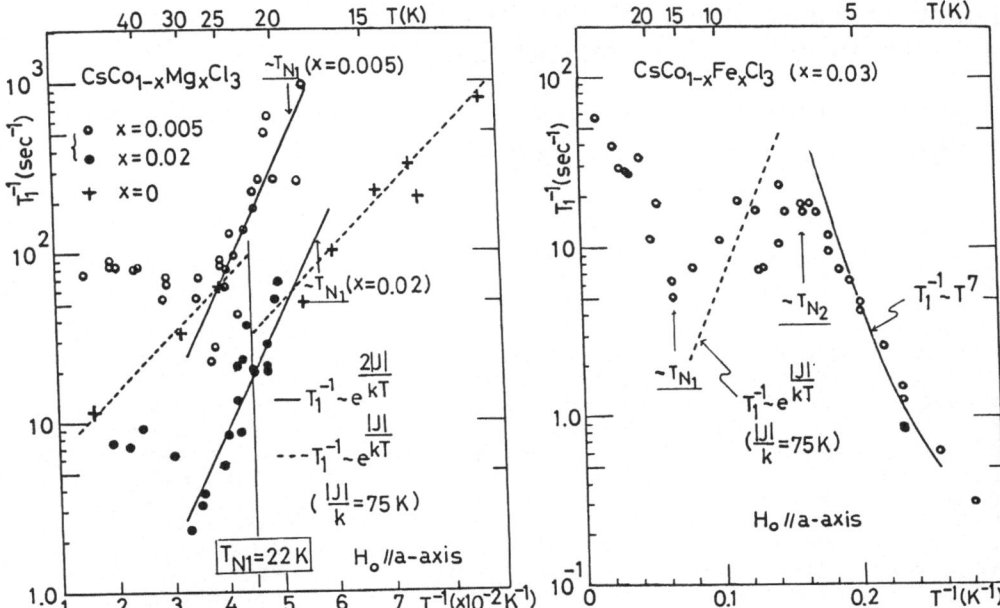

Fig.1. T_1^{-1} of ^{133}Cs vs T^{-1} in Mg^{2+}–doped $CsCoCl_3$ for $H_0//a$-axis and $\omega_N=6.5$ and 5.0MHz. The dashed and solid lines represent (1) and (2) in the text.

Fig.2. T_1^{-1} of ^{133}Cs vs T^{-1} in Fe^{2+}–doped $CsCoCl_3$ for $H_0//a$-axis and $\omega_N=6.5$ and 5.0MHz. The dashed line represents (1) in the text.

$$T_1^{-1} \sim \Delta^{-1} \sim (V_0\lambda)^{-1}\exp(2|J|/kT). \qquad (n_s\lambda \ll 1) \qquad (2)$$

The solid lines in Fig.1 represent the theoretical curve of (2) with $|J|/k=75K$ which are fitted independently to the data for x=0.005 and 0.02. The theoretical prediction agrees qualitatively with the experimental results for 20~30K, thus giving evidence for the diffusive soliton. It is rather surprising that the incoherent model is appropriate for 20~30K even for the sample of x=0.005. If it is assumed that the soliton is scattered only at the impurity sites, we find $n_s\lambda\gg1$ in the relevant temperature range, which is fulfilled for the coherent model. As for the reason for this, we suppose that even a small amount of the non-magnetic impurity has an important effect to reduce λ appreciably. This may be related to the partial destruction of the frustration caused by the impurity, which disturbs the soliton propagation. The reason for the fact that T_1^{-1} turns to be almost constant in the high temperatures is uncertain at the present. As is seen in Fig.2, the temperature dependence of T_1^{-1} of Fe^{2+}–doped compound for the intermediate phase is more gentle than the prediction of (1) with $|J|/k=75K$. It is added that the temperature dependences of T_1 for both compounds below about 7K are interpreted qualitatively in terms of the two–phonon process via the magnon–phonon interaction described by the equation $T_1^{-1}\sim T^7$ (solid line in Fig.2).

In conclusion, experimental evidence for the diffusive motion of the soliton has been presented in $CsCo_{1-x}Mg_xCl_3$. The authors wish to thank Mr.N.Fujiwara for the experimental contribution.

References

1. J.P.Boucher, H.Benner, F.Devreux, L.P.Regnault, J.Rossat–Mignod, C.Dupas, J.P.Renard, J.Bouillot, and W.G.Stirling: Phys.Rev.Lett. 48, 431 (1982)
2. K.Adachi: J.Phys.Soc.Jpn. 50, 3904 (1981)
3. J.P.Boucher, L.P.Regnault, J.Rossat–Mignod, Y.Henry, J.Bouillot, and W.G.Stirling: Phys.Rev. B31, 3015 (1985)
4. Y.Ajiro et al.: to be published
5. M.Mekata, T.Tatsumi, T.Nakashima, K.Adachi, and Y.Ajiro: J.Phys.Soc.Jpn. 56, 4544 (1987)

Soliton-Mediated Magnetic Phase Transition in the Frustrated Antiferromagnet CsCoCl₃

H. Kikuchi and Y. Ajiro

Faculty of Science, Kyoto University, Kyoto 606, Japan

Hexagonal $CsCoCl_3$ is a quasi-one-dimensional (1D) $S=1/2$ Ising antiferromagnet with intrachain interaction $J=-75K$. With decreasing temperature, a 3D magnetic transition occurs due to the weak antiferromagnetic interchain interaction as usual. However, a frustration effect due to triangular arrangement of the constituent magnetic chains in the c-plane gives rise to unusual successive phase transitions at $T_{N1}=21K$ and $T_{N2}=9K$. The spin frustration model identifies the intermediate-temperature phase between T_{N1} and T_{N2} as a partially disordered (PD) phase in which 2/3 of magnetic chains order antiferromagnetically with each other, leaving the rest uncorrelated, and the low-temperature phase below T_{N2} as a ferrimagnetic (FR) phase in which all magnetic chains order in a collinear ferrimagnetic arrangement. While the model has succeeded in explaining the various experimental results, there remain several basic questions; Whether the PD phase is stable or not? How can 1/3 of chains behave paramagnetically? What happens at T_{N2}? We have carried out magnetic resonance experiments (NMR, ESR) in order to solve these problems from a microscopic point of view.

1. NMR

NMR measurements were performed at 5 MHz using a pulsed spectrometer. As shown in Fig. 1, the angular dependence of the resonance fields of ^{133}Cs at 4.2K can be well explained by the calculated dipolar fields at the Cs nucleus for the FR configuration shown in Fig. 2(a), confirming that the low temperature phase is the FR phase. Interestingly, the NMR spectrum specific to the FR phase remarkably changes in shape as the temperature approaches T_{N2}, as shown in Fig. 3. For $H_o//a$-axis, three peaks with almost equal intensity were observed at 4.2K in agreement with the model. With increasing temperature, however, the central peak diminishes faster than the outer two peaks without change of the resonance field. This curious behavior evidences a disordering process of 1/3 of the chains. When the central spin in Fig. 2(a) reverses its direction with time, transition from (a) to (b), we notice that there are two different sites, site 1 feels a time-dependent $h_{//}$ and site 2 feels a time-independent $h_{//}$, where $h_{//}$ is the internal field parallel to the external field. Supposing that the rate of spin-flipping increases with increasing temperature, the signal originated from site 1 corresponding to the central peak becomes invisible due to strong fluctuations of the internal field. The outer two peaks also decrease in intensity due to indirect effects of the fluctuations without a change of the resonance field. Spin-flipping is caused by the propagation of domain-wall solitons as evidenced by NMR spin-lattice relaxation measurements. The effect of spin fluctuation on the relaxation rate is strongest when the spin-flipping rate is comparable to the NMR frequency, and thus the relaxation rate shows a pronounced anomaly at T_{N2}. As the soliton density increases exponentially with increasing temperature, the spin-flipping rate also increases exponentially. When the spin-flipping rate becomes faster than the NMR observation time, the signals reappearat the averaged positions between (a) and (b) owing to the motional averaging effect, resulting in the spin configuration shown in Fig. 2(c) corresponding to the PD phase in which the central spin now has no time-averaged moment.

Figure 4 shows the angular dependence of the resonance fields at 15K in the intermediate phase. The solid lines are the calculated internal fields based on

Springer Series in Synergetics, Vol. 43   **Cooperative Dynamics in Complex Physical Systems**
Editor: H. Takayama   © Springer-Verlag Berlin, Heidelberg 1989

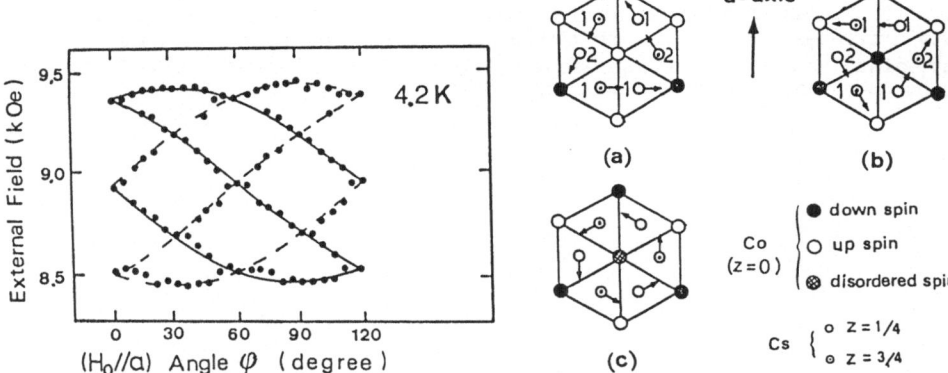

Fig. 1 Angular dependence of the resonance fields at 4.2K

Fig. 2 Direction of internal fields at Cs nuclei in FR phase (a) and (b), and in PD phase (c)

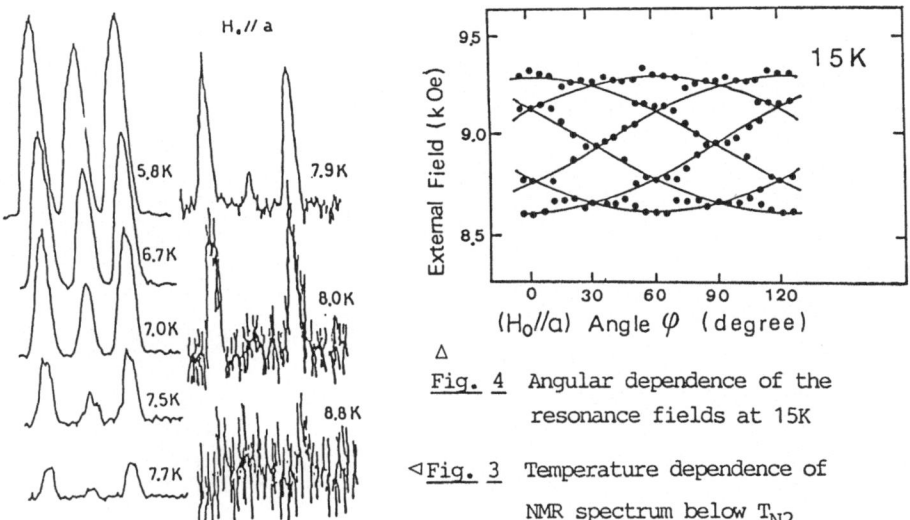

Fig. 4 Angular dependence of the resonance fields at 15K

◁Fig. 3 Temperature dependence of NMR spectrum below T_{N2}

the PD phase and are in good agreement with the experimental results. We conclude that the intermediate phase is certainly the PD phase, at least on the time scale of NMR and in the temperature region not very close to T_{N1}.

2. ESR

ESR measurements give direct evidence for the existence of moving solitons. The intensity of ESR signal from solitons increases exponentially since the thermally excited soliton density increases exponentially with increasing temperature. The linewidth of the ESR signal also increases exponentially with increasing temperature above T_{N2}. The temperature dependence can be explained in terms of lifetime broadening due to collisions among solitons, evidencing that the solitons propagate in the PD chains above T_{N2}.

3. Conclusion

The low-temperature phase below T_{N2} is the FR phase. As T_{N2} is approached from below, spins in 1/3 of magnetic chains reverse their direction very slowly through the soliton movement. Thus, the character of the phase transition at T_{N2} is dynamical. The spins have no measurable time-averaged moment as the soliton density increases with increasing temperature, resulting in the PD phase in the intermediate temperature phase between T_{N2} and T_{N1}. The PD phase is stable at least on the time scale of NMR up to near T_{N1}.

Computer Experiment on Sublattice Dynamics of Antiferromagnetic Triangular Lattice Systems with $1/f^n$ Fluctuations: Proposal of SSA for the Study of Frustrated Systems

S. Ozawa[1], S. Kamata[2], T. Kobayashi[2], and T. Haseda[2]

[1]Ibaraki University, Faculty of Engineering, 4-12-1 Nakanarusawa, Hitachi 316, Japan
[2]Toin University of Yokohama, Faculty of Engineering,
  1614 Kurogane, Midoriku, Yokohama 227, Japan

1. Successive Spin Aggregation (SSA) in Frustrated Systems - General Method

This paper proposes SSA computer experiment for the study of frustrated systems. SSA consists of the following processes : (1) spins are successively deposited on a non-magnetic substrate of a defined lattice, (2) the deposited spins move to lower energy sites on the substrate and form spin aggregates, and (3) spin flipping occurs for obtaining the lowest energy of the aggregates. Although we can consider SSA as a kinetic model of actual phenomena e.g. adsorption of electrons on helium surface, we rather intend to use it here as a tool for the study of obtaining microscopic understanding of propagation of order in frustrated systems.

2. Deposition of Ising Spins onto Triangular lattice

In the spin deposition process of SSA, let us assume that spins are wise enough that they find the lowest energy site on the substrate and settle there. If we have a number of lowest energy sites, we pick one out of them by casting a dice. Figure 1 shows spin aggregates grown from a seed on 60x60 triangular lattice with AF-JNN and with or without F-JNNN interactions. In Fig. 1(a), complete A and B sublattices grow forming a honeycomb structure, and C sublattice begins to grow after they cover the whole area of substrate. While in Fig. 1(b), the pattern is something like that of

(a) JNNN/JNN=-0.1   (b) JNNN/JNN=0

Fig. 1 Spin aggregate successively grown from a seed marked by circle, # : + spin, - : - spin.

DLA and has no single domain sublattice. It is seen from these figures that the existence of JNNN removes the degeneracy of ground state of the system and is essential for the occurrence of the single domain sublattice. The ground state spin structure in frustrated systems clearly comes into sight by the SSA experiment.

3. Coagulation of Randomly Deposited Ising Spins

We firstly set a given number of Ising spins at random positions on the triangular lattice, and then allow them to move sequentially into neighboring lowest energy sites. Figure 2 shows an example of this experiment, where JNN=10 and JNNN=-5. The honeycomb structure develops again, although in this case the spin notices the energy at neighboring sites alone. The occurrence of honeycomb structure is the result of the frustrated interactions of triangular lattice. However, the honeycomb structure in itself is not frustrated and stable at finite temperatures. Addition of spins into the center sites of honeycomb starts the frustration of aggregates. Computer experiments of spin addition into complete A and B sublattices at finite temperatures show that the complete sublattices begin to be destroyed with increasing the number of added spins on C sublattice. These results of SSA suggest the mechanism of sublattice switching.

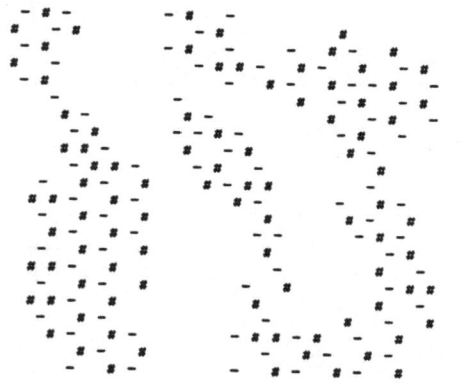

Fig.2 Coagulated spin aggregates, JNN=10 and JNNN=-5.

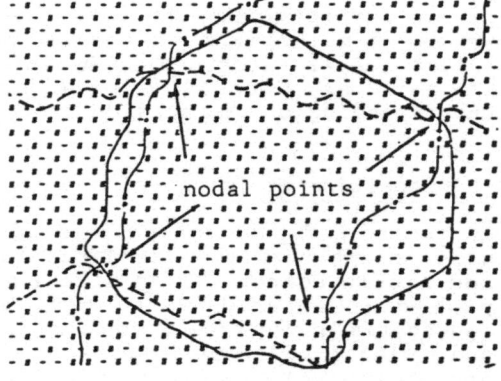

Fig. 3 SSA pattern of domain boundaries, JNN=10 and JNNN=-2.

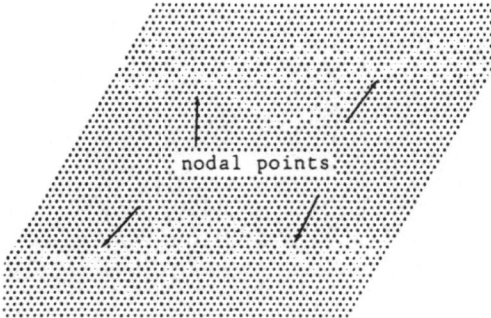

Fig. 4 MC pattern of domain boundaries, JNN=10 and JNNN=-1.

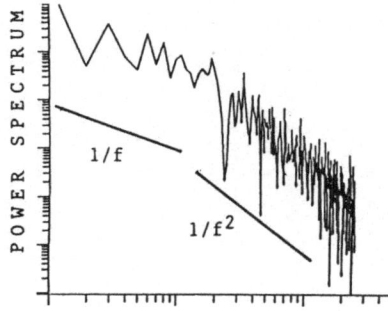

Fig. 5 Fourier analysis of the MC magnetization experiment.

4. Spin Flipping, Domain Movement and $1/f^n$ Fluctuation

We arrange $\pm$ Ising spins randomly on the whole lattice sites, and flip spins only when the energy of the system is lowered. Figure 3 shows the domain boundaries formed in this process. It has been found from SSA spin flipping experiments that (1) a bubble shaped domain is easily swept out, (2) a stripe shaped domain stands still, (3) domain boundaries coincide at one point. In the usual Monte Carlo experiment (see Fig. 4), the domain boundaries move with $1/f^n$ fluctuations (see Fig. 5). However, the rules of overall movement of domain boundaries of MC patterns are the same as those given above.

A New Approach to the Frustrated Two-Dimensional Ising Models

T. Chikyu[1] *and S. Miyashita*[2]

[1]Department of Physics, Faculty of Science, University of Tokyo,
  Hongo, Bunkyo-ku, Tokyo 113, Japan
[2]College of Liberal Arts, Kyoto University, Kyoto 606, Japan

In studying spin glass phenomena, it is important to extract mechanisms for the characteristic properties, e.g., the divergence to $-\infty$ of the non-linear susceptibility χ_2 at $T \to T_c - 0$ [1] and so on , which are different from those in the pure ferromagnetic or antiferromagnetic case. Here we consider two factors, that is, randomness and frustration.

We have an interest in how the behavior of non-linear susceptibility χ_2 is modified from the Mattis model [2] by the effects of frustration. In order to evaluate these effects, we study the following simple two-dimensional Ising model and analyze it using exact expressions for correlation functions. The Hamiltonian is

$$H = - \sum_{<i,j>} J_{ij} \varepsilon_i \varepsilon_j \sigma_i^z \sigma_j^z - h \sum_i \sigma_i^z, \tag{1}$$

where J_{ij}'s are interactions regularly distributed with a 3×3 unit cell as shown in Fig.1. The first sum is taken over nearest pairs and the second sum is taken over all spins. ε_i and ε_j take values 1 or -1 and the simultaneous transformations ε_i to $-\varepsilon_i$ and σ_i to $-\sigma_i$ correspond to the local gauge transformation in the model.

To begin with, we calculate exactly two-spin-correlation functions $< \sigma_i \sigma_j >$ for $\varepsilon_i = 1$ for all spins (a regularly frustrated model shown in Fig.1). Here we assign a coordinate to each spin as $\sigma_{(ij)}^{(k)}$, where (ij) is a coordinate of a unit cell and (k) is the one in the unit cell (see Fig.1). Noting that free energy and correlation functions in infinite systems can be expressed as some kind of determinant [3,4], we analyze the present model by calculating the exact expressions directly. We have found that the present model has a phase transition at $T_c \simeq 1.358J$. The spins except the spins $\sigma_{(ij)}^{(5)}$ exhibit symmetry breaking ,while $< \sigma_{(ij)}^{(5)} >= 0$ at all temperatures. In other

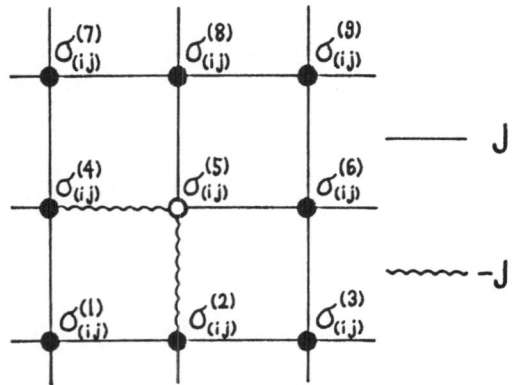

Fig.1 A 3×3 unit cell of our model.

Springer Series in Synergetics, Vol. 43   **Cooperative Dynamics in Complex Physical Systems**
Editor: H. Takayama   © Springer-Verlag Berlin, Heidelberg 1989

words, the spins except $\sigma_{(ij)}^{(5)}$ form the *skeleton* lattice and the spins $\sigma_{(ij)}^{(5)}$ are *floating* spins. Another point to stress is that the correlation functions $< \sigma_{(ij)}^{(k)} \sigma_{(ij)}^{(5)} >$ except $k = 3$ and 7 are very large near T_c, although they are zero at $T = 0$. These results suggest that the fluctuation of the floating spins $\sigma_{(ij)}^{(5)}$ has an important role for the fluctuation near T_c.

At the next stage, we calculate the susceptibility χ_0 and the non-linear susceptibility χ_2 of the present model by taking an average over $\varepsilon_i = \pm 1$ as is done in the Mattis model. χ_0 and χ_2 can be expressed using $< \sigma_i >$ and $< \sigma_i \sigma_j >$ of the regular system as follows:

$$\chi_0 = \frac{1}{N \cdot k_B T} \cdot \sum_i (1 - < \sigma_i >^2) \quad , \tag{2}$$

$$\chi_2 = \frac{1}{N \cdot (k_B T)^3} \cdot [\sum_i (-1 + 4 < \sigma_i >^2 - 3 < \sigma_i >^4)$$

$$+ \sum_{i \neq j} (12 < \sigma_i \sigma_j >< \sigma_i >< \sigma_j > -3 < \sigma_i \sigma_j >^2 -9 < \sigma_i >^2 < \sigma_j >^2)], \tag{3}$$

where the values $< \sigma_i >$ depend on the location in a unit cell. Using the correlation functions calculated previously, we obtained the following results.

χ_0 has a cusp at $T = T_c$ and χ_2 diverges to $+\infty$ at $T \to T_c - 0$ as in the Mattis model. It should be noted that the signs of χ_2 are minus at $T = 0.6J$ and $0.8J$. An indication of the divergence of χ_2 appears at $T \simeq 1.2J$ and the correlation length $\xi \simeq 1.6$ at that temperature, where the unit of the length is a distance between two nearest-neighbor spins. The results suggest that when the correlation length ξ is about a half period of the distribution of the frustrated plaquettes (the linear size of a unit cell of the present model is 3), each spin begins to feel that the whole systems has a regular property. On getting closer to T_c, the divergence of χ_2 shows behavior qualitively similar to the Mattis model.

Here we discuss the behavior of χ_2 near T_c more quantitively and compare it with the Mattis model. Assuming the scaling properties for the correlation functions near T_c, we can obtain the critical behavior of χ_2 near T_c in a way similar to the case in the Mattis model:

$$\chi_2 \simeq \chi_2^{(0)} \cdot t^{-1.5}, \qquad t = | (T_c - T)/T_c | \quad , \tag{4}$$

where $\chi_2^{(0)}$ is a constant depending on the model. Our data for χ_2 are well fitted in the form (4) and

$$\chi_2 \simeq 0.17 \cdot t^{-1.5}. \tag{5}$$

OGUCHI and ISHIKAWA have obtained $\chi_2^{(0)}$ for the Mattis model on the basis of the asymptotic form of the correlation functions as $\chi_2^{(0)} \simeq 24.5$ [5]. Though the quantity χ_2 of our model has the same sign and critical exponent near T_c as that of the Mattis model, the coefficient of singularity of χ_2 is very small compared with that of the Mattis model. We think that the small coefficient comes from frustration and large fluctuation of the floating spins.

We will report the results for other models with frustration in the near future.

The authors would like to thank Professor M. Suzuki for his encouragement. This study is partially financed by the Research Fund of the Ministry of Education, Science and Culture.

1. M. Suzuki: Prog. Theor. Phys. **58** (1977) 1151.
2. D. C. Mattis: Phys. Lett. **56A** (1976) 421.
3. E. W. Montroll, R. B. Potts and J. C. Ward: J. Math. Phys. **4** (1963) 308.
4. T. T. Wu: Phys. Rev. **149** (1966) 380.
5. T. Oguchi and T. Ishikawa: J.P.S.J. **50** (1981) 2180.

New Type of Phase Transition in an Antiferromagnetic Ising Model on a Stacked Triangular Lattice by the Interface Method

K. Mitsubo, Gang Sun, and Y. Ueno

Department of Physics, Tokyo Institute of Technology,
Oh-okayama, Meguro-ku, Tokyo 152, Japan

The antiferromagnetic (AF) Ising model on a stacked triangular lattice (a hexagonal loose packed one) has been of considerable interest as a model of compound $CsCoCl_3$ which exhibits extraordinary phase transitions, but not well elucidated yet[1]. In the triangular layers besides AF couplings between nearest neighbors (nn) $-J_1(<0)$, it has ferromagnetic (F) couplings J_2 between next nn, and F couplings J_0 between the layers. In this article we briefly give preliminary results and further study will be given elsewhere.

We recently developed the interface method with the use of Monte Carlo simulations which first enables us to fully make size-dependent analyses of the interface free energy ΔF[2]. This has been successfully applied to 3d AF Potts models and proved to have many advantages. We assume $\Delta F \sim L^{a(T)}$ for large L. This *stiffness exponent* has the following properties: $a<0$ for $T>T_c$, $a=0$ at $T=T_c$ and $a>0$ for $T<T_c$ only if the order is of *long range*. Figure 1 shows them in the 2d triangular antiferromagnet with $J_2/J_1=0.15$, where the region of $a=0$ clearly indicates the Kosterlitz-Thouless phase.

For the present model of $J_0/J_1=1$ and $J_2/J_1=0.1$ we calculate ΔF under the two kinds of boundary conditions to investigate the responses to some F and AF stresses. For the former we fix all the spins on the boundaries (which are the top and bottom layers) up or down and for the latter (of the sine type), we do them such that the spins on the A sublattice are up, those on the B sublattice are down and the remains are free. Figures 3 and 4 show the temperature dependences of ΔF with size $L=12,15,18,21,24$ on the F and AF

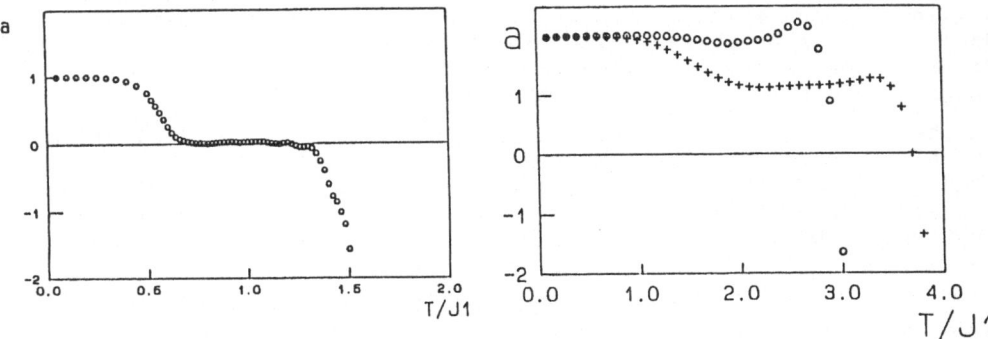

Fig.1(left) and Fig.2(right) Temperature dependences of the stiffness exponents in the 2d and 3d triangular frustrated Ising models, respectively.

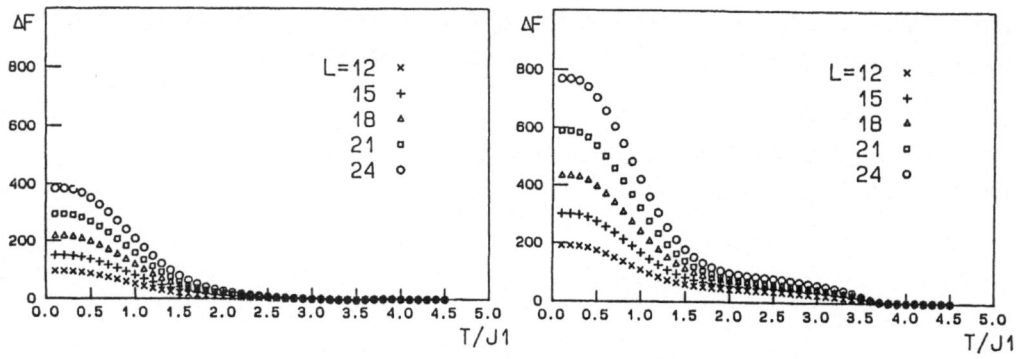

Fig.3(left) and Fig.4(right)   Temperature dependences of the
interface free energies of the 3d triangular Ising model of
sizes under F and AF boundary conditions, respectively.

boundary conditions, respectively. Figure 2 shows those of the
stiffness exponent both for the F(o) and AF(+) boundary conditions.
The former values near the critical point are less accurate because
ΔF shows very small values there(see Fig.3). These results lead to
the following three continuous phase transitions.

(1) The first transition of the AF components at $T_{AF1} \sim 3.7$ (in unit
of J_1). Since there are two AF components in the present model one
naturally considers that this transition is in the 3d XY universality
class, but both a $\sim 1.25(>1)$ and the divergent behavior of specific
heat calculated here deny it. Very similar behaviors to these are
seen in the 3d three-state AF Potts model[2], which suggests the
discreteness of spin variables still remaining.
(2) The transition of the F component at $T_F \sim 2.9$. This may be the
usual transition of the Ising universality class into the ordered
state with $a=2$, though we do not find any anomaly in specific heat
which is attributed to very weak contributions (as suggested from
Fig.3). The intermediate state is very remarkable: it is stiff to a
F stress while it is soft to an AF stress.
(3) The second transition of the AF components at $T_{AF2} \sim 2.1$. The
ordered state becomes stiff from soft in the AF nature as T is
lowered, which may be also seen in 3d F clock models. This is a new
type of phase transitions in the following senses. i) The stiffness
of the same component changes between two long-range ordered states.
ii) There is no symmetry change from the usual symmetry arguments.
However it can be characterized in the structure of phase space: the
distribution is completely anisotropic for the phase below T_{AF2} and
(perhaps almost) isotropic for the one above T_{AF2}.

    The above results together with those calculated on fluctuations
which are consistent with those of the previous Monte Carlo
studies[1] deny the existence of the partially-disordered phase[3]
suggested in experiments[4], which suggests that other interactions
must be allowed for, for example, those with XY character.

1. F. Matsubara: J. Phys. Soc. Jpn. 53, 4373 (1984) and references
therein.
2. Y. Ueno, G. Sun and I. Ono: submitting to J. Phys. Soc. Jpn.
3. M. Mekata: J. Phys. Soc. Jpn. 42, 76 (1977)
4. Y. Ajiro : private communication.

Spin Dynamics of an Ising-like $S = 1/2$ Antiferromagnet on Finite Chains and Triangular Lattices

F. Matsubara, S. Inawashiro, and H. Ohhara

Department of Applied Physics, Tohoku University, Sendai 980, Japan

We propose a method of studying dynamics of an Ising-like model, and apply it to an antiferromagnetic model on a linear chain and on a triangular lattice. In the former, dynamics of solitons are discussed and, in the latter, stability of a three sublattice spin structure occurring in the classical case is examined.

We briefly mention the method used here. We start with the model of S = 1/2 described by

$$H = 2J \sum_{\langle ij \rangle} (S_i^z S_j^z + \varepsilon (S_i^x S_j^x + S_i^y S_j^y)) \tag{1}$$

where the sum runs over all nearest neighbor pairs and $\varepsilon \ll 1$. The space-time correlation function at a temperature T can be approximately calculated by using a Monte Carlo method of random sampling of states[1,2]. That is

$$\langle S_i^\eta S_j^\eta (t) \rangle = \sum_{k=1}^{M} \langle k | S_i^\eta e^{iHt} S_j^\eta e^{-iHt} e^{-\beta H} | k \rangle / \sum \langle k | e^{-\beta H} | k \rangle \tag{2}$$

where $|k\rangle = \sum_{l=1}^{2^N} C_l^k |l\rangle$ with $\{ |l\rangle \}$ being a complete set of Ising states and N is the number of spins. The coefficient C_l^k is a random number uniformly distributed between -1 and 1. Here the unit $\hbar = 1$ is used, and $\beta = 1/kT$ and $\eta = x$, y or z. A procedure of calculating (2) is as follows. (i) We take a state $|k\rangle$ and operate $\exp(-\beta H/2)$ to it yielding a state at 2T, $|\tilde{k}\rangle$. (ii) Once $|\tilde{k}\rangle$ is given, we can successively obtain time-sequences of two states $|\tilde{k}(n)\rangle$ and $\langle \tilde{kS}(n)|$, i.e., $|\tilde{k}(n+1)\rangle = \exp(-iH\Delta t)|\tilde{k}(n)\rangle$ with $|\tilde{k}(0)\rangle = |\tilde{k}\rangle$ and $\langle \tilde{kS}(n+1)| = \langle \tilde{kS}(n)|\exp(iH\Delta t)$ with $\langle \tilde{kS}(0)| = \langle \tilde{k}|S_i^\eta$. The time-sequences of the correlation functions, for the initial $|k\rangle$, are calculated by $\langle \tilde{kS}(n)|S_j^\eta|\tilde{k}(n)\rangle$. (iii) Repeating this M times starting with different initial state $|k\rangle$, we can obtain the time-sequenses of the averaged correlation functions of different two spins. The dynamical structure factor $S_{\eta\eta}(q,\omega)$ is a Fourier transform of the correlation functions. Since the model treated here is of Ising-like and $|\tilde{k}(n)\rangle$ is also expressed in terms of a linear combination of the Ising states, we can use an expansion method based on the Ising states for calculating $\exp(-iH\Delta t)|\tilde{k}(n)\rangle$. The approximation (2) together with this expansion reduces much both CPU time and memory size of computer.

We firstly apply the method to an Ising-like antiferromagnet on a linear chain whose dynamics is governed by propagation of domain-walls (or solitons)[3]. Theoretical studies have predicted that a steep double maximum at $\omega = \pm \Omega_q$ ($= 4\varepsilon J |\sin(q)|$) results in the structure factor $S_{zz}(q,\omega)$, whereas experimental observations of it on $CeCoCl_3$ and $CeCoBr_3$ have revealed the occurrence of shoulders (or very broad maxima) around $\pm \Omega_q$ [4,5]. Our purpose in studying the

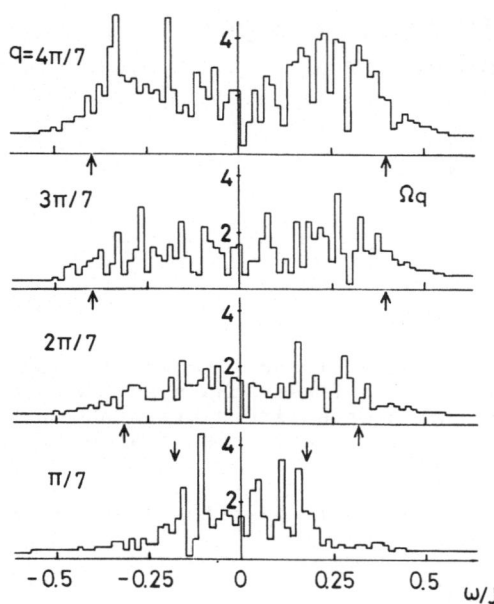

Fig.1 $S_{zz}(q,\omega)$ of the model of $\varepsilon = 0.1$ on the linear chain with N = 14 spins at kT/J = 0.5. Here we use M = 5, $J\Delta t = 0.2$ and $Jt_{max} = 400$.

model is to examine whether the experimentally observed profile of $S_{zz}(q,\omega)$ is actually found in our model or not. A typical result is presented in Fig. 1. We see that $S_{zz}(q,\omega)$ has finite values for $|\omega| \lesssim \Omega_q$ and exhibits a very broad double peak at the edges. These results are in agreement with the experimental observations. To make clear the mechanism responsible for the broadening of the peak, we make the following calculations. (i) Time evolution of a state describing a single-soliton by the use of our method. It is shown that the state is stable although a considerable mixing between the single-soliton state and higher excited states occurs. (ii) Diagonalization technique of the matrix in a subspace E_2 consisting of all Ising states with lowest and next lowest excitation energies. $S_{zz}(q,\omega)$ calculated in terms of a set of eigenstates with lower energies exhibits a clear double peak with their centers inside the edges, whereas that calculated in terms of the other eigen states exhibits very broad maxima. The contribution of the latter to total $S_{zz}(q,\omega)$ is shown to be not small at temperatures of interest. We conclude that the broadening of the steep peak mainly owes to multi-soliton states rather than relaxation of single-soliton states.

Next we apply the method to an Ising-like antiferromagnet on the triangular lattice. We examine the following problems: (i) Whether a three sublattice spin structure occurring in a classical model is stable against quantum fluctuation or not. (ii) Whether a special state called as 'RVB state [6]' really occurs in the model or not. We calculate correlations of sublattice magnetizations as well as pair correlations of spins and find a strong fluctuation of the sublattice magnetization revealing instability of the sublattice structure. We also investigate microscopically a change in the spin structure and find that it is characterized by a change of clusters with the sublattice structure rather than drift of "singlet pairs".

References

1. M. Imada and M. Takahashi: J. Phys. Soc. Jpn. 55, 3354 (1986)
2. F. Matsubara and S.Inawashiro: Solid State Commun. 67, 229 (1988)

3. J. Villan: Physica 79B, 1 (1975)
4. S. E. Nagler, W. J. L. Buyers, R. L. Armstrong and B. Briat: Phys. Rev. 28, 3873 (1983)
5. J.P.Boucher,L.P.Regnault,J.Rossat-Mignod, Y.Henry, J.Bouillot and W.G.Stirling: Phys. Rev. 31, 3015 (1985)
6. P. W. Anderson: Mater. Res. Bull. 8 153 (1973)

Higher-Order Commensurate, Incommensurate and Liquid Phases of an Atomic Monolayer System on a Periodic Substrate

H. Mitani

Institute for Solid State Physics, University of Tokyo, Roppongi,
Minato-ku, Tokyo 106, Japan

1. THE MODEL

We study a grand canonical system of a two-dimensional(2D) competitive model where a monolayer of atoms is located on a graphite-type substrate potential. The Hamiltonian is as follows:

$$H = \frac{2A}{21} \sum_{<i,j>} \frac{1}{|\vec{x}_i - \vec{x}_j|^5}$$

$$+ \frac{E_s}{6} \sum_{i}^{M} (3 - \cos(\vec{g}_1 \cdot \vec{x}_i) - \cos(\vec{g}_2 \cdot \vec{x}_i) - \cos(\vec{g}_3 \cdot \vec{x}_i)) , \tag{1}$$

where $\vec{x}_i$ ($i=1,\cdots,M$) is the 2D location of the atoms and $\vec{g}_j$ ($j=1,2,3$) are the reciprocal vectors of the substrate lattice. The ratio of the substrate to the interatomic interaction will be represented by a factor $\lambda_0 = 2E_s a^5/5A$, where a is the substrate lattice parameter.

The purpose of this study is to obtain the phase diagram in the μ-T-λ_0 (chemical potential-temperature-substrate strength) space by distinguishing between higher-order commensurate(C) phases and an incommensurate(IC) phase. The competitive spin systems on which a lot of work has been done are essentially the same as the limiting case, $\lambda_0 = \infty$, of the present system.

The study is done in two steps; first for the ground state phase diagram (on μ-λ_0), and then for the finite temperature phase diagram (mainly on μ-T).

2. GROUND STATE PHASE DIAGRAM[1]

The essential point of this procedure is to obtain the minimum value of the grand canonical "Hamiltonian" H - μN. We ignore the possibility of a stripe discommensuration. Then the average locations of the atoms form a triangular lattice, which can be indicated by the distance (b) of the nearest neighbor (n.n.) atoms of the lattice and the rotation angle (ϕ) between the lattice and the host lattice.

We choose those C-structures which contain less than 57 atoms (see Fig.1). From calculating the energy $E(b_c,\phi_c)$ for every C-structure, it is found that $E(b,\phi)$ has a downward cusp at every (b_c,ϕ_c). Thus a C-structure (b_c,ϕ_c) has the minimum value of $E(b_c,\phi_c) - \mu N$ for a given chemical potential. The structures with the minimum grand potential are represented by filled circles. The system is locked in each of the C-structures, which condense on the line shown in Fig.1. This line has been previously derived from an experimental rule and a Ginzburg-Landau-type theory[2]. We can therefore conclude that the

Springer Series in Synergetics, Vol. 43   **Cooperative Dynamics in Complex Physical Systems**
Editor: H. Takayama   © Springer-Verlag Berlin, Heidelberg 1989

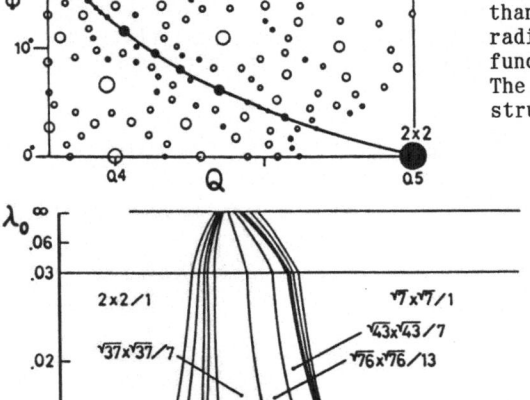

Fig.1 The distribution of C-structures in the $Q(\equiv a/b)$-ϕ plane. The circles represent C-structures which contain less than 57 atoms in their unit cell. The radii are chosen to be a decreasing function of the number of atoms.
The filled circles represent the C-structures assumed by the system.

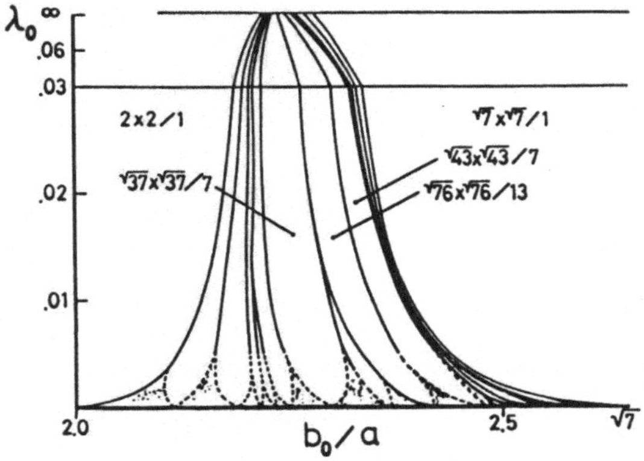

Fig.2 The phase diagram of the ground state in the the b_0(or μ) - λ_0 plane.

$\sqrt{n}$ x $\sqrt{n/m}$ represents a C-structure whose unit cell contains n graphite hexagons and m atoms.

transition is devil's staircase-like. In Fig.2, the phase diagram is shown in the b_0(or μ)-λ_0 plane where b_0 is the natural distance between n.n. atoms when the substrate is absent: $b_0 = (A/\mu)^{1/5}$.

3. FINITE TEMPERATURE PHASE DIAGRAM[3]

Here we use the grand-canonical Monte-Carlo technique. We take a rhombic system with a periodic boundary condition, where the size of the system is $\sqrt{148}$x$\sqrt{148}$ on the unit of the graphite lattice. This area is compatible with the 2x2/1 and $\sqrt{37}$x$\sqrt{37}$/7 structures that are forming.
   In Fig.3, the phase diagram is shown in the μ-T plane. It is seen that 2x2/1 and $\sqrt{37}$x$\sqrt{37}$/7-locked regions occur. Moreover, in spite of the existence of one excess graphite hexagon area (1% mis-match), the $\sqrt{7}$x$\sqrt{7}$/1 structure also appears. These C-regions reach the liquid region out of the IC region. We can expect that other C-phases of higher-order melt inside the IC region because

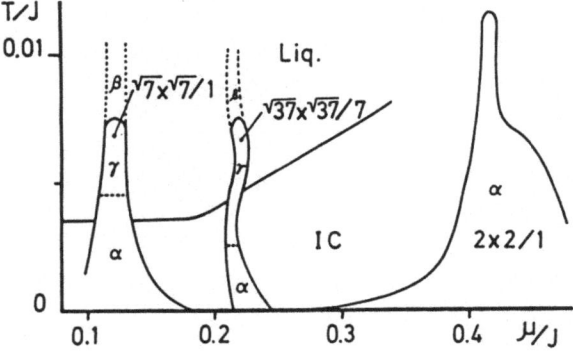

Fig.3 The phase diagram in the μ - T plane for λ_0= 0.0685, where J=2A/21a^5.

For the phases α,β,γ, see the text. The results at T=0 are taken from the previous chapter (see Fig.2).

of their commensurability energies being lower than the liq.-IC phase transition temperature.

The C-region has two phases denoted by α and γ in Fig.3. In the α-phase, all the atoms are in the same hexagons as for the ground state, but in the γ-phase, some fluctuations are present. The regions denoted by β belong essentially to the liquid phase, but C-states appear in the fluctuations. These regions should be considered to be a finite size effect.

1) H.Mitani and K.Niizeki, J.Phys.C21(1988)1895-1903
2) M.Mori, S.C.Moss, Y.M.Yan and H.Zabel, Phys.Rev.B25(1982)1287-1296;
   Y.Yamada and I.Naiki, J.Phys.Soc.Jpn.51(1982)2174-2180
3) H.Mitani, to be submitted to J.Phys.C

Critical Phenomena
of a Quantum Heisenberg-like Antiferromagnet

M. Matsuura[1], *H. Kageyama*[1], *and K. Koyama*[2]

[1]Faculty of Engineering Science, Osaka University, Toyonaka 560, Japan
[2]Department of Electrical Engineering and Electronics,
  Toyohashi University of Technology, Toyohashi 440, Japan

Ordering of two-dimensional Heisenberg (2DH) spin systems has been a most interesting problem from the viewpoint of cooperativity. Any phase transition associated with long range order was predicted not to occur down to 0 K[1]. So the phase
transition with spontaneous magnetization found in the actual 2DH-like magnets examined so far has been attributed to a non-negligible interplane interaction and/or
Ising(I)-type anisotropy. Apart from the ideal behaviour, the phase transition of
2DH-like antiferromagnets, especially of S=1/2 if any, is interesting because the
critical phenomena could be modified by large quantum fluctuations[2] and a new aspect of cooperativity may be found.

Among these, layer structure compounds $Cu(HCOO)_2 \cdot 4H_2O$ (CuF4H) and $Cu(HCOO)_2 \cdot 2H_2O$
$\cdot 2CO(NH_2)_2$ (CuFUH) are attractive candidates because the expected quantum fluctuations have been suggested from the observed large spin contraction[3,4] of ~50 %.
The transition temperature T_N of these compounds is roughly half the intraplane exchange interaction. The associated entropy change at T_N is extraordinarily small
in spite of the high T_N[5]. As shown in Fig.1, the negligible heat capacity anomaly makes a remarkable contrast to the divergent susceptibility[5]. A new type of
phase transition has thus been suspected to occur in these compounds.

A detailed examination of spontaneous staggered moment L_0 in the deuterated compounds CuF4D and CuFUD has revealed a critical exponent $\beta=0.22$[6]. From comparison
with other 2DH-like magnets and from the difference between CuF4D and CuFUD, this
novel exponent, quite different from values for both 2DI and 3D universality classes, was concluded to be an essential value which characterizes the phase transition
of a 2DH-like antiferromagnet with an extraordinarily weak I-type anisotropy.

For further information, the field-induced magnetization has recently been examined in the neighbourhood of T_N[7]. Owing to the staggered field effect, which is
expected for a crystallographic two-sublattice (C_2) system[8], the critical isotherm of field-induced staggered moment L can be derived in this case by analyzing

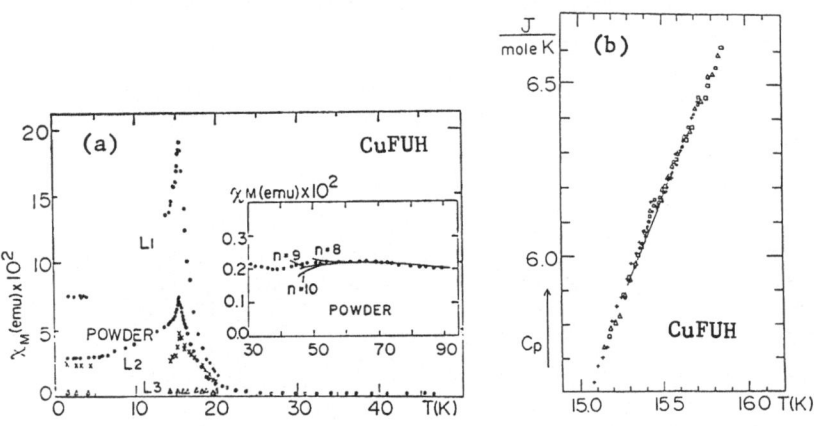

Fig. 1. Temperature dependence of (a) susceptibility and (b) heat capacity near
T_N [5]

NMR line shifts(from the value for free nucleus). The present situation, which is different from a simple antiferromagnet, is briefly as follows. The NMR line shift is generally determined by both uniform and staggered moments through the hyperfine interaction with magnetic ions[8](dipole-dipole interaction with the neighbouring Cu ions in the present nucleus, proton in the formate bond). Since the staggered field is the conjugate field of the order parameter of the antiferromagnet, the field-induced staggered moment is much larger than the uniform one in the neighbourhood of T_N, as long as the external field intensity is much smaller than the exchange field. As a result, the NMR line shift is essentially determined by the staggered moment and the contribution of the uniform one is negligible. It is experimentally confirmed for CuF4D[9], and the situation is quite the same for CuFUD. Since the effective staggered field intensity H_s is proportional to the external one H[8], the external field dependence of the NMR line shift ΔH_0 or ΔH_0-H curve directly shows the L-H_s relationship, except for the proportionality constant.

Figure 2a is the obtained critical isotherm plotted on a logarithmic scale. The derived critical exponent δ is 5.8. A deviation in the low field region is explained as due to the fact that the direction of effective staggered field (crystallographic b axis in this case) is different from the easy direction (in the ac plane)[7]. As in the case of β, again it is quite different from the values for both 2DI and 3D universality classes and numerically similar to the theoretical value predicted for the SO_3 symmetry group of order parameters[10]. However, if the present exponents are attributed to the SO_3 group, the crossover phenomenon in the L_0-$\varepsilon(=1-T/T_N)$ curve of CuF4D cannot be well explained[6]. Unlike the case of β, unfortunately, the obtained result cannot be compared with those for other 2DH-like antiferromagnets, because this is the first successful derivation of δ.

Now, if we try to apply the scaling law, we get another critical exponent for staggered susceptibility as $\gamma=1.1$. The divergent part of susceptibility in Fig.1a can be replotted on a logarithmic scale as shown in Fig.2b. The exponent γ in the temperature range $\varepsilon>0.1$ is nearly equal to 1.75 for the 2DI universality class, while for $\varepsilon<0.1$ an apparent rounding makes it difficult to determine the exponent. In any case, further experimental investigation is necessary for the full understanding of the present novel critical phenomena.

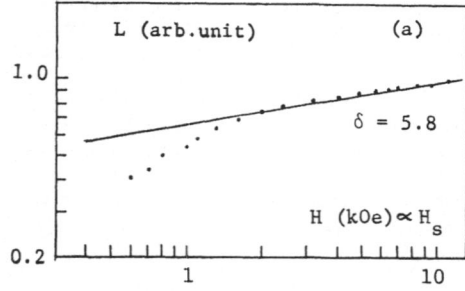 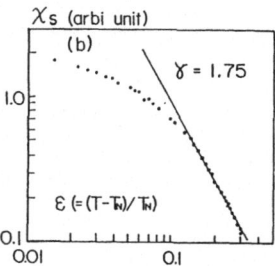

Fig. 2. (a) Critical isotherm of staggered moment for CuFUD, (b) a logarithmic plot of susceptibility near T_N of CuFUH

References

1. N.D. Mermin and H. Wagner: Phys. Rev. Lett. 17 1133 (1966).
2. J. Frölich and H.E. Lieb: Phys. Rev. Lett. 38 440 (1977).
3. A. Dupas and J.P. Renard: Phys. Lett. 33A 470 (1970).
4. K. Koyama: Thesis, Faculty of Engineering Sci. of Osaka Univ. (1984).
5. Y. Yamamoto, M. Matsuura and T. Haseda: J. Phys. Soc. Jpn. 40 1300 (1976).
6. K. Koyama and M. Matsuura: J. Phys. Soc. Jpn. 54 4085 (1985).
   K. Koyama, H. Nobumasa and M. Matsuura: J. Phys. Soc. Jpn. 56 1553 (1987).
7. M. Matsuura and K. Koyama: J. Phys. Soc. Jpn. to be published.
8. M. Matsuura and Y. Ajiro: J. Phys. Soc. Jpn. 41 44 (1976).
9. Y. Ajiro, K. Enomoto, N. Terata and M. Matsuura: Solid State Commun. 20 1151 (1976).
10. K. Kawamura: J. Phys. Soc. Jpn. 54 3220 (1985).

Ground-State Long-Range Order
of the Two-Dimensional Quantum Antiferromagnet

Y. Ozeki, H. Nishimori, and Y. Tomita

Department of Physics, Tokyo Institute of Technology,
Oh-okayama, Meguro-ku, Tokyo 152, Japan

The two-dimensional quantum antiferromagnet is of active current interest. One of the main unsettled problems is whether magnetic long-range order exists or not in the ground state of the quantum antiferromagnet on the square lattice. In the present contribution we treat the *XYZ* model,

$$H = \sum_{\alpha \in \Lambda} \sum_{\delta} (J_x S_\alpha^x S_{\alpha+\delta}^x + J_y S_\alpha^y S_{\alpha+\delta}^y + J_z S_\alpha^z S_{\alpha+\delta}^z) , \qquad (J_x, J_y, J_z \geq 0) \qquad (1)$$

and derive the parameter region in which long-range order certainly exists. We also show that long-range order along the non-principal axis (e.g. *z*-axis in the *XY*-like region) exists in the *XXZ* model ($J_x = J_y = 1$, $J_z = \Delta$) under certain conditions.

The method we use is originally due to DYSON, LIEB and SIMON[1]. The DLS theory has been applied to the ground state of two-dimensional system by many authors[2-8]. From these references, the sufficient condition for the existence of Néel-type long-range order in the direction i (=x or y or z) is written as

$$-\langle S_\alpha^i S_{\alpha+\delta}^i \rangle > \int^{(+)} \frac{d^\nu p}{2\nu(2\pi)^\nu} (B_p^{(i)} C_p^{(i)})^{1/2} (-\sum_{m=1}^\nu \cos p_m) , \qquad (2)$$

where (+) means that the integral is restricted to the region where the integrand is positive. ν is the dimensionality (=2), p is the wave number, and

$$B_p^{(i)} = \frac{1}{2J_i \sum(1+\cos p_m)} , \qquad C_p^{(i)} = \langle [S_p^i, [H, S_{-p}^i]] \rangle . \qquad (3)$$

Let us first discuss the parameter region $J_x \geq J_y \geq J_z$, in which the appropriate long-range order is that of the spin component $i=x$. After some calculations, the condition (2) is reduced to

$$-2\langle xx \rangle \sqrt{J_x} > \sqrt{-J_y \langle yy \rangle - J_z \langle zz \rangle} \; \Gamma_\nu , \qquad (4)$$

where $\langle jj \rangle$ (j=x, y, z) is an abbreviation of $\langle S_\alpha^j S_{\alpha+\delta}^j \rangle$ and

$$\Gamma_\nu = \int^{(+)} \frac{d^\nu p}{(2\pi)^\nu} \sqrt{\frac{\nu - \sum \cos p_m}{\nu + \sum \cos p_m}} (-\frac{1}{\nu} \sum_m \cos p_m) . \qquad (5)$$

Further simplification comes from the inequality $-\langle xx \rangle \geq -\langle yy \rangle \geq -\langle zz \rangle \geq 0$. We estimate a lower bound of $-\langle xx \rangle$ by applying the variational principle to the Hamiltonian (1) with the Néel-state wave function in the x-direction:

$$J_x \langle xx \rangle + J_y \langle yy \rangle + J_z \langle zz \rangle \leq -S^2 J_x. \tag{6}$$

Consequently, we derive the final condition for the existence of long-range order as

$$2SJ_x > \Gamma_\nu \sqrt{(J_x + J_y + J_z)(J_y + J_z)} \; . \tag{7}$$

In two dimensions, Γ_2 is 0.646, which leads to the existence of long-range order if $S \geq 1$ for any value of J_x, J_y and J_z in the present parameter region. In the case of $S = 1/2$ in two dimensions, the inequality (7) is satisfied only if $0 \leq (J_y + J_z)/J_x < 1.13$. The sufficient condition in the other parameter regions (such as $J_y \geq J_x \geq J_z$) can be obtained simply by permuting x, y and z in (7).

If we restrict ourselves to the *XXZ* model, the result of KUBO and KISHI[7] (who proved the existence of long-range order for $\Delta < 0.13$ and $\Delta > 1.78$) is reproduced. The spin-1/2 isotropic model $\Delta = 1$ escapes the best estimate of the ground-state energy as suggested by KENNEDY *et al.*[6]. A different approach is required to settle this case. We have also proved that *XXZ* model on the two-dimensional hexagonal lattice has ground-state long-range order if $S \geq 1$, or if $S = 1/2$ and $\Delta > 2.55$.

A natural question arises: what happens to long-range order of the component for which J_i has not the largest value in the Hamiltonian (1)? We investigate this problem, restricting ourselves to the *XXZ* model for simplicity. If the system is *XY*-like ($0 \leq \Delta \leq 1$), the long-range order of our interest is related to $\langle S_p^z S_p^z \rangle$. Using (4), we find that long-range order in the z-direction exists if

$$-16 \langle xx \rangle < \Delta \Gamma_\nu^2 + 8S^2 - \Gamma_\nu \sqrt{\Delta(\Delta \Gamma_\nu^2 + 16S^2)} \; . \tag{8}$$

If $\Delta = 1$, since the left hand side is bounded from above by $16S(S + 1/2\nu)/3$, (8) is satisfied if $S \geq 3/2$ when $\nu = 2$. Therefore, if the left hand side is continuous in the neighborhood of $\Delta = 1$ in the thermodynamic limit, (8) is satisfied by sufficiently large S in a range $\Delta_c(S) < \Delta$ with $\Delta_c(S) < 1$. A similar argument can be given on the long-range order in the x-direction in the Ising-like region ($\Delta \geq 1$).

REFERENCES
1. F.J. Dyson, E.H. Lieb, B. Simon: J. Stat. Phys. 18, 335 (1978)
2. E.J. Neves, J.F. Perez: Phys. Lett. 114A, 331 (1986)
3. I. Affleck, T. Kennedy, E.H. Lieb, H. Tasaki: Commun. Math. Phys. 115, 477 (1988)
4. K. Kubo: Phys. Rev. Lett. 61, 110 (1988)
5. H. Nishimori, K. Kubo, Y. Ozeki, Y. Tomita, T. Kishi: preprint
6. T. Kennedy, E.H. Lieb, B.S. Shastry: preprint
7. K. Kubo, T. Kishi: preprint
8. Y. Ozeki, H. Nishimori, Y. Tomita: preprint

Random Interfaces and the Physics of Microemulsions

S.A. Safran

Exxon Research and Engineering, Corporate Research, Annandale, NJ 08801, USA

1. Introduction

The structures of mixtures of amphiphiles, water and oil can sometimes show only microscopic correlations between the components (e.g. a three component solution) and sometimes exhibit long-range ordered structures (e.g. lyotropic liquid crystals). The term microemulsion in its most general use, connotes a thermodynamically stable, fluid, oil-water-surfactant mixture [1]. In practice, microemulsions are taken to consist of structures with intermediate-range correlations. The oil and water regions are fairly well separated and the surfactant molecules are organized as monolayers at the internal water-oil interfaces. There are long-range correlations between the oil and water molecules in that they are separated on length scales of the order of hundreds of Angstroms. In addition, there are long range correlations among the surfactant molecules which self-assemble into a monolayer film at the set of internal water-oil interfaces. In this respect, microemulsions are different from three-component solutions. However, the set of interfaces which comprise the microemulsion do not show long range order comparable to that found in lyotropic liquid crystals, where there exists a periodic array of surfactant bilayers separating adjacent water or oil regions.

It is precisely for these reasons that the understanding of microemulsions is both interesting and difficult: interesting, because the structure can be idealized as a set of interfaces, and difficult, because there is no long-range order of these interfaces. In this summary article, the structures and phase behavior of microemulsions are discussed from the interfacial point of view. The details of the approach and of the results can be found in the references [2, 3, 4, 5, 6]. Although many workers have implicitly used this "phenomenological" approach to analyze globular (mostly spherical) microemulsions [7, 8, 9], the application to random, or bicontinuous systems was initiated by Talmon and Prager [10] and further developed by de Gennes and co-workers [11, 12], and Widom [13]. Most recently, this approach was extended to include the thermal fluctuations of the surfactant film to model both the thermodynamics and spatial correlations of bicontinuous microemulsions [2, 3, 4,

6]. One regards the oil and water as continuum liquids; the interfacial surfactant layer is treated as a flexible sheet. For systems where the surfactants pack in a condensed liquid state, the dominant energy is the bending or curvature energy of the monolayer [14]. While this energy is minimized by a droplet or domain with a given curvature (determined by the molecular details of the surfactant packing at the interface), the entropy tends to randomize the structure. It is this competition that gives rise to the rich phase behavior and structures that are observed. The model predicts phase behavior, scattering structure factors, and microstructure that are consistent with experiment.

An alternative approach to the interfacial model is based on the construction of microscopic lattice models in which a cell contains only a small number of molecules [16, 17, 18]. This point of view which focuses on the microscopic interactions between the water, oil and surfactant molecules is well suited to describe three-component solutions and their relation to microemulsions. The microscopic approach, though of undoubted fundamental interest, may be more difficult to implement than the phenomenological one. In particular, a microscopic model of microemulsions must produce structural organization on a length scale much larger than that of the molecular (or lattice) size. In contrast, the interfacial point of view presented here is tailored to focus on the microstructure and its relation to the phase behavior of microemulsions. This is because the length scales of interest for the problems of microstructure and scattering are in the range of tens to hundreds of Angstroms for good microemulsions [15]. These length scales are best treated by the continuum theory presented here; microscopic theories such as the lattice models referenced above, would have to include correlations between thousands of molecules to accurately model the interfaces at these length scales. In contrast, the interfacial models presented here assume that the strong correlations needed to self-assemble the surfactant molecules at the internal water-oil interfaces are always present.

2. Random Microemulsions vs. Lamellar Structures

In many cases, real microemulsion systems show lamellar, long-range order at high surfactant volume fractions where the structure consists of monolayers of surfactant separating oil and water regions which are arranged in a one-dimensionally ordered array [19]. However, at small values of the surfactant volume fraction, ϕ_s, there is often a single phase which is isotropic and obviously non-lamellar. For such systems (which are referred to as balanced, with no spontaneous curvature of the surfactant monolayer towards either the water or oil), NMR and conductivity measurements indicate that for phases with comparable volumes of oil and water, both the oil and the water diffuse with essentially their bulk diffusivities [1]. These structures were termed *bicontinuous* by Scriven [20] who presented examples of periodic, ordered, bicontinuous structures which may describe some systems at larger values of ϕ_s. In this review, the term random is used to describe microemulsions which are isotropic, non-periodic, and which do not have a particular globular component; they may or may not be bicontinuous, depending on the volume fractions of water and oil.

De Gennes and Taupin [11] suggested that these random microemulsions are related to the ordered, lamellar phases which can occur at larger values of ϕ_s. Although the bicontinuous systems have a random structure, their characteristic length scale is related to the persistence length of an infinite surfactant monolayer. The lamellar persistence length, ξ_K, is the typical distance over which the normal to the film is decorrelated by thermal fluctuations. For surfactant layers which can be modeled as *incompressible*, two-dimensional liquids, the restoring force to random fluctuations of the monolayer is provided by the curvature or bending energy. For the case of balanced films with no spontaneous curvature towards either the water or the oil, this energy is proportional to the integral of the square of the mean curvature. The proportionality constant is the bending modulus, K. The two-dimensional nature of the monolayer results in a decorrelation of the normal to the layer in a distance $\xi_K \sim \exp(4\pi K/aT)$, where a is a parameter that is discussed in Refs. [21, 22]. At length scales shorter than ξ_K, the monolayer is essentially flat, while at longer length scales, the layer randomly fluctuates in space [23]. These fluctuations are limited by the steric interaction of the layer - both with itself and with other fluctuating monolayers in the system [24, 25].

Jouffroy et al. attempted to derive the phase diagram of microemulsions by assuming that the microemulsions were always characterized by the length ξ_K. However, only two-phase equilibria and not the characteristic three-phase coexistence of microemulsion, water, and oil were predicted. In a series of papers by the Exxon group [2, 3, 4], it was shown that the observed phase behavior can be obtained by including the effect of thermal fluctuations in the free energy, rather than by postulating that the length scale of the microemulsion - for all values of the concentrations - is the persistence length.
By allowing the length scale of the random microemulsion to be determined by the conservation of surfactant, an idea introduced by Talmon and Prager [10] (who did not include the bending energy in the expression for the free energy) and extended by Widom [13] (who did not include the effects of thermal fluctuations on the free energy), Refs. [2, 3, 4] were able to derive phase diagrams which showed the regions of stability of the lamellar and bicontinuous phases as well as the multi-phase equilibria. The length scale over which the single phase,

bicontinuous, microemulsion is stable is of order the persistence length, but not identical to it. The connection of the microemulsion length scale to the persistence length arises by considering the effects of thermal fluctuations on the bending constant of the layer. Following the work of Helfrich [21] and others [22], a length-scale dependent bending modulus is used to estimate the free energy cost of bends in the random structure. Bends which occur on length scales comparable to the molecular scale cost the bare (molecular) bending energy, proportional to K_0, while bends which occur on larger length scales take advantage of the spontaneous undulations of the layer induced by thermal fluctuations. The free energy cost of these bends is reduced from K_0 in an approximate manner by allowing the bending modulus to vary with length scale. This reduction of the bare bending modulus by thermal fluctuations allows the microemulsion free energy to be lower than that of a phase with long-range lamellar order.

The theory predicts a phase diagram in qualitative agreement with experiment. Under the assumption that all of the surfactant is packed incompressibly at the water-oil interface, the length scale, ζ, of a random microemulsion is related to the volume fractions of water (ϕ_w), oil (ϕ_0), and surfactant (ϕ_s) by the relation $\zeta \sim \phi_0 \phi_w / \phi_s$. At high surfactant volume fractions, where $\zeta << \zeta_K$, the lamellar, ordered phase is stable. At lower surfactant volume fractions where $\zeta \sim \zeta_K$, the bending modulus is sufficiently reduced by thermal fluctuations that the microemulsion free energy is lower than that of the lamellar phase. At very small values of ϕ_s, however, the single-phase microemulsion is unstable to phase separation; the microemulsion coexists with phases that are nearly all water or oil. For equal volume fractions of water and oil, the smallest value of the surfactant volume fraction at which the single-phase, random, microemulsion is stable is denoted as ϕ_m and is known as the "middle-phase" microemulsion.

The model has been used to predict the generic phase behavior [2, 3, 4], scattering cross-section [6], and microstructure [6] of microemulsions. For non-ionic surfactants, the predicted behavior is in semi-quantitative agreement with experimental results [4]. For short-chain surfactants (corresponding to a small value of the bare bending modulus, K_0), the value of the surfactant volume fraction in the middle phase, ϕ_m, is large, but there is no lamellar phase, even for high surfactant concentration. On the contrary, for longer surfactant molecules (corresponding to a higher value of the bare bending modulus, K_0), the value of ϕ_m is much smaller and the lamellar phase boundary is nearer to the microemulsion phase. The observed [19] linearity of these phase boundaries with the $\log(\phi_s)$ is in agreement with the predictions that ϕ_m is inversely proportional to the persistence length - $\phi_m \sim \exp(-4\pi K_0 / \alpha T)$ - and suggests that the model correctly describes the competition between the lamellar and microemulsion phases [4].

64

3. Unresolved Issues

In this section, some of the outstanding problems relating to the interfacial description of microemulsions, the bending free energy, and the structures are discussed. These topics are currently under study or are areas for future investigations.

While the interfacial description presented here is well-suited to describe microemulsions where the characteristic length scale is much larger than a molecular size, it is not designed to describe systems which are closer to surfactant-water-oil solutions. On the other hand, the microscopic lattice models are simple to deal with if correlations only involve several molecules, but are not optimized to describe long-wavelength undulations and structures. Experiments indicate, however, that there is a continuum of behavior which spans the small scale to large scale regime. Thus, a problem that is worthy of further study is the construction of a bridge between the continuum theory outlined here and the more microscopic approach. Such a bridge will be able to treat micelles, microemulsions, vesicles, and ordered phases, on the same footing.

It should be noted that the continuum model presented here has not completely analyzed the implications of the bending energy on the phase diagram and structure. Most notably, the role of the saddle-splay [14,9] in stabilizing random structures with zero mean curvature has not been discussed. Huse and Leibler [26] have discussed the stability of such a state which they term the "plumbers's nightmare"; a detailed treatment of its structure, phase behavior as a function of concentration, and the effects of the spontaneous curvature deserve further inquiry.

Another such area is the dependence of the phase diagram for a random microemulsion on the chemical potential of the surfactant in the dilute, mostly water and mostly oil phases [3]. The deep reason for this sensitivity is not apparent. Another puzzle is the fact that the microemulsion model presented here only yields the three-phase equilibrium for a limited range of values of the renormalization parameter a, defined above. For values of a that are larger than $3/2$, there is no free energy minima on the boundaries of the phase diagram (i.e. at values of $\phi = \phi_s/2$, $(1-\phi) = \phi_s/2$) and there is only a two-phase equilibrium between two microemulsions. In Ref. [3], this problem was overcome by noting that the microemulsion free energy does not quantitatively account for the energy of the micelles present in the nearly all water and nearly all oil phases. An additional parameter, f*, which represents the free energy of the micellar state, was introduced. For large enough values of |f*|, the free energy minima at small values of ϕ and $(1-\phi)$ are restored. The phase diagram then also exhibits the three-phase equilibrium, in agreement with experiment.

The model of random microemulsions reviewed above predicts that in the limit of small oil or water volume fractions, the structure is that of nearly isolated, dilute domains of oil in water or water in oil, respectively - even for balanced systems where there is no energetic tendency for the surfactant film to bend towards either the water or the oil. However, an alternative structure for the system in the dilute limit may be that of connected sheets of thin, surfactant-coated oil or water layers in a water or oil continuum. This random, sheet-like structure could have substantially lower bending energy than the dilute drop-like system and almost comparable entropy (due to the undulations of the sheets). Such a structure has, in fact, been proposed for the L_3 phase of lyotropic systems and some microemulsions [5]. It would be of great interest to construct a model which can describe both limits of comparable water and oil volume fractions (random domains) and of dilute systems (sheet-like structure) in a unified manner. In addition, very few structural measurements have been performed on balanced systems where both ϕ and the spontaneous curvature are small. This is because these experiments are performed as salinity scans for systems where the overall volume fractions of oil and water are equal. Salinity scans of the case of small oil or water volume fractions may indicate whether the structure of the single phase, dilute microemulsion at the optimal salinity (where the system is balanced) is sheet-like or a dilute, droplet-like dispersion. Another question that deserves further study is the nature of the transition between the sheet-like and vesicle phases [5, 26].

Acknowledgements

The work summarized here is a result of a collaboration with D. Roux, S. Milner, M. Cates, and D. Andelman. The author also is grateful to C. Safinya and J. S. Huang for their comments.

References

1. For a general survey see (a) Surfactants in Solution, ed. K. Mittal and B. Lindman, (Plenum, N.Y., 1984), and ibid 1987; (b) Physics of Complex and Supermolecular Fluids, ed. S.A. Safran and N.A. Clark (Wiley, N.Y., 1987).
2. S.A. Safran, D. Roux, M. Cates, D. Andelman, Phys. Rev. Lett. **57**, 491 (1986), and in Surfactants in Solution: Modern Aspects, ed. K. Mittal, (Plenum, N.Y., in press).
3. David Andelman, M. Cates, D. Roux, and S. A. Safran, J. Chem. Phys. **87**, 7229 (1987).
4. D. Andelman, S. A. Safran, D. Roux, and M. Cates in Langmuir, in press.
5. M. E. Cates, D. Roux, D. Andelman, S. T. Milner, and S. A. Safran, Europhysics Lett. **5**, 733 (1988).
6. S. T. Milner, S. A. Safran, D. Andelman, M. Cates, and D. Roux, J. de Phys. (Paris) **49**, 1065 (1988).
7. A. Calje, W.G.M. Agerof, A. Vrij in Micellization, Solubilization, and Microemulsions, ed. K. Mittal, (Plenum, N.Y.) 1977, p. 779; R. Ober and C. Taupin, J. Phys. Chem. **84**, 2418 (1980); A.M. Cazabat and D. Langevin, J.

Chem. Phys. **74**, 3148 (1981); D. Roux, A.M. Bellocq, P. Bothorel in Ref. 1a, p. 1843; J.S. Huang, S.A. Safran, M.W. Kim, G.S. Grest, M. Kotlarchyk, N. Quirke, Phys. Rev. Lett. **53**, 592 (1983); M. Kotlarchyk, S. H. Chen, J.S. Huang, M.W. Kim, Phys. Rev. A **29**, 2054 (1984).

8.   C. Huh, J. Coll. Interface Sci. **97**, 201 (1984) and **71** (1979); S.A. Safran and L.A. Turkevich, Phys. Rev. Lett. **50**, 1930 (1983); S.A. Safran, L.A. Turkevich, P.A. Pincus, J. Phys. (Paris) Lett. **45**, L69 (1984).

9.   S.A. Safran, L.A. Turkevich, P.A. Pincus, J. Phys. (Paris) Lett. **45**, L69 (1984).

10. Y. Talmon and S. Prager, J. Chem. Phys. **69**, 2984 (1978) and **76**, 1535 (1982).

11. P. G. de Gennes and C. Taupin, J. Phys. Chem. **86**, 2294 (1982).

12. J. Jouffroy, P. Levinson, P.G. de Gennes, J. Phys. (Paris) **43**, 1241 (1982).

13. B. Widom, J. Chem. Phys. **81**, 1030 (1984).

14. W. Helfrich, Z. Naturforsch. **28a**, 693 (1973).

15. W. Jahn and R. Strey, J. Phys. Chem. **92**, 2294 (1988).

16. S. Alexander, J. de Phys. Lett. (Paris), **39**, 1 (1978).

17. J.C. Wheeler and B. Widom, J. Am. Chem. Soc. **90**, 3064 (1968); B. Widom, J. Chem. Phys. **84**, 6943 (1986).

18. M. Schick and W. H. Shih, Phys. Rev. B **34**, 1797 (1986) and Phys. Rev. Lett. **59**, 1205 (1987); K. Chen, C. Ebner, C. Jayaprakash, R. Pandit, J. Phys. C **20**, L361 (1987).

19. M. Kahlweit, R. Strey, P. Firman, and D. Haase, Langmuir, **1**, 281 (1985); Ang. Chem., Int. Ed. Engl. **24**, 654 (1985); J. PHys. Chem. **91**, 1553 (1987); J. Phys. Chem. **90**, 671 (1986); D. H. Smith, J. Coll. Int. Sci. **108**, 471 (1985).

20. L.E. Scriven, in Micellization, Solubilization, and Microemulsions, ed. K. Mittal, (Plenum, N.Y., 1977), p.877.

21. W. Helfrich, J. Phys. (Paris) **46**, 1263 (1985); ibid. **48**, 285 (1987).

22. L. Peliti and S. Leibler, Phys. Rev. Lett. **54**, 1690 (1985); D. Foerster, Phys. Lett. **114A**, 115 (1986); H. Kleinert, Phys. Lett. **114A**, 263 (1986).

23. Y. Kantor, M. Kardar, and D. R. Nelson, Phys. Rev. Lett. **57**, 263 (1986).

24. W. Helfrich, Z. Naturforsch. **33a**, 305 (1978).

25. C. R. Safinya, D. Roux, G. Smith, S. K. Sinha, P. Dimon, N. Clark, A. M. Bellocq, Phys. Rev. Lett. **57**, 2718 (1986).

26. D. Huse and S. Leibler, J. Phys. (Paris), **49**, 605 (1988).

Dynamical Aspects of Volume Phase Transition and Pattern Formation in Polymer Gels

S. Hirotsu, A. Kaneki, K. Iwasaki, and I. Yamamoto

Department of Physics, College of Science, Tokyo Institute of Technology, Ohokayama, Meguroku, Tokyo 152, Japan

Some polymer gels are known to undergo a volume phase transition, i.e. a phase transition between two gel phases with different polymer concentrations. Most of the previous studies on this subject have been confined to equilibrium properties of homogeneous gels. Recent experimental and theoretical studies have revealed a number of interesting problems related to nonequilibrium states, anisotropic deformation, and inhomogeneity inside gels. These include spinodal decomposition [1], pattern formation on surface of swelling gels [2-5], formation of domain walls during 1st-order transition [6,7], effect of anisotropic stresses on the the volume transition [8], and so on. We will be concerned here with two of the topics, i.e. evolution of patterns during swelling and the nucleation of new phase at the 1st-order transition.

Tanaka et al.[2] observed that regular honeycomb-like patterns develop on surface of gel while it undergoes extensive swelling. It should be pointed out here that their experiment was not to do with a bulk discontinuous transition but with a continuous swelling from a concentrated, nonequilibrium state to a swollen equilibrium state. Thus the pattern they observed was due to continuous swelling and not due to discontinuous transition. We want to make this distinction clear because there is no reason to assume that the kinetics of continuous and discontinuous swelling (or shrinking) follow the same law.

We have studied the evolution of surface pattern during continuous swelling. Samples used were acrylamide(AA)/water and N-isopropylacrylamide(NIPA)/water gels containing acrylic acid. As-prepared gels had a concentration $\phi = 0.05 \sim 0.08$. They were dried in air to $\phi \simeq 0.6 \sim 1.0$. As soon as the concentrated gel was put into distilled water, it began to swell and a fine pattern appeared on its surface. The characteristic size R (i.e. the mean length of the edges of polygons) of the pattern was measured with a video-microscope as a function of time. The results are shown in Fig.1. The actual values of R ranged from ~0.05 to ~1mm. It is seen that R fits the power law $R \sim t^{\alpha}$ very well. However, the value of α depends on the initial concentration of gel, ϕ_I. For $\phi_I \simeq 0.6 \sim 0.8$ we have $\alpha = 0.45 \pm 0.1$, while for $\phi_I = 1$ we have $\alpha = 0.95 \pm 0.1$. The uncertainty of α comes from that of R which in turn is due to irregular shape of polygons. As long as ϕ_I is smaller than about 0.8, the value of α does not depend on the precise value of ϕ_I. Moreover, the values of α obtained on NIPA gels were almost identical to those obtained on AA gels.

In the existing model, R should be proportional to the thickness of the swollen layer in the surface region. If this layer grows via diffusion, its thickness at time t will be proportional to $\sqrt{Dt}$, where D is the diffusion constant of the network. Thus α is predicted to be 0.5. The present result for $\phi_I \simeq 0.6 \sim 0.8$ is consistent with this prediction, while for $\phi_I = 1$ α is almost twice the predicted value. The fact that $\alpha \simeq 1$ in the latter case shows that the surface layer grows with a nearly constant speed, which in turn shows that the swelling must be dominated by a mechanism other than diffusion. In this context it would be important to note that for a network with ionizable groups the swelling from the state $\phi_I = 1$ requires a special consideration as compared with the swelling from any state with $\phi_I < 1$ because in the vicinity of $\phi_I=1$ ionic osmotic pressure increases with increasing volume (or vice versa) leading to negative network modulus. The anomalous value of α obtained above must be related to this instability at the interface between swollen and unswollen regions. Similarly, if swelling involves some kind of reaction at the interface, α will deviate from the value predicted for diffusion-dominated swelling mechanism.

Springer Series in Synergetics, Vol. 43 **Cooperative Dynamics in Complex Physical Systems**
Editor: H. Takayama   © Springer-Verlag Berlin, Heidelberg 1989

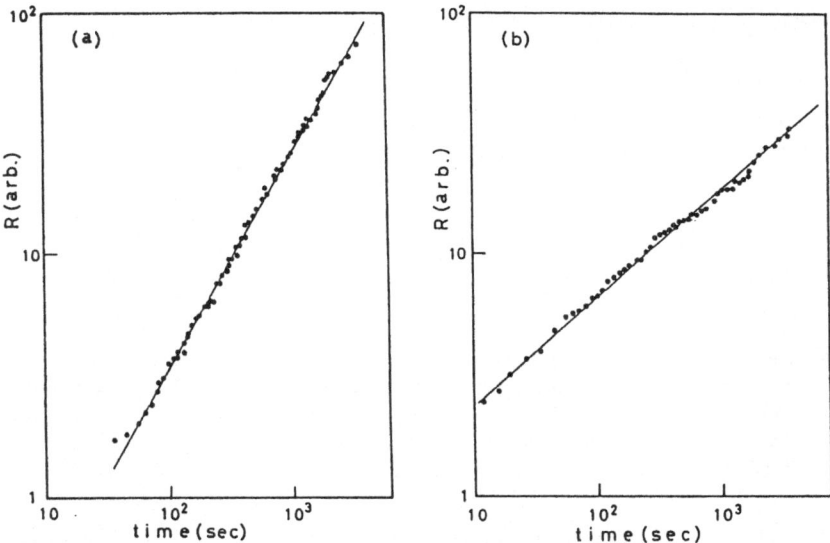

Fig. 1 Evolution of the characteristic length of the pattern observed on acrylamide-acrylic acid copolymer gels during continuous swelling. (a) ϕ_I =1, the line represents R $\sim$ t$^\alpha$ with α = 0.95, (b) $\phi_I \sim$ 0.6, the line corresponds to α = 0.46

We have also studied a nucleation and growth of the new phase at the discontinuous transition [9] in neutral NIPA/water gel around 34°C. The qualitative features observed are as follows: (1) The process of the transition depends strongly on shape and size of a sample. In a thin rod with a diameter less than a few mm, a nucleation always occurs at edges of the rod, leading to a "sausage-shaped" domain [6]. The transition proceeds via a movement of the domain wall along the rod axis. (2) In a thicker rods, on the other hand, a nucleus of new phase cannot be formed even at edges. At discontinuous shrinking of thick rods, the whole sample undergoes a spinodal decomposition at a temperature a little higher than the coexistence temperature determined on a thin rod. On the other hand, the discontinuous swelling transition seems to proceed via a pattern formation on surface. The pattern is similar to that observed in a continuous swelling, though its evolution law has not yet been investigated in detail. (3) At spinodal decomposition many tiny bubbles appear at surface, which can be identified as phase-separated swollen regions. At the same time, the interior of gel becomes strongly opaque.

A number of anomalous features [10] observed at the discontinuous transition of gels show that the kinetics of transition is dominated by the elasticity of network and that it is quite different from that of the usual phase separation.

References

1. S. Hirotsu and A. Kaneki: Proc. Int. Symp. Dynamics of Ordering Processes in Condensed Matter, Kyoto, 1987 (Plenum Press, 1988).
2. T. Tanaka et al.: Nature 325 796 (1987).
3. A. Onuki: J.Phys.Soc.Jpn. 57 703 (1988).
4. K. Sekimoto and K. Kawasaki: J.Phys.Soc.Jpn. 56 2997 (1988).
5   T. Hwa and M. Karder: Phys.Rev.Letters 61 106 (1988).
6. S. Hirotsu: J.Chem.Phys. 88 427 (1988).
7. K. Sekimoto and K. Kawasaki: J.Phys.Soc.Jpn. 57 No.8 (1988).
8. A. Onki: J.Phys.Soc.Jpn. 57 1868 (1988).
9. S. Hirotsu: J.Phys.Soc.Jpn. 56 233 (1986).
10. S. Hirotsu: to be published.

Coherent-Anomaly Method in Polymer Physics

Xiao Hu and M. Suzuki

Department of Physics, Faculty of Science,
University of Tokyo, Hongo, Bunkyo-ku, Tokyo 113, Japan

A coherent-anomaly method (CAM) in cooperative phenomena has been proposed by SUZUKI [1,2,3]. Here in this article, we report an application of this new theory to self-avoiding walk problem (SAW)[4]. SAW has been introduced as a model of linear polymer chain in good solvents, and takes into account the *excluded volume* effect. We investigate one of the most interesting aspects of SAW, namely the asymptotic behavior of the total number of N-step SAWs.

As an approximation of SAW, we consider finite-order-restricted walk (FORW). In an R-order-restricted-walk (R-ORW), no visited site can be revisited within R steps. It is obvious that when R appraoches infinity we obtain a SAW. As R-ORW is a Markov process, it is straightforward to obtain the following asymptotic formula for the total number of R-ORWs:

$$C_{FORW}(N;R) \sim \lambda(R)^N; \qquad N \to \infty, \tag{1}$$

with $\lambda(R)$ denoting the maximum eigenvalue of some transition matrix which governs the Markov process R-ORW [4,5]. The above asymptotic behavior does not change no matter how large a finite R is taken. It is essentially a classical one, namely it has the same behavior of a random walk without any memory. Constructing such a kind of classical approximation is the starting point of the CAM theory[1,2,3].

On the other hand, as a larger R is taken more walks with self-intersections are excluded. DOMB et al. [5] investigated one effect of this fact, namely the R-dependence of the maximum eigenvalue $\lambda(R)$. Their data for the f.c.c. lattice are $\lambda(3) \simeq 10.6569$, $\lambda(4) \simeq 10.4891$, $\lambda(5) \simeq 10.3923$, from which we can obtain[4,5]

$$\lambda(R) \simeq 10.005 \times (1 + 0.1935/R). \tag{2}$$

DOMB et al. [5] tried to extract the non-Markov property of SAW only from this R-dependence of $\lambda(R)$. Unfortunately there are some serious ambiguities in their argument, because another important aspect has been disregarded.

In a general viewpoint of the CAM, having obtained the approximate asymptotic formulae (1) with (2) means that we have already constructed a cannonical classical approximation series[1,2,3]. Then, following the prescription of the CAM, we should investigate the coefficient of the most singular part in the approximate asymptotic formula (1). For this purpose we rewrite (1) as follows

$$C_{FORW}(N;R) \simeq \beta(R) \times \lambda(R)^N; \qquad N \to \infty. \tag{3}$$

It is straightforward to estimate $\beta(R)$ numerically, and our results are $\beta(3) \simeq 1.16018$, $\beta(4) \simeq 1.21459$, $\beta(5) \simeq 1.25854$. Investigating the R-dependence of $\beta(R)$, we can find systematic power-law variance of the coefficient $\beta(R)$ with respects to R:

$$\beta(R) \simeq 0.9747 \times R^{0.1587}, \tag{4}$$

for the f.c.c. lattice.

Springer Series in Synergetics, Vol. 43   **Cooperative Dynamics in Complex Physical Systems**
Editor: H. Takayama   © Springer-Verlag Berlin, Heidelberg 1989

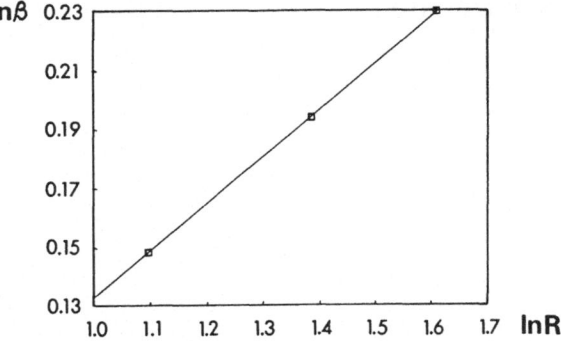

Fig.1 The R-dependence of the coefficient β

Combining (3) with (2) and (4), we arrive at

$$C_{FORW}(N;R) \simeq 0.9747 \times R^{0.1587} \times (1 + 0.1935/R) \times 10.005^N. \tag{5}$$

Now it is ready to derive the true asymptotic formula for SAW. We make analytical continuation of N and R. Increasing R the total-number curve (5) approaches the true one for SAW uniformly, and the *fixed curve* of (5) with R as a parameter should give the true one for SAW. The condition that an approximate curve has reached the *fixed curve* can be formalized from the fact that there should be no variance of the curve on the parameter R in the following differential equation

$$\partial C_{FORW}(N;R)/\partial R = 0, \tag{6}$$

and the *fixed curve*, namely the envelope curve is given by

$$C_{ENVE}(N) = C_{FORW}(N;R)|_{R=R(N)}, \tag{7}$$

with the function $R = R(N)$ determined by (6). Then our conclusion is that the envelope function of the differential equations about the approximate asymptotic formulae for FORW-approximants should be the true asymptotic formula for SAW.

The differential equation (6) can be solved easily, and numerically we have

$$C_{SAW}(N) = C_{ENVE}(N) \simeq 1.179 \times N^{0.1587} \times 10.005^N, \tag{8}$$

for the f.c.c. lattice.

Our estimate of the universal exponent $g \simeq 0.1587$ differs from the expected value $g = 1/6$ only by 5%. Another new aspect of our theory is that we can also determine the coefficient in the asymptotic formula for SAW systematically.

The present authors thank Dr. S.Miyashita and Dr. M.Katori for useful discussions. This work is partially financed by the Scientific Research Fund of the Ministry of Education, Science and Culture.

References

1. M.Suzuki: J.Phys.Soc.Jpn.**55** (1986) 4205; Phys.Lett.**116A** (1986) 375.
2. M.Suzuki, M.Katori and X.Hu: J.Phys.Soc.Jpn. **56** (1987) 3092.
3. M.Katori and M.Suzuki: J.Phys.Soc.Jpn. **57** (1988) 807.
4. X.Hu and M.Suzuki: Physica **A150** (1988) 310.
5. C.Domb and F.T.Hioe: J.Phys. **C3** (1970) 2223.

Computer Modelling of Pattern Formation in Gels

K. Sekimoto, N. Suematsu, and Kyozi Kawasaki

Department of Physics, Kyushu University 33, Fukuoka 812, Japan

Pattern formation in gels has some peculiar nature due to the nonlinear elastic deformation of gel-materials. Examples are the surface patterns of swelling gels [1-5] and diffusive phase boundary between swollen phase and shrunken phase in gel rods [6,7]. The domain patterns of coexisting phases in bulk gels, however, have not been studied with emphasis on their peculiarities. The existing theories of phase coexistence in gels either describe gels essentially as multi-phase fluids separated by semi-permeable membranes [8], which is a valid description for isotropic states, or restrict to the uniaxial deformation where a parallelism to fluid phase separation again exists [9].

The present paper is the first attempt by computer simulation to elucidate the nature of phase coexistence in gels fully taking into account the nonlinear elastic deformations. We have numerically studied stable states of a lattice version of two-dimensional model of gel with free energy F having the following general form

$$F= \int f(tr(^{t}M(\vec{x}) \cdot M(\vec{x})), \, det(M(\vec{x})) \; d^{2}x$$

where the 2×2 matrix $M(\vec{x})$ describes local deformation through $d\vec{X}(\vec{x}) = M(\vec{x}) \cdot d\vec{x}$

with $\vec{X}(\vec{x})$ the position of the material point which was at $\vec{x}$ in a uniform reference state. Figure 1 shows diffusive phase boundary [6,7] between the swollen phase and the shrunken phase in a two-dimensional rod with both top and bottom boundaries allowed to move freely. We have also confirmed the predicted proportionality between the rod diameter and the thickness of phase boundary [7]. Figures 2(a) and 2(b) show, respectively, the random initial state of bulk system and its relaxed stable state with the periodic boundary conditions both in horizontal and perpendicular directions. We clearly see that the rather compact and nearly isotropic domains of swollen phase are sharply bounded by a highly deformed (especially stretched) percolating domain of shrunken phase. We may call this a bubble structure or a sponge-like structure. We expect that a similar structure might be observed experimentally. The uniformity of compact nonpercolating domains can be explained by our very recent analytical study [10].

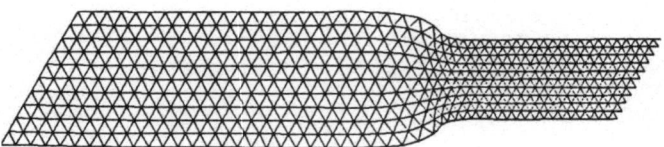

Fig.1 Diffusive phase boundary between coexisting phases under the free boundary conditions on both the top and bottom surfaces.

Springer Series in Synergetics, Vol. 43   **Cooperative Dynamics in Complex Physical Systems**
Editor: H. Takayama   © Springer-Verlag Berlin, Heidelberg 1989

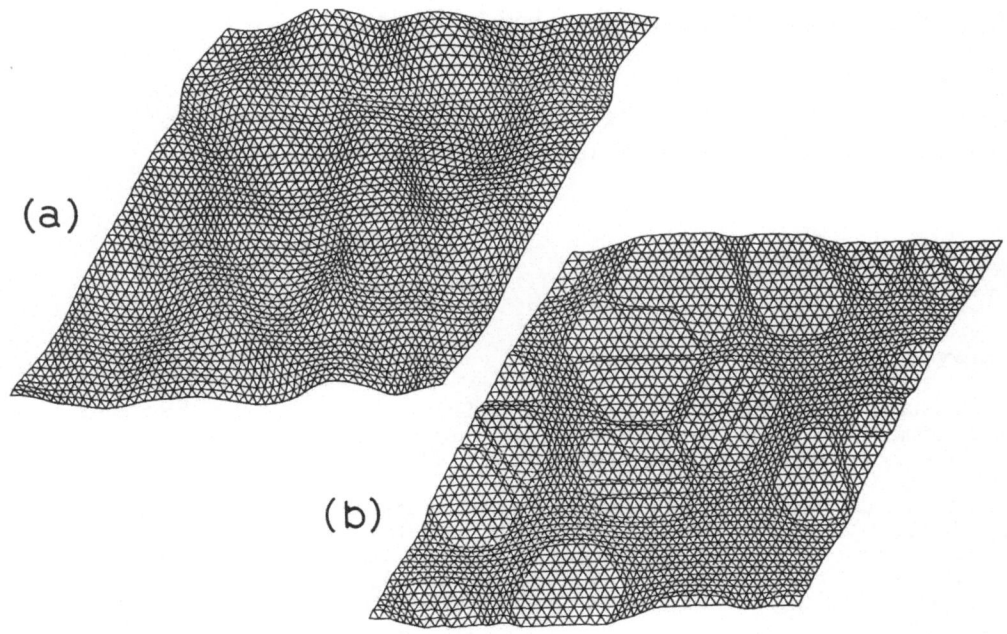

(a)

(b)

Fig.2 Phase coexistence in bulk gel under the periodic boundary condition. (a)
A random initial state and (b) the relaxed stable state.

Acknowledgements

We are grateful to S. Hirotsu for informing us of experimental details. All of
our computations were done at the Computer Center of the Institute of Plasma
Physics, Nagoya University.

References

1. T. Tanaka, S. T. Sun, Y. Hirokawa, S. Katayama, J. Kucera, Y. Hirose, and
   T. Amiya, Nature 325 (1987) 796, and the references cited therein
2. K. Sekimoto and K. Kawasaki, J. Phys. Soc. Japan 56 (1987) 2997
3. A. Onuki, J.Phys. Soc. Japan 57 (1988) 703
4. K. Sekimoto and K. Kawasaki, J. Phys. Soc. Japan 57 (1988) 2594
5. T. Hwa and M. Kardar, Phys. Rev. Lett. 61 (1988) 106
6. S. Hirotsu, J. Chem. Phys. 88 (1988) 427
7. K. Sekimoto and K. Kawasaki, J. Phys. Soc. Japan 57 (1988) 2591
8. B. Erman and P. J. Flory, Macromolecules, 19 (1986) 2342
9. A. Onuki, J. Phys. Soc. Japan 57 (1988) 1868
10. K. Sekimoto and K. Kawasaki, submitted to Physica A

Molecular Theory of Thermoreversible Gelation

F. Tanaka

Department of Physics, Faculty of General Education,
Tokyo University of Agriculture and Technology, Fuchu, Tokyo 183, Japan

1. INTRODUCTION

The formation of gel network by physical crosslinking can be brought about by a
change of material conditions such as stereoregularity of a polymer chain, content of associative side groups, or polarizabity of the solvent.

Since the discovery[1] that chemically inactive atactic polystyrene (at-PS)
can exhibit gelation in a series of solvents, there has been a growing interest
in the phenomena of physical gelation.   Recent experimental studies[2] have
furnished strong evidence that the gelation of at-PS is thermally reversible,
and also that the gel melting temperature agrees with the gel freezing temperature.   The current understanding of the experimental observations is based on
the cooccurrence of sol-gel transition and two-phase separation on the temperature-concentration plane.

The purpose of this presentation is to derive the observed phase diagram from a
molecular point of view, and to study the interference between the two intrinsically different phase transitions: gelation and phase separation.

2. THERMODYNAMICS

Besides the usual solvent-polymer contact interaction, we assume the existence
of specific associative interaction capable of forming physical crosslinks
between the polymer segments.   Although the origin has not been clarified, the
binding energy is expected to be of the order of thermal energy, so that bonding
-unbonding equilibrium is easily attained. In thermal equilibrium intermolecular
crosslinking yields polydisperse molecular aggregates, whose distribution is
completely specified by the multiple-equilibria conditions

$$\mu_m / m = \text{independent of } m, \tag{1}$$

where m is the number of chains in an aggregate, and μ_m its chemical
potential. The distribution is thermally controlled and strongly dependent on
the solute concentration. Lattice theory of polydisperse solution, incorpolated
with the condition (1), yields: $\phi_m = K_m \phi_1^m$ for the volume concentration
ϕ_m of the m-clusters. K_m is the association constant and can be expressed
in terms of the free energy gain δ_m of a chain when participating in an m-
cluster from isolation.

3. GELATION vs PHASE SEPARATION

The total solute concentration is given by an infinite sum:

$$\phi = \sum_{m=1}^{\infty} K_m \phi_1^m . \tag{2}$$

Springer Series in Synergetics, Vol. 43   **Cooperative Dynamics in Complex Physical Systems**
Editor: H. Takayama   © Springer-Verlag Berlin, Heidelberg 1989

The sum takes a finite value at the radius of convergence $\phi_1{}^*$ of the series, while it is infinite above $\phi_1{}^*$, indicating the appearence of a macroscopic net-work. Above the threshold of the total concentration ϕ^*, the infinite sum remains constant since it can not accommodate the contribution from the macronet-work. The excess quantity $\phi - \phi^*$ is simply absorbed into this macronetwork. This is the gel formation.
The two-phase equilibrium condition reduces to a single one: $\mu_1(\phi') = \mu_1(\phi'')$ for the isolated chains by the help of eq.(2), where ϕ' and ϕ'' are the upper and the lower concentration of the coexisting phases.

4. COMPARISON WITH THE EXPERIMENTS

For the quantitative evaluation of physical quantities, the energy gain must be specified. We introduce a simple model for the internal structure of an aggregate. It is an acyclic tree constructed with m chains, each carrying f identical functional units. The calculated phase diagram based on this model is shown in Fig.1 and compared with the experimental data on at-PS/CS2 system. The temperature is measured by the dimensionless deviation τ from the theta temperature. Although the observed binodals are almost symmetric, the calculated one is highly asymmetric, showing characteristics of polymer solutions.

The osmotic pressure and the number-averaged mean cluster size are shown in Fig.2 as a function of the concentration. The temperature is varied from curve to curve. The pressure shows a kink at the gelation point, indicating that the gelation is a second-order phase transition.

The possibility of appearence of new critical phenomena, such as tricritical point (TCP) or critical end point (CEP) as a result of interference between gelation and phase separation is examined.

We also show one possible mechanism of segment association. In order for the polymer segment to be tightly bound, the large side groups must be on the same side of the backbone chain. Local isotacticity is required for this. Under such assumption the functionality f of a chain is evaluated.

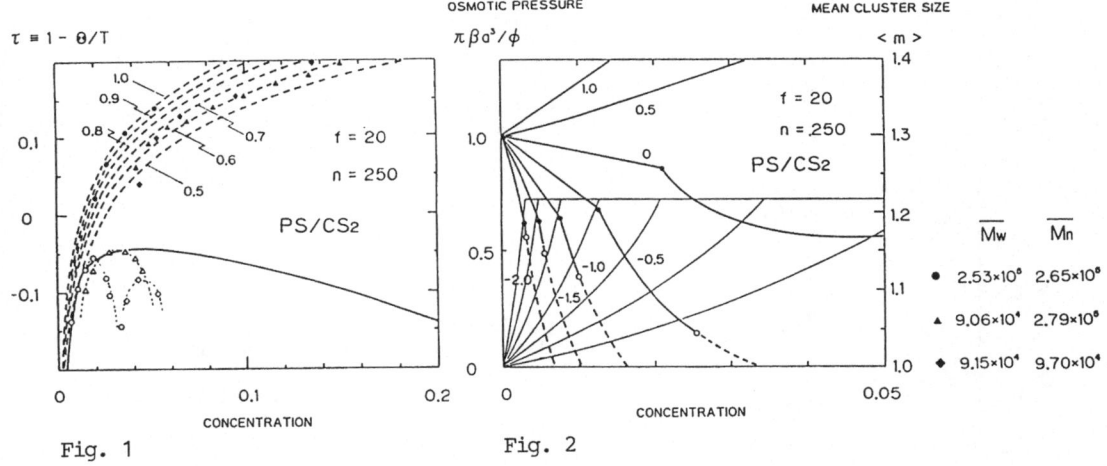

Fig. 1   Fig. 2

[1] S.Wellinhoff, J.Shaw and E.Baer, Macromolecules 12, 932 (1979)
[2] T.M.Tan, A.Moet, A.Hiltner and E.Baer, Macromolecules 16, 28 (1983)

Electrostatic Effects and Counterion Condensation in Gels

Y.Y. Suzuki and T. Tanaka

Department of Physics and Center for Materials Science & Engineering,
Massachusetts Institute of Technology, Cambridge, MA 02139, USA

A universal property of gels is that they undergo a volume phase transition when external conditions are varied. The equilibrium phase properties of these gels are well characterized by the classical Flory-Huggins theory [1]. However, the theory is not yet quantitative in the case of ionic gels [2]. Hence, we now consider the electrostatic effects on the equilibrium phase properties of ionic gels in an attempt to clarify the current discrepancies.

The exponent ν which characterizes the configuration of a single polymer chain is defined by

$$r_{chain} \propto n^{\nu},$$

where r_{chain} is the end-to-end distance of the chain consisting of n monomers. The value of the exponent ν ranges from 1/2 for Gaussian chain configuration to 1 for fully extended chains. In gels, a polymer chain exists between two crosslinks within the network. Assuming a Gaussian distribution for r_{chain}, the size of the bulk gel R_{bulk} is

$$R_{bulk} \propto \left(\frac{N}{n}\right)^{1/2} r_{chain} \propto n^{\nu'} \propto n_{BIS}^{-\nu'} , \qquad \nu' = \nu - \frac{1}{2} ,$$

where N is the total number of monomers in the gel, which is constant. Then, the average number of monomers for a chain separated by crosslinks, n, is inversely proportional to the number of crosslinks n_{BIS}. In Fig. 1 we show the experimental dependence of the size of gels as the number of crosslinks are varied. The exponents, $\nu = 0.58$ and $\nu = 0.85$ reflect a Flory configuration for nonionic gels and a rodlike configuration for ionic gels, respectively.

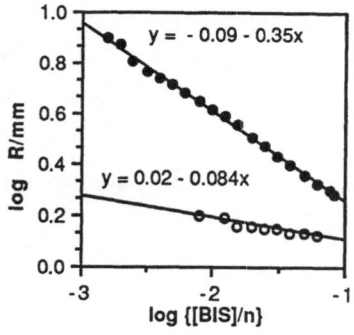

Fig. 1: Equilibrium sizes of acrylamide-sodiumacrylate ($n_{ion}/n=0.14$) gels (●) and acrylamide (nonionic) gels (O) are exhibited as a function of bisacrylamide (crosslinker) concentration in the pre-gel solution. The best fit values are $\nu'=0.35$ for highly ionized gel, and $\nu'=0.084$ for nonionic gel.

Consider now the effect of electrostatic interactions on bulk gels with a fixed crosslink number but whose ionic content varies. In terms of the end-to-end distance of a polymer chain within gels with respect to increasing number of ionized monomers n_{ion}, an exponent β is defined by

$$r_{chain} \propto n_{ion}^{\beta} .$$

Springer Series in Synergetics, Vol. 43   **Cooperative Dynamics in Complex Physical Systems**
Editor: H. Takayama   © Springer-Verlag Berlin, Heidelberg 1989

In the Flory-Huggins theory, an extensive force arises only from the osmotic pressure of freely moving counterions. Coulomb interactions were assumed negligible because the charge groups were effectively screened. However, the experimental data indicate that the assumption is not valid; the polymer chain configuration is extended due to repulsion among neighboring charges. Thus, an electrostatic term must be included in the free energy for ionic gels. Since long range interactions are screened and only neighboring charges interact, the electrostatic free energy of gels is proportional to n_{ion}. Moreover, due to the presence of electrostatic interactions, the Flory configuration of the polymer chain is now modified. Based on scaling arguments, the polymer chain is considered to break up a series of "blobs" under a strong stretching force where the elastic restoring force becomes nonlinear [3]. The free energy is thus given by

$$F = F_{electro} + F_{elastic} \approx n_{ion} \frac{e_0^2}{\varepsilon_0 R} + kT \left(\frac{R}{R_F} \right)^{5/2} .$$

Minimizing this free energy, we obtain $\beta = 2/7$. Fig.2 shows the dependence of the volume of gels on the amount of n_{ion} incorporated within the gels. The exponent $\beta = 0.27$ suggests that electrostatic interaction term is dominant in this region. Beyond a certain amount of charge groups, however, there appears a plateau where the size of gel is constant regardless of the increase in ionic content. This indicates the existence of counterion condensation in gels which is briefly discussed below.

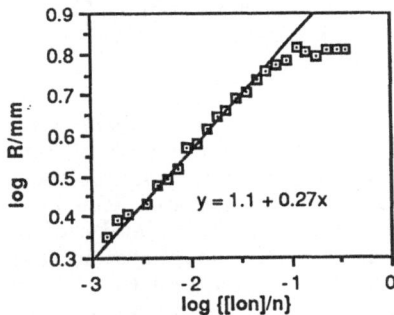

Fig. 2: Equilibrium sizes of acrylamide-sodiumacrylate gels are shown as a function of n_{ion}/n. The experiment exhibits $\beta = 0.27$ for data points excluding the counterion condensation region. The counterion condensation starts at $n_{ion}/n \approx 10^{-1}$, which is approximately equivalent to the Bjerrum length, as predicted by theory.

The distribution of counterions and the average potential around polymers can be expressed by the Poisson-Boltzmann equation(PBE). The solution of PBE for rod-like polymers has a singularity at the Bjerrum length $e_0^2/\varepsilon_0 kT$. For low charge polymers the counterion distribution is nearly uniform. For highly charged polymers the counterions tend to "condense" near the polymer when the average interval between the charge groups becomes less than the Bjerrum length ($d=7\text{Å}$ in water), which results in a fixed apparent charge density at $1/d$ in spite of further increase in the number of charge groups. This phenomenon is called "counterion condensation" which is analogous to the gas-liquid condensation [4]. Hence the equilibrium size of the gels does not increase beyond this point.

From simple size measurements of swollen bulk gels, we observed that the configuration of polymer chains within gels is modified due to electrostatic interactions, as predicted. In addition, the experiment also confirmed the observation of counterion condensation when the interionic distance in ionic gels becomes less than the Bjerrum length. Based on these studies, we believe that electrostatic effects are significant in ionic gels.

1. P. J. Flory, In Principle of Polymer Chemistry (Cornell University, Ithaca, 1963)
2. S. Hirotsu, Y. Hirokawa, T. Tanaka, J. Chem. Phys., 87 1392 (1987)
3. P. G. de Gennes, Scaling Concepts in Polymer Physics (Cornell University, Ithaca, 1979)
4. F. Oosawa, In Polyelectrolytes (Marcel Dekker, New York, 1971)

Role of Fluctuations in Fluid Mechanics and Dendritic Solidification

H.E. Stanley

Center for Polymer Studies and Department of Physics, Boston University, Boston, MA 02215, USA

Our purpose is to review certain recent advances in understanding the role of fluctuations in fluid mechanics and dendritic solidification; many of these represent joint work of the author and J. Nittmann. If one understands completely the simple Ising model, then one understands virtually all systems near their critical points—although the detailed descriptions of many such systems requires a suitably-chosen variant of the Ising model (such as the XY or Heisenberg model). By analogy, we shall argue here that if one understands completely the simple diffusion limited aggregation (DLA) model or the closely-related dielectric breakdown model (DBM), then one understands the role of fluctuations in a range of fluid mechanical systems, as well as in dendritic solidification. The detailed descriptions of some such systems requires suitably-chosen variants, such as DBM with anisotropy and noise reduction.

The theme I'll develop is that recent work on relatively simple *non-deterministic* models has some utility for describing experimentally-observed phenomena in fluid mechanics and dendritic growth. I'll first make the case that we can approach these experimental subjects of classic difficulty with the same spirit that has been used in recent years to approach problems associated with phase transitions and critical phenomena. This approach is to carefully choose a microscopic model system that captures the essential physics underlying the phenomena at hand, and then study this model until we understand "how the model works." Then we reconsider the phenomena at hand, to see if an understanding of the model leads to an understanding of the phenomena. Sometimes the original model is not enough, and a variant is needed, and we shall see that this is the case here also. Fortunately, however, we shall see that the same underlying physics is common to the model and its variants (cf. Table 1).

We begin, then, with the classic Ising model.

The Ising Model and Its Variants

The first time I heard a lecture on the Ising model, the speaker apologized for having what was termed "the Ising disease" (an appellation attributed to Montroll). The Ising model was proposed 67 years ago[1] and its solution for a one-dimensional lattice occurred 62 years ago.[2] However, at that time no one knew that the Ising model describes a wide range of materials near their critical points. Over 1000 papers have been published on this model, but only since 1977 have we known that if one understands the Ising model thoroughly, one understands the essential physics of virtually all 3-dimensional materials systems near thermal critical points. This is because other systems are simply variants of the Ising model. For example, most systems are related to special cases of the n-vector model, which in turn is a simple Ising model in which the spin variable s has not one component but rather n separate components s_j: $\mathbf{s} \equiv (s_1, s_2, \ldots, s_n)$.

Springer Series in Synergetics, Vol. 43   **Cooperative Dynamics in Complex Physical Systems**
Editor: H. Takayama   © Springer-Verlag Berlin, Heidelberg 1989

Table 1

A "Rosetta stone" connecting the physics underlying (a) an electrical problem (dielectric breakdown), (b) a fluid mechanics problem (viscous fingering), and (c) a diffusion problem (dendritic solidification).

(a) electrical	(b) fluid mechanics	(c) dendritic solidification
electrostatic potential: $\phi(r,t)$	pressure: $P(r,t)$	concentration: $c(r,t)$
electric field: $E \propto -\nabla\phi(r,t)$	velocity: $v \propto -\nabla P(r,t)$	growth rate: $v \propto -\nabla c(r,t)$
conservation: $\nabla \cdot E = 0$	$\nabla \cdot v = 0$	$\nabla \cdot v = 0$
Laplace Equation: $\nabla^2 \phi = 0$	$\nabla^2 P = 0$	$\nabla^2 c = 0$

The Ising model solves the puzzle of how it is that nearest-neighbor interactions of **microscopic** length scale 1Å "propagate" their effect cooperatively to give rise to a correlation length ξ_T of **macroscopic** length scale near the critical point (Fig. 1a). In fact, ξ_T increases without limit as the coupling $K \equiv J/kT$ increases to a critical value $K_c \equiv J/kT_c$,

$$\xi_T \sim A\left(\frac{K - K_c}{K_c}\right)^{-\nu_T}. \tag{1a}$$

The "amplitude" A has a numerical value on the order of the lattice constant a_o. A snapshot of an Ising system shows that there are fluctuations on all length scales from a_o ($\cong 1$Å) to ξ_T (which can be from $10^2 - 10^4$Å in a typical experiment).

Attempts to simplify the essential problem of propagation of order from one spin to its neighbors by making mean-field type of truncations (such as the Weiss approximation, the Bethe approximation, and the Kasteleyn-van Kranendonk constant coupling approximation) fail to describe 3-dimensional systems near their critical points.

To describe the specific heat near the λ point of ^{4}He, one finds that the Ising model is not appropriate. This is because the order parameter in ^{4}He is not a one-dimensional variable with only two values (up or down), but rather a two-dimensional object with an amplitude and a phase. Accordingly, the Ising model has to be replaced by a "variant" for which the one-dimensional Ising spins are replaced by two-dimensional XY spins.

Random Site Percolation on a Lattice, and Its Variants

In its simplest form, one randomly occupies a fraction p of the sites of a d-dimensional lattice (the case $d = 1$ is shown schematically in Fig. 1b). Again, phenomena occur-

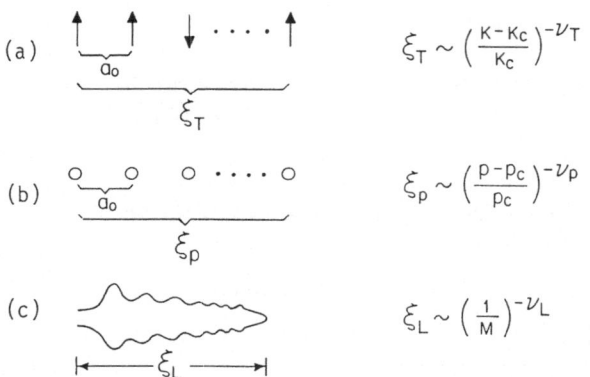

$$\xi_T \sim \left(\frac{K - K_c}{K_c}\right)^{-\nu_T}$$

$$\xi_p \sim \left(\frac{p - p_c}{p_c}\right)^{-\nu_p}$$

$$\xi_L \sim \left(\frac{1}{M}\right)^{-\nu_L}$$

Fig. 1: Schematic illustration of the analogy between (a) the Ising model, which has fluctuations in spin orientation *on all length scales* from the microscopic scale of the lattice constant a_o up to the macroscopic scale of the thermal correlation length ξ_T, (b) percolation, which has fluctuations in characteristic size of clusters *on all length scales* from a_o up to the diameter of the largest cluster—the pair connectedness length ξ_p, and (c) the DLA/DBM problem, whose clusters have fluctuations *on all length scales* from the microscopic length $d_o = \gamma/L$ (γ is the surface tension and L the latent heat) up to the diameter of the cluster ξ_L. Also shown, on the right side, is the analogy between the scaling behavior of the three length scales ξ_T, ξ_p, and ξ_L.

ring on the local 1Å scale of a lattice constant are "amplified" near the percolation threshold $p = p_c$ to a macroscopic length ξ_p.

Here p plays the role of the coupling constant K of the Ising model. When p is small, the characteristic length scale is comparable to 1Å. However when p approaches p_c, there occur phenomena on all scales ranging from a_o to ξ_p, where ξ_p increases without limit as $p \to p_c$

$$\xi_p \sim A\left(\frac{p - p_c}{p_c}\right)^{-\nu_p}. \tag{1b}$$

Again, the amplitude A is roughly 1Å.

It is by now a well-known piece of "magic" that each phenomenon of thermal critical phenomena has a corresponding analog in percolation, so that the percolation problem is sometimes called a geometric or "connectivity" critical phenomenon. Any connectivity problem can be understood by starting with pure random percolation and then adding interactions, or whatever. Thus, e.g., we understand why the critical exponents describing the divergence to infinity of various geometrical quantities (such as ξ_p) are the same regardless of whether the elements interact or are non-interacting.[3,4] This has been predicted on theoretical grounds and confirmed by detailed numerical simulations. Similarly, the same connectivity exponents are found regardless of whether the elements are constrained to the sites of a lattice or are free to be anywhere in a continuum (see, e.g., Gawlinski and Stanley[5] for $d = 2$, and Geiger and Stanley[4,6] for $d = 3$).

The Laplace Equation and Its Variants

Is there some lesson to be learned for fluid mechanics from our experience with thermal and geometric critical phenomena? We don't know the answer to this question, but J. Nittmann and I have been exploring this possibility in recent months.

Just as variations in the Ising and percolation problems were found to be sufficient to describe a rich range of thermal and geometric critical phenomena, so we have found that variants of the original Laplace equation are useful in describing puzzling patterns in fluids mechanics and dendritic growth.

In the Ising model, we place a spin on each pixel (site) of a lattice. In percolation we allow each pixel to be occupied or empty. In fluid mechanics, we assign a number—call it ϕ—to each pixel. Generally we shall understand ϕ to be the pressure at this region of space.

The spins in an Ising model interact with their neighbors. Hence the state of one Ising pixel depends on the state of all the other pixels in the system—up to a length scale given by the thermal correlation length ξ_T. The "global" correlation between distant pixels in an Ising simulation arises from the fact that neighboring pixels at i and j have a "local" exchange interaction J_{ij}. Similarly, the correlation in connectivity between distant pixels in the percolation problem arises from the "propagation" of local connectivity between neighboring pixels. In fluid mechanics, the pressure on each pixel is correlated with the pressure at every other pixel because the pressure obeys the Laplace equation.

One can calculate an equilibrium Ising configuration by "passing through the system with a computer" and flipping each spin with a probability related to the Boltzmann factor. Similarly, one can calculate the pressure at each pixel by "passing through the system" and re-adjusting the pressure on each pixel in accord with the Laplace equation.* If we were to arbitrarily flip the configuration of a single pixel in the Ising problem (from $+1$ to -1), we would significantly influence the equilibrium configuration of the system out to a length scale on the order of ξ_T. Similarly, if we were to arbitrarily impose a given pressure on a single point of a system obeying the Laplace equation, we would drastically change the resulting pattern out to a length scale that we shall call ξ_L.

Does ξ_L obey a "scaling form" analogous to Eqs. (1a) and (1b) obeyed by the functions ξ_T and ξ_p for the Ising model and percolation? We believe that the answer to this question is "yes," although our ideas on this subject remain somewhat tentative and subject to revision.

The best way to see the fluctuations inherent in structures grown according to the Laplace equation is to first introduce some specific models. There are two models that were at once thought to be fully equivalent, although it is now recognized that the actual patterns produced by each have a different "susceptibility to lattice anisotropy."[7] The first of these models is diffusion limited aggregation (DLA). Here one releases a random walker from a large circle surrounding a seed particle placed at the origin. When the random walker touches a perimeter site of the seed, it "sticks" (i.e., the perimeter site becomes a cluster site), and we have a cluster of mass = 2. A second random walker is then released. This process continues until a large cluster is formed. Initially the "mass" M of clusters was typically 10^3 to 10^4. However, it has become possible to make very fast algorithms, and the largest cluster to date has a mass of 4×10^6.[8]

The dielectric breakdown model (DBM) differs from DLA in that nothing happens until the random walker touches a cluster site, at which time the perimeter site it was just on at the previous step is transformed into a cluster site. Not surprisingly, this tiny local change in boundary conditions does not affect the "critical exponents" of this problem—DLA and DBM have the same value of the fractal dimension d_f

---

* There is an intimate connection between the diffusion equation and the random walk problem (see, e.g., Chandrasekhar.[9])

describing how the cluster mass depends on cluster diameter L: $M \sim L^{d_f}$.† In both thermal critical phenomena (or percolation) the length L introduced when we have a finite system size scales the same as the correlation lengths ξ_T (or ξ_p). Hence for DLA we expect that there will be fluctuations on length scales up to ξ_L, where ξ_L itself increases with the cluster mass according to

$$\xi_L \sim A\left(\frac{1}{M}\right)^{-\nu_L} \qquad [\nu_L = 1/d_f].\qquad (1c)$$

Here the amplitude A is again on the order of 1Å. Note that (1c) is analogous to (1a) and (1b) if we think of $M \to \infty$ as being analogous to $K \to K_c$. This reasoning is common in polymer physics, where we relate the radius of gyration R_g of a polymer to the mass through an equation of the form of (1c), $R_g \sim (1/M)^{-1/d_f}$. Note that $\nu_L = 1/d_f$ plays the role of the critical exponents ν_T and ν_p of (1a) and (1b). Suppose we test this idea, qualitatively, by examining the largest DLA clusters in detail. We find that indeed there are fluctuations in mass on length scales less than, say, the width W of the side branches. If one makes a log-log plot of W against mass M, one finds the same slope $1/d_f$ that one finds when one plots the diameter against M.

Evidence for Similarity of Viscous Fingering Patterns and Laplace Equation (DLA/DBM) Patterns

In the remainder of this talk, we'll describe in some detail the sorts of results we obtain from variants of the Laplace equation. First, it is necessary to describe the simplest system that produces patterns resembling interesting objects found in nature. Consider, e.g., the classic Saffman-Taylor viscous fingering problem. Here one injects a low-viscosity fluid into a medium filled with high viscosity fluid. In the limit that the viscosity ratio between the high and low viscosity fluids can be taken to be zero, we can assume that the pressure everywhere inside the low viscosity fluid is a constant: $P(i) = 1$ for $i \in$ [cluster of pixels occupied by low-viscosity fluid]. The pressure everywhere else in the system will have a value given by the solution of the Laplace equation, (2). This problem is modelled by the dielectric breakdown model or DBM[10] or diffusion-limited aggregation model or DLA.[11] These two models have in common that both are solutions to the Laplace equation for the case in which the pressure is zero at infinity and $P = 1$ on an object called the cluster.

Daccord has made accurate measurements on the fractal dimension of viscous fingers in both lateral[12] and radial[13] geometries (Fig. 2). He reduced the length scale normally imposed by surface tension by using liquids with zero interfacial tension— the two fluids were water and a viscous aqueous solution of polysaccharide (Fig. 3). He found that the resulting patterns are indeed fractal, with a fractal dimension identical to that of DLA/DBM (Fig. 4). Måløy et al[14] found analogous behavior

---

† The difference in boundary conditions *does* affect the rate at which the asymptotic behavior shows up.[7] For example, for DLA the screening will be more severe: as soon as a random walker steps on a perimeter site, the walker is stopped and the perimeter site becomes a cluster site. However for the DBM a random walker is free to walk on perimeter sites with impunity: only when the walker steps on a **cluster** site does the walker stop walking. Hence in the DBM the walkers can better penetrate the fjords of the system, so in overall appearance DBM clusters appear to have thicker branches and to be more "compact." The critical exponent $\nu_L = 1/d_f$ is not changed since it depends not on the density but on the *rate* at which the density decreases as the mass increases.

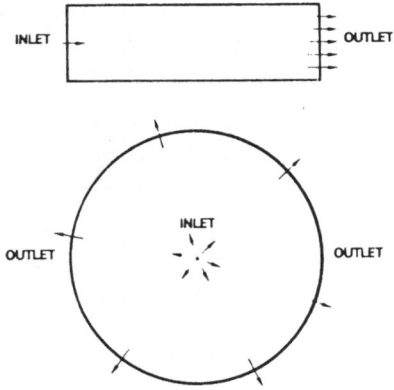

Fig. 2: Schematic illustration of the lateral and radial Hele-Shaw cells. Shown are top views. The spacing between the plates is typically 1 mm or less. From Daccord et al.[13]

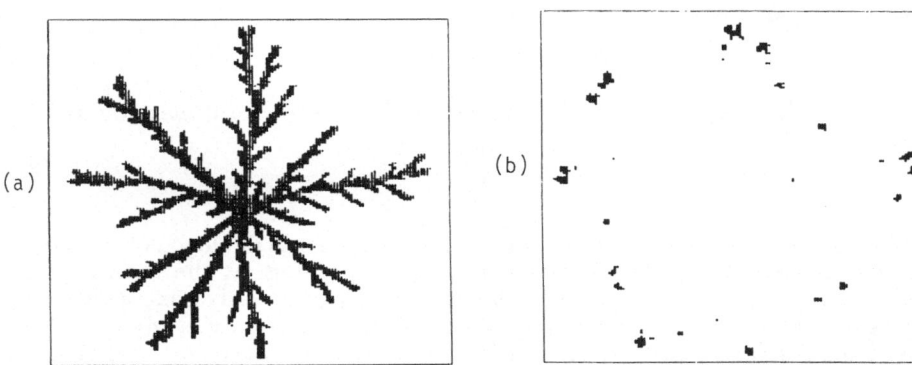

(a)   (b)

Fig. 3: The growth region of a radial viscous finger, a typical experimental pattern for which DLA is the appropriate model. The finger at time $t = t_o$ is shown in (a), while (b) displays the difference between the pattern at $t = t + \Delta t$ and $t = t$, obtained experimentally by simply subtracting the images of the same finger photographed at slightly different times. After Daccord et al.[13]

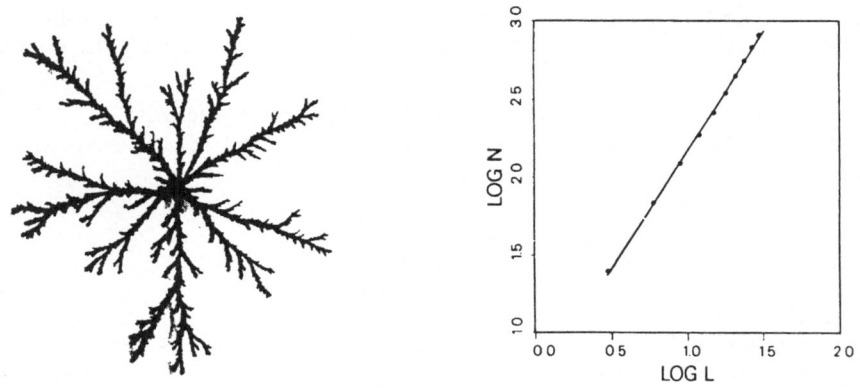

Fig. 4: Analysis of the fractal dimension typical of a radial viscous finger by the sandbox method (N is the number of occupied pixels in a $L \times L$ sandbox whose center is on an occupied pixel). The slope of the straight line shown is $d_f = 1.70 \pm 0.05$, while for DLA d_f is believed to be about 1.71 (from Daccord et al.[13]).

where the cell itself introduced the randomness: he accomplished this by placing glass beads inside the cell at random. Chen and Wilkinson[15] imposed the randomness by studying viscous fingering inside a network of glass tubes whose diameter L was randomly chosen from a probability distribution $\pi(L)$.

Not only is the fractal dimension the same for the fluid mechanics problem and for the Laplace patterns, but *so also are the multifractal properties the same.* Multifractals arise when one defines some quantity on all the pixel sites. Perhaps the simplest example is that of a charged needle: if we assign to every pixel a number equal to the electric field, then the set $\{E_i\}$ of field values for the perimeter sites of the needle form a multifractal set. The distribution $n(E)$ giving the number of perimeter pixels with electric field E is characterized, like all distribution functions, by its moments

$$Z(q) = \sum_E n(E)E^q. \tag{2}$$

As might be anticipated for a self-similar system, these moments scale with the mass M (or with the diameter L)

$$Z(q) \sim M^{\sigma(q)} \sim L^{-\tau(q)}. \tag{3a}$$

Since $M \sim L^{d_f}$, the exponents σ and τ are related by the fractal dimension d_f,

$$\sigma = \frac{\tau}{d_f}. \tag{3b}$$

For thermal and geometric critical phenomena, exponents analogous to the $\sigma(q)$ and $\tau(q)$ can be defined by considering a large L × L system at the critical point $[K = K_c]$ (or $p = p_c$). One finds that the ratio of two successive exponents is a constant "gap," so that there is no new information obtained by studying higher moments of the distribution. Connected with this simplicity is the fact that there is only one independent exponent in finite size scaling at the critical point (a second exponent arises if we wish to relate quantities that describe the approach to the critical point).

In percolation, these exponents have geometric interpretations:

(i) $y_h = d_f$, the fractal dimension of the incipient infinite cluster (the largest cluster found in a box of edge L at $p = p_c$), and

(ii) $y_T = d_{red}$, the fractal dimension of the red bonds that occur inside the largest spanning cluster (red bonds are singly connected bonds: when cut, the cluster falls into two pieces).

Relation (i) was noted by Stanley[16] while (ii) was proved by Coniglio.[17]

In the case of the moments Z_q, there is an infinite hierarchy of exponents in the sense that the ratio $\tau(q+1)/\tau(q)$ depends on q:

$$\frac{\tau(q+1)}{\tau(q)} = D(q). \tag{4}$$

For the case of a long thin needle, the exponent $D(q)$ sticks at the value 3/2 for small q, but for q above a critical value $q = q_c$, $D(q)$ becomes "unstuck" and varies continuously with q.

The same considerations apply to the fluid mechanics problem. Here the analog of the electric field $E \propto \nabla V$ is the growth probability $p_i \propto \nabla P$, where the index i runs over all perimeter sites i. Thus p_i is the probability that site i is the next to be added to the cluster. If we think of random walkers (Fig. 5), then p_i is the hit

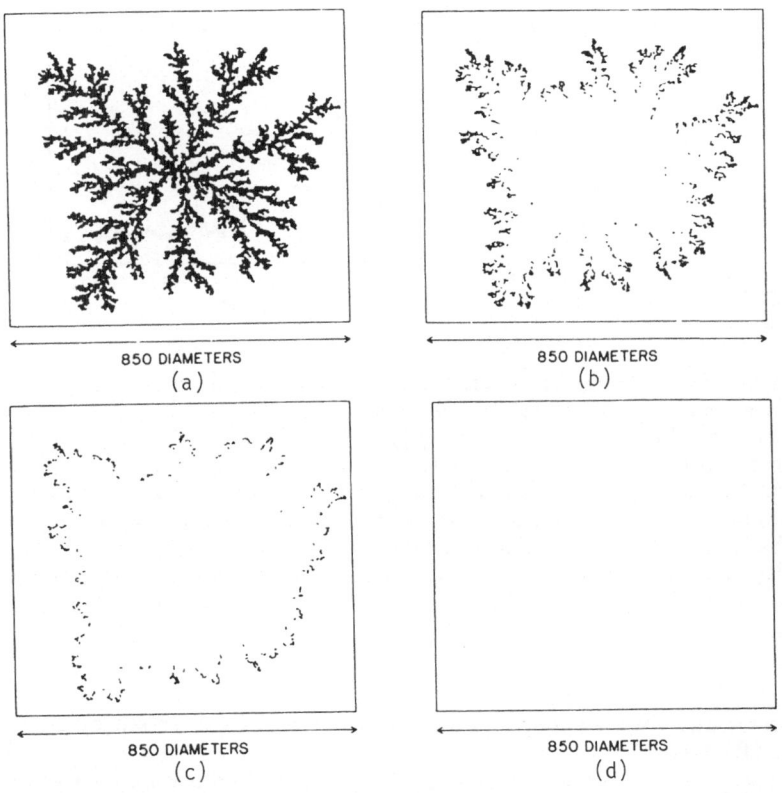

850 DIAMETERS
(a)

850 DIAMETERS
(b)

850 DIAMETERS
(c)

850 DIAMETERS
(d)

Fig. 5: This figure illustrates the harmonic measure for a 50,000 particle off-lattice 2d DLA aggregate. Figure 5a shows the cluster. Figure 5b shows all 6803 perimeter sites which have been contacted by at least one of 10^6 random walkers (following off-lattice trajectories). Figure 5c shows all of those perimeter sites which have been contacted 50 or more times and Fig. 5d shows those sites which have been contacted 2500 or more times. The maximum number of contacts for any perimeter site was 8197 so that $p_{max} = 8.2 \times 10^{-3}$. After Meakin et al.[19]

probability (the probability that site i is the next to be hit by a random walker). Clearly the set p_i plays a vital role in determining the dynamics of growth, since if we know all the p_i for every perimeter site i at a given time t, then we can predict (in a statistical sense) the state of the system at time $t + 1$.

Recently, considerable attention has focussed on the question of how a DLA aggregate grows. Such growth phenomena are **completely** characterized by assigning to each perimeter site i the number p_i, the probability that site i is the next to grow.

Theoretical evidence has been advanced recently to suggest that the numbers p_i form a multifractal set: this set cannot be characterized by a single exponent (as in the case of the DLA aggregate itself) but rather an infinite hierarchy of exponents is required. The physical basis for this fact is that the hottest tips of a DLA aggregate grow much faster than the deep fjords (which hardly grow at all); hence the **rate of change** of the p_i differs greatly when i is a tip perimeter site than when i is a fjord perimeter site.

Although there have been theoretical calculations of the multifractality of DLA,[18-20] there had been no experimental tests of these predictions. We have recently carried

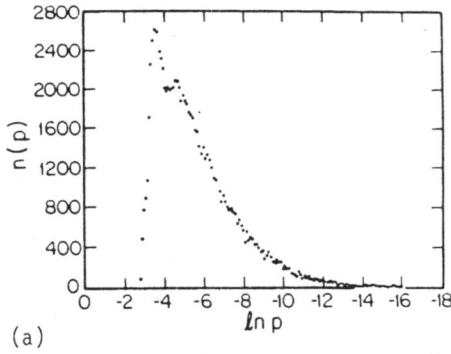
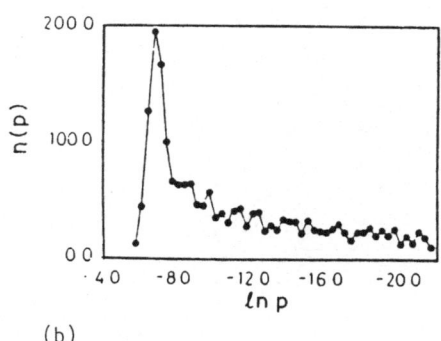

Fig. 6: Comparison between the distribution functions $n(p)$ for simulated (a) and "experimental" (b) viscous fingering patterns. Here $n(p)\delta p$ is the number of perimeter sites with growth probabilities in the range $[p, p + \delta p]$. The simulated patterns and their growth probabilities were obtained using the dielectric breakdown model. The growth probabilities for the experimental patterns were obtained by numerically solving the Laplace equation in the vicinity of a digitized representation of the pattern with absorbing boundary conditions on the sites occupied by the pattern. Similar results were obtained for large α (corresponding to the "tips") by directly subtracting two successive experimental patterns. After Amitrano et al[40] and Nittmann et al.[41]

out the first such tests, and found experimental confirmation of the broad outlines of the theory of multifractals.[21]

There are many experimental realizations of DLA, and for the present work we will focus upon two-dimensional fractal viscous fingers since it is possible to study the real-time growth using a movie camera and to digitize precisely the observed time development of the DLA fractal. By subtracting two successive "snapshots" we can obtain an accurate estimate of the appropriate normalized growth probability p_i for each perimeter site of the finger (Fig. 3).

We first calculated the distribution function $n(p)$, where $n(p)dp$ is the number of perimeter sites with p_i in the range $[p_i, p_i + dp_i]$. This curve has a long tail extending to the extremely small values of p_i for perimeter sites deep inside fjords. We found good agreement between the experimental $n(p)$ for viscous fingers (Fig. 6b) and the corresponding theoretical $n(p)$ calculated for DLA (Fig. 6a).

We next formed the moments $Z_q = \Sigma(p_i)^q$ which are characterized by the hierarchy of exponents τ_q defined through $Z_q = L^{-\tau_q}$, where L is a characteristic linear dimension. The experimental results (Fig. 7b) show that when q is large, τ_q is linear in q but for q small there is downward curvature in τ_q, showing that the fjords are characterized by different growth rates than the tips. It is conventional to also calculate the Legendre transform with respect to q of τ_q: $-f(\alpha) = \tau(q) - q\alpha$ where $\alpha = d\tau/dq$. Downward curvature in $\tau(q)$ corresponds to upward curvature in $-f(\alpha)$ [Fig. 8a]. The experimental data of Figs. 7b and 8b compare favorably with the theoretical DLA model calculations shown in Figs. 7a and 8a.

"Dendritic Solidification": Variants of the Fluid Mechanical Models

By analogy with the Ising model and its variants, we can modify DLA/DBM to describe other fluid mechanical phenomena. One of the most intriguing of these concerns a variation of the viscous fingering phenomenon in which there is present

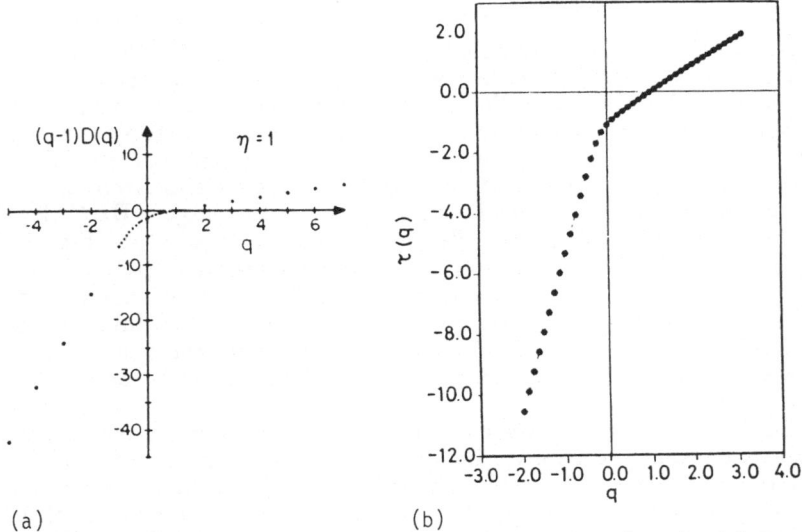

(a)   (b)

Fig. 7: Comparison of the critical exponents $\tau(q) = (q-1)D(q)$ for the (a) theoretical and (b) "experimental" viscous fingering patterns. In both cases, $\tau(q)$ was obtained numerically (see caption to Fig. 6). After Amitrano et al.[40] and Nittmann et al.[41]

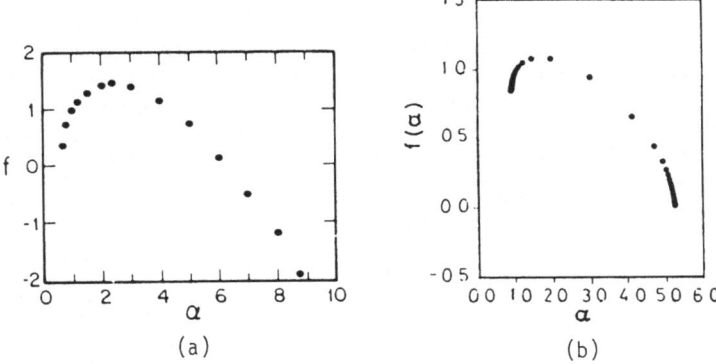

(a)   (b)

Fig. 8: Comparison between (a) theoretical and (b) "experimental" plots of the function $f(\alpha)$. After Amitrano et al.[40] and Nittmann et al.[41]

anisotropy. Ben Jacob et al[22] imposed this anisotropy from by scratching a lattice of lines on their Hele-Shaw cell. They found patterns that strongly resemble snow crystals! If viscous fingers are described by DLA, then can the Ben Jacob patterns be described by DLA with imposed anisotropy?

Nittmann and Stanley[23] attempted to answer this question—specifically, they attempted to reproduce the Ben Jacob patterns with suitably modified DLA. A scratch in a Hele-Shaw cell means that the plate spacing b is increased along certain directions, and the permeability coefficient k relating growth velocity to ∇P is proportional to b^2 ($k \propto b^2$). Hence Nittmann and Stanley calculated DLA patterns for the case in which there was imposed a periodic variation in the k. It is significant that their simulations reproduce snow crystal type patterns, just like the experiments. These simulations relied for their efficacy on the presence of noise reduction.

Noise Reduction

The original DLA and DBM models are prototypes of completely chaotic systems. No discernable pattern emerges. If there is a weak anisotropy, we expect that the resulting pattern reflects this anisotropy. For example, if the simulations are carried out on a lattice, then the presence of the lattice imposes a weak anisotropy (e.g., on a square lattice, it is more likely that particles attach to the westernmost tip if they approach from the west than from the north or south). This weak anisotropy is not visually apparent unless large clusters are grown. However, the largest DLA clusters made[8] with mass about 4 million sites, clearly display the anisotropy (Fig. 9). Unfortunately, no one can afford the computer resources to make such "mega-DLA" clusters each time we wish to model a new phenomenon. Noise reduction is a computational trick that seems to have the property that it speeds up the attainment of this asymptotic limit. In the absence of noise reduction, a perimeter site becomes a cluster site whenever it is chosen (e.g., whenever a random walker lands on that site).

"Noise reduction" means that we associate a counter with each perimeter site; each time that site is chosen, the counter increments by one. The perimeter site becomes a cluster site only after the counter reaches a pre-determined threshold value termed s.[23-25] When $s = 1$, we recover the original noisy DLA. Growth is dominated by the stochastic randomness in the arrival of random walkers. If s is very large, then growth is determined by the actual probability distribution.

For example, suppose we start with a large disc as a seed particle (instead of a single site). The growth probability at all points on the disc surface will be equal, assuming a continuum.

By the D'Arcy growth law this disc should evolve in time into a larger disc. On the other hand, for ordinary DLA ($s = 1$), as soon as a random walker touches a single perimeter site on the disc, this site will become part of the cluster and the disc will lose its circular symmetry. The growth probabilities will all be re-calculated, and the perimeter sites close by the one that just grew will have higher growth probabilities. Thus the disc with a single site added to it will be more likely to grow in the direction of that single site. At a later time we will almost certainly not find a cluster with circular symmetry.

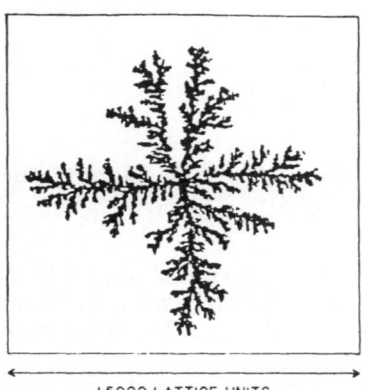

15000 LATTICE UNITS

(a)

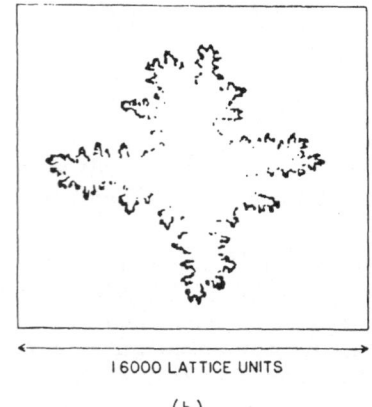

16000 LATTICE UNITS

(b)

Fig. 9: A huge DLA cluster with a mass of 4 million sites grown on a square lattice. Shown is only the last 5% of the growth. In reality, there is structure on all scales less than the width W of the 4 arms. Moreover, W scales with cluster mass as $W \sim (1/M)^{-1/d_f}$, just in the same way as the quantity ξ_L defined in Eq. (1c). The spontaneous appearance of side branches is reminiscent of experimental dendritic growth patterns such as those shown in Fig. 13. After Meakin.[8]

Clearly if s is very large, then the initial growth will preserve an almost circular structure. This is because before the first site is added to the circular seed, all the perimeter sites will acquire large numbers in their counters ($s - 1$, $s - 2$, etc.). After the first site is added, these additional perimeter sites will be very close to the threshold for growing while the new perimeter sites that were born when the first cluster site is added will all have counters initialized at zero. A typical cluster grown in this fashion is shown in Fig. 10; actually this cluster is grown on a square lattice with first and second neighbor interactions, not on a continuum. However, Meakin et al.[26] have found an almost identical pattern for the continuum case.

At first sight, there is little economy in computational speed, since one needs "s times as many" random walkers to reach a given cluster size. Thus to grow a cluster with merely 4000 sites with $s = 1000$ requires almost as much time as to generate a mega-DLA with 4,000,000 sites and $s = 1$. Fortunately, there is a way around this problem. Instead of using random walkers to solve the Laplace equation (to sample the growth probabilities p_i on each perimeter site), we can directly solve the Laplace equation numerically. This is the approach used when the dielectric breakdown model was first proposed (Fig. 11). Whether one calculates the growth probabilities by sending in random walkers or by solving the Laplace equation is immaterial: the difference between DLA and DBM is the boundary conditions, not the method of calculation.

The advantage of the Laplace equation approach when s is large is obvious: one need re-solve the Laplace equation only after a site is actually added to the cluster. In between adding sites, one simply chooses random numbers weighted by the growth probabilities of each perimeter site. This is a relatively rapid procedure for the computer, compared with its counterpart of sending random walkers.

"Snow Crystals"

Of course, real dendritic growth patterns (such as snow crystals) do not occur in an environment with periodic fluctuations in $k(x,y)$. Rather, the *global* asymmetry of the pattern arises from the *local* asymmetry of the constituent water molecules. Can this local asymmetry give rise to global asymmetry? Buka et al.[27] replaced the Ben Jacob experiment (isotropic fluid, anisotropic cell) by the reverse: isotropic cell but anisotropic fluid!

To accomplish this, they used a nematic liquid crystal for the high viscosity fluid. Thus the analog of the water molecules in a snow crystal are the rod-shaped anisotropic molecules of a nematic. This experiment shows that the underlying anisotropy can as well be in the fluid as in the environment.

Snow crystal formation is thought to involve mainly the aggregation of tiny ice particles and droplets of supercooled water. To the extent that snow crystals grow by accreting water molecules previously in the vapor or liquid phase, the growth rate is thought to be limited by the diffusion away from the growing snow crystal of the latent heat released by these phase changes. Under conditions of small Peclet number, the diffusion equation describing the space and time dependence of the temperature field $T(\mathbf{r}, t)$ reduces to the Laplace equation. Thus a reasonable starting point is DLA, independent of whether we wish to focus on particle aggregation, heat diffusion, or both.

DLA reflects well the randomness inherent in a wide range of growth processes, including colloidal aggregation, it fails to describe dendritic solidification. While the deterministic models of snow crystals produce patterns that are much too "symmetric," the DLA approach suffers from the opposite problem: DLA patterns are too "noisy." That DLA is too noisy has long been recognized as a defect of this otherwise physically appealing model. Recently, an approach has been proposed[21] that retains the "good" features of DLA and at the same time produces patterns that resemble real (random) snow crystals.

Firstly, we introduce[21] controlled amounts of noise reduction of the same sort used previously for both DLA and for DBM. It is believed that noise-reduced DLA is in the same universality class as ordinary DLA—i.e., it has the same fractal dimension d_f, the only difference being an increase in the characteristic local length scale W. One advantage of setting $s > 1$ is that the asymptotic ("mass" $= \infty$) behavior shows up much sooner than if $s = 1$. We do not explicitly introduce anisotropy—the only anisotropy present is the six-fold anisotropy arising from the underlying triangular lattice.

The patterns obtained[21] have the same general features for all values of s greater than about $s = 100$—the effect of increasing s seems mainly to be that of increasing the width W of the fingers and side branches. The fjords between the 6 main branches contain much empty space. Some snow crystals have such wide "bays" but some do not. A better model would seem to require some tunable parameter that enables the complete range of snow crystal morphologies to be generated. We have found one

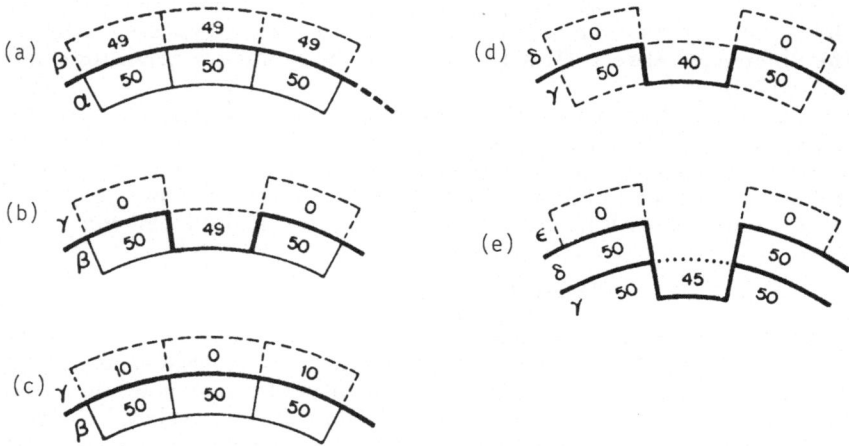

Fig. 10: Schematic illustration of the difference between an outward ('positive') and an inward ('negative') interface fluctuation. A positive fluctuation tends to be damped out rather quickly, as mass quickly attaches to the side of the extra site that is added. On the other hand, a negative fluctuation grows, in the sense that mass accumulates on both sides of the tiny notch. The notch itself has a lower and lower probability of being filled in, as it becomes the end of a longer and longer fjord. This is the underlying mechanism for the tip-splitting phenomenon when no interfacial tension is present. a shows the advancing front (row α) of a cluster with $s = 50$. The heavy line separates the cluster sites (all of which were chosen 50 times) from the perimeter sites (all of which have counters registering less than 50). In a, no fluctuations in the counters of these three sites have occurred yet, and all three perimeter counters register 49. b shows a negative fluctuation, in which the central perimeter site is chosen slightly less frequently than the two on either side; the latter now register 50, and so they become cluster sites in row β. The perimeter site left in the notch between these two new cluster sites grows much less quickly because it is shielded by the two new cluster sites. For the sake of concreteness, let us assume it is chosen 10 times less frequently. Hence by the time the notch site is chosen one more time, the two perimeter sites at the tips have been chosen 10 times (c). The interface is once again smooth (row γ), as it was before, except that the counters on the three perimeter sites differ. After 40 new counts per counter, the situation in d arises. Now we have a notch whose counter lags behind by 10, instead of by 1 as in b. Thus the original fluctuation has been amplified. After Nittmann and Stanley.[23]

such parameter, η, that has the desired effect of reducing the difference in the ratio of the growth probabilities between the tips and fjords. Specifically, we relate by the rule $p_i \propto (\nabla \phi)^\eta$ the growth probability p_i (the probability that perimeter site i is the next to grow) to the potential ϕ (e.g., ϕ may be the temperature $T(\mathbf{r})$ at point $\mathbf{r}$, or the probability that a tiny ice particle is at point $\mathbf{r}$). Our model is thus the analog for DLA of the "η model."

We used η to tune the balance between tip growth and fjord growth and found growth patterns that resemble better the wide range of snow crystal morphologies that have been experimentally observed.[21] To what does the case $\eta \neq 1$ correspond? For $\eta = k$ (k = positive integer), we have a model[28] in which a site grows only if it is chosen k times in succession ($k = 1$ is pure DLA). It is possible that we have a situation not altogether different from the classic n-vector model of isotropically-interacting n-dimensional classical spins: this model makes physical sense only if n is a positive integer, yet its study for other values of n has led to rich insights—particularly the cases $n = 0$ (the dilute polymer chain limit), $n = \infty$ (the spherical model) and $n = -2$ (the mean field limit). Similarly, the Q-state Potts model makes physical sense only if Q is an integer above 1, yet the cases $Q = 0$ (random resistor network), $Q = 1$ (percolation) and $Q = 3/2$ (a spin glass model) are of great interest.

The fractal dimension d_f is believed independent of the value of the noise reduction parameter s (s renormalizes the cluster mass). We confirmed this belief. However, we found d_f does depend on η. The most reliable estimates were obtained by first calculating estimates of d_f for a sequence of increasing cluster masses, and then extrapolating this sequence to infinite cluster mass. Our values for d_f agreed remarkably well with values we obtained by digitizing photographs of experimentally observed snow crystals. Of course this preliminary study[21] does not completely "solve" the snow crystal problem:

(i) The initial seed of a snow crystal is almost certainly hexagonal (i.e., quasi-2-dimen- sional), since this is the local geometry that water molecules take when they form hexagonal ice I_h. Are DBM-type considerations (small growth probability near the center of a plate-like structure) sufficient to explain why a snow crystal remains quasi-2-dimensional as it continues growing? Why does its thickness remain less than its width? It is perhaps appropriate to mention that no adequate explanation has yet been advanced for why a snow crystal remains quasi-2-dimensional throughout its growth, despite the fact that the "assembly plant" is certainly 3-dimensional. Intuition on this subject stems from experience not only from critical phenomena but also from recent theoretical and experimental work on pattern formation, where it was found that even minute amounts of anisotropy are sufficient to stabilize structures of lower effective dimension.

(ii) What are the microscopic mechanisms that give rise to the feature that real snow crystals contain branches (and side branches) which are much more than one molecule thick? Is noise reduction relevant, or is noise reduction merely a "computational trick" that allows one to see the asymptotic form of a DLA cluster using reasonable masses? (E.g., on a square lattice, the same cross-like pattern for a mass of 5,000 sites seen in noise-reduced DLA with a noise-reduction parameter of $s = 500$ is also seen in ordinary "noisy" DLA ($s = 1$) provided the mass is allowed to increase to roughly 5,000,000 sites!) We know that DLA is obtained even if the incoming random walkers have a sticking probability that is less than one. Hence we anticipate that DLA might possibly describe a modest range of phenomena with structural re-arrangement. What is the actual sticking probability for newly arriving water molecules in real snow crystals? Is a value of the sticking probability less than unity sufficient to account for the fact that the arms and sidebranches of real snow crystals have macroscopic thickness.

(iii) Are those real snow crystals which possess relatively compact cores with ramified dressing on their surfaces products of different environments of assembly, or did melting and structural re-arrangement take place after formation? Can one

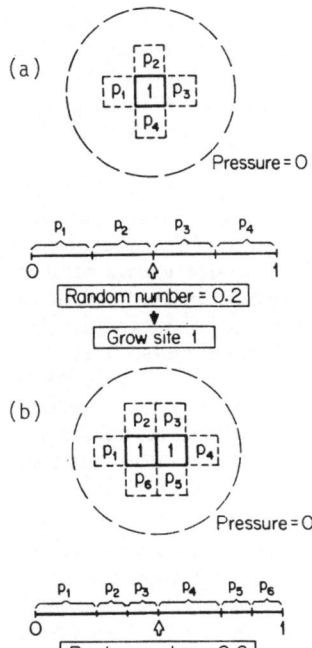

Fig. 11: Schematic illustration of the first steps in the generation of a DLA cluster by solving directly the Laplace equation on a square lattice.

mimic the effect of the changing environments in which a given snow crystal is actually assembled? Do these correspond to varying parameters such as η or γ *in the course of the growth process*? To study this effect, we generated patterns with values of η and γ that change during the growth process–e.g., we might choose $\eta \ll 1$ for an initial fraction f of the growth (thereby creating a hexagonal core), and $\eta = 1$ thereafter (thereby creating a ramified exterior portion).

(iv) Does the presence in the clouds of a wind whose direction and speed varies randomly (both in time and in space, with characteristic time scales and length scales that are microscopic) imply that the actual trajectories of water molecules and water droplets might more resemble those of some extremely "pathological" path than those of a conventional DLA type random walk? We know that the random walk trajectories of DLA correspond exactly to the present electrostatic growth model, the DBM with DLA boundary conditions. What are the trajectories in "real space" corresponding to a choice of the η parameter below unity? One can speculate that a Lévy flight with tunable fractal dimension may be related to the path of a real ice particle buffeted around in a cloud.

(v) How significant, in practice, is the role played by diffusion of latent heat away from the growing aggregate in determining the actual structure of a snow crystal? We know that this phenomenon is of paramount importance in dendritic growth of crystals from a liquid phase How significant is the role played by the capillary length $d_o = \gamma/L$ in vapor phase deposition of water molecules onto a growing snow crystal?

(Here L is the latent heat.) An ideal model might encompass *both* the diffusion of heat away from the snow crystal *and* the aggregation of particles toward the snow crystal?

(vi) Are real snow crystals sometimes fractal objects? This intriguing question has been the object of considerable discussion in recent years. Our growth patterns are fractal, for all positive values of η. We found[21] that the fractal dimension d_f is independent of the value of the noise reduction parameter s (s seems to mainly renormalize the cluster mass), but d_f does depend on η. We also found that these values for d_f agreed well with values we obtained by digitizing the corresponding photographs of experimentally observed snow crystals (Fig. 12).

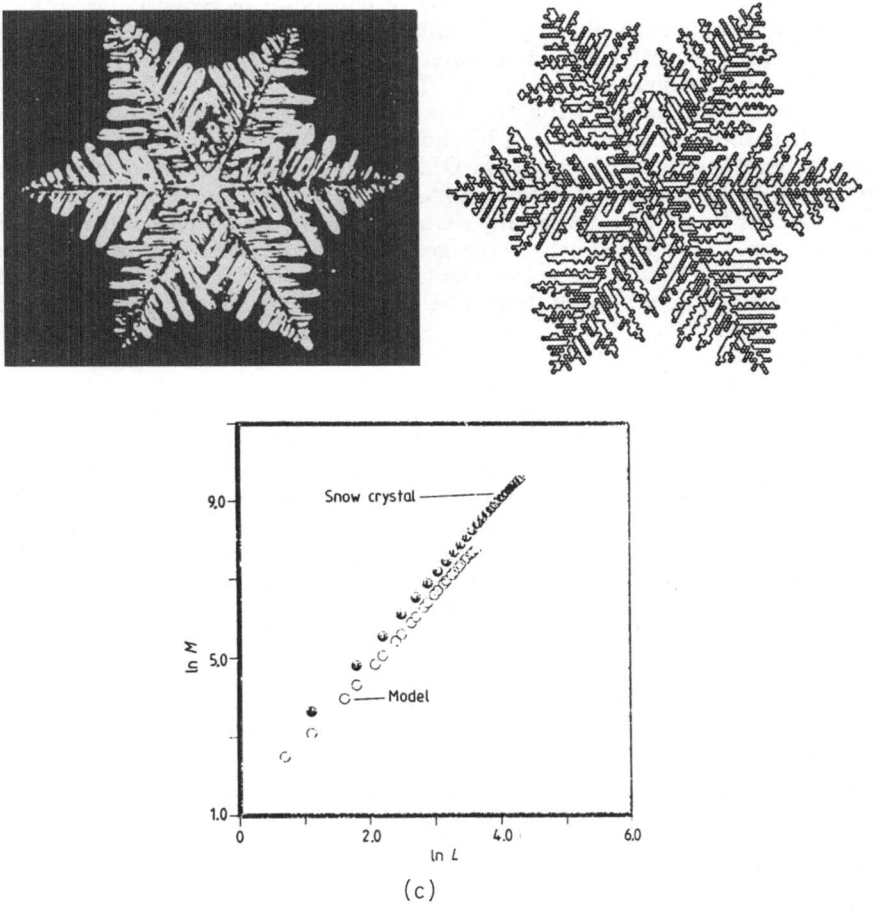

Fig. 12: (a) A typical snow crystal from the collection of 2453 photographs assembled in Bentley and Humphreys.[44] Other experimental examples may be found in Nakaya[42] and LaChapelle.[43] (b) A DLA simulation with noise reduction parameter of $s = 200$ and non-linearity parameter $\eta = 0.5$. (c) Comparison between the fractal dimensions of (a) and (b) obtained by plotting the number of pixels inside an $L \times L$ sandbox logarithmically against L. The same slope, $d_f = 1.85 \pm 0.06$, is found for both. The experimental data extend to larger values of L, since the digitizer used to analyze the experimental photograph has 20,000 pixels while the cluster has only 4000 sites. After Nittmann and Stanley.[21]

Dendritic Growth of NH_4Br

Dendritic crystal growth has been a field of immense recent progress, both experimentally and theoretically. In particular, Dougherty et al.[29] have recently made a detailed analysis of stroboscopic photographs, taken at 20 second intervals, of dendritic crystals of NH_4Br (Fig. 13a). They have found three surprising results: (i) the sidebranches are non-periodic at any distance from the tip, with random variations in both phase and amplitude, (ii) sidebranches on opposite sides of the dendrite are essentially uncorrelated, and (iii) the rms sidebranch amplitude is an exponential function of distance from the tip, with no apparent onset threshold distance. Some of these results are apparently at variance with predictions from recent theories.[30-32]

How can we understand these new experimental facts? Many existing models reflect the essential physical laws underlying the growth phenomena, but fail to find a tractable mechanism to incorporate the effects of noise on the growth. Growth of a dendrite from solution is controlled by the diffusion of solute towards the growing dendrite. In the limit of small Peclet number, the diffusion equation reduces to the Laplace equation (as mentioned above). The Laplace equation for a moving interface (the growing dendrite) brings to mind the diffusion limited aggregation model (DLA). Growth patterns produced by the various DLA simulation algorithms do *not* resemble dendritic growth patterns: DLA patterns are much too chaotic in appearance. We shall discuss here a related model[33] whose asymptotic structure does resemble the patterns found experimentally—both in broad qualitative features and in quantitative detail. The picture that emerges is one of Laplacian growth, where noise arises from the fact that there are concentration fluctuations in the vicinity of the growing dendrite (these are estimated to be roughly $\pm 10^5$ NH_4Br molecules per cubic micron).

Our starting point is the observation that minute amounts of anisotropy become magnified as the mass of a cluster increases. In fact, even the weak anisotropy of the underlying lattice structure can become so amplified that clusters of 4,000,000 particles take on a cross-like appearance (cf. Fig. 1 of Ref. 8). A real dendrite has a mass of roughly 10^{16} particles; it is impossible to generate clusters of this size on a computer, since even clusters of size 10^6 require hundreds of hours on the fastest available computers. Fortunately, there is a computational trick—termed *noise reduction*—that speeds the convergence of the pattern toward its asymptotic "infinite mass" limit. The patterns we obtained with noise-reduced DLA resemble Fig. 1 of Dougherty et al.[29], reproduced in Fig. 13a.

A typical result[33] for a mass of 4000 particles is shown in Fig. 13b. After each 333 particles are added, a contour is drawn:

(i) It is apparent from the "stroboscopic" representation of Fig. 13b that the distance between successive tip positions is a decreasing function of the mass; in fact, we find that $log\ x_{tip}$ is linear in $log\ M$ with slope 2/3. This result is consistent with the belief that $d_f = 1.5$ for DLA with anisotropy.

(ii) The tip is remarkably parabolic: specifically, when we form $(y_c - y_o)^2$ (where y_c is the contour, and y_o is the centerline of the dendrite) and plot this on linear graph paper as a function of $x - x_{tip}$, we obtain a straight line with an R value of 0.997.

(iii) The sidebranches are non-periodic at any distance from the tip, with random variations in both phase and amplitude. To demonstrate this, we have analyzed our simulations in exactly the same mathematical fashion as Dougherty et al. analyzed the experimental dendrite patterns.

An open theoretical question concerns the microscopic origin of the sidebranching phenomenon. One current hypothesis predicts that the sidebranch amplitude would be periodic and the two sides of the dendrite should have correlated sidebranching. Dougherty et al.[29] noted that their experimental data are not consistent

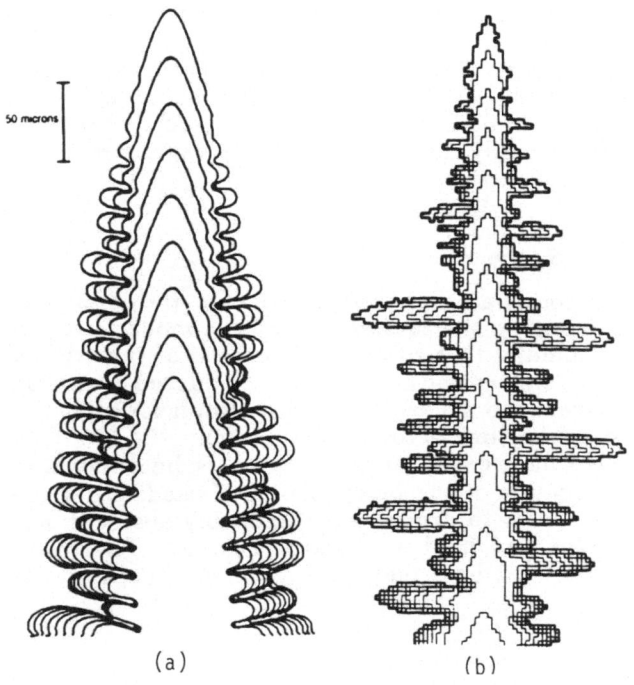

(a)   (b)

Fig. 13: (a) Experimental pattern of dendritic growth, measured for NH$_4$Br by Dougherty et al.[29] (b) DLA simulation with noise reduction parameter $s = 200$ (after Nittmann and Stanley[33]).

with this hypothesis, and we can make similar remarks for the present model. A second hypothesis views sidebranching as a result of the noise arising from concentration fluctuations. To test this hypothesis, Dougherty et al[29] plot the sidebranch amplitude as a function of $x - x_{tip}$, the distance from the tip. They found that the sidebranch amplitude decreases as the distance variable $x_{tip} - x$ decreases, and shows no sign of a threshold distance below which the amplitude is zero.

Moreover, they found that close to the tip the sidebranch amplitude is roughly linear on semi-log paper. If we plot y_c, the amplitude, which should scale roughly as the square root of the area under the peak if the peak maintains its shape as a function of $x - x_{tip}$; we find exactly the same exponential growth of sidebranch amplitude with distance from the tip.

In summary, we have developed a model in which noise reduction is used to tune the effect of noise, and cubic anisotropy is introduced through the use of an underlying square lattice. The resulting patterns obtained strongly resemble the experimental patterns of Dougherty et al. both in their *qualitative* appearance and in the same degree of *quantitative* detail studied experimentally. Sidebranching arises from the fact that an approximately flat interface in the DLA problem grows trees (which resemble "bumps" in the presence of noise reduction); these compete for the incoming flux of random walkers. If one tree gets ahead, it has a further advantage for the next random walker and so gets ahead still more. Thus some sidebranches grow while others do not. The characteristic spacing λ between sidebranches scales with the dendrite mass with the same exponent 2/3 that characterizes the growth of dendrite length x_{tip}. Moreover, the patterns we obtain are reasonably independent of details of the simulation in that similar patterns are obtained when we vary the surface tension parameter σ over a modest range; we can also alter the boundary conditions of the model with some latitude and even allow for non-linearity in the growth process ($\eta \neq 1$).

The significance of the present findings is that the essential physics embodied in the DLA model—previously used to describe fluid-fluid displacement phenomena ("viscous fingering")—seems sufficient to describe the highly uncorrelated (almost random) dendritic growth patterns recently discovered from the experiments and quantitative analysis of Dougherty et al.

Summary

We have argued that it is worth exploring all the consequences of a straightforward physical model. Our optimism is based on the success of the Ising model and percolation in the past. We must be mindful that substantial variants of the original model may be called for. In our case, e.g., anisotropy must be introduced or else the pattern bears absolutely no resemblance to dendritic growth. Also, noise reduction must be introduced or else the computer time becomes prohibitive.

This modest work perhaps raises more questions than it answers, but it nonetheless might stimulate further investigation of the basic physics of random systems that must be better understood in order to explain experimentally-observed non-symmetric dendritic growth patterns and fluid mechanics patterns. The reader interested in more details than provided here may consult recent books on the subject.[34-39]

Acknowledgements

I'd like to thank the Organizing Committee for inviting me. The results I report here are largely due to interactions with A. Coniglio, G. Daccord, P. Meakin, E. Touboul, T. A. Witten and, most especially, J. Nittmann.

[1] LENZ, W., 1920, Phys. Z., **21**, 613-5.

[2] ISING, E., 1925, Ann. Physik, **31**, 253-258.

[3] KERTÉSZ, J., STAUFFER, D., and CONIGLIO, A., 1985, Ann. Israel Phys. Soc. (Adler, Deutscher and Zallen, eds), pp. 121-148.

[4] GEIGER, A., and STANLEY, H. E., 1982b, Phys. Rev. Lett., **49**, 1895.

[5] GAWLINSKI, E.T., and STANLEY, H. E., 1981, J. Phys. A., **14**, L291-L299.

[6] GEIGER, A., and STANLEY, H. E., 1982a, Phys. Rev. Lett., **49**, 1749.

[7] BALL, R., 1986, Physica, 140A, 62.

[8] MEAKIN, P., 1986a, J. Theor. Biol., **118**, 101.

[9] CHANDRASEKHAR, S., 1943, Rev. Mod. Phys., **15**, 1.

[10] NIEMEYER, L., PIETRONERO, L., and WEISMANN, H. J., 1984, Phys. Rev. Lett., **52**, 1033.

[11] WITTEN, T. A., and SANDER, L. M., 1981, Phys. Rev. Lett., **47**, 1400.

[12] NITTMANN, J., DACCORD, G., and STANLEY, H. E., Nature **314**, 141 (1985).

[13] DACCORD, G., NITTMANN, J., and STANLEY, H. E., 1986, Phys. Rev. Lett., **56**, 336.

[14] MÅLØY, K. J., FEDER, J., and JØSSANG, T., 1985, Phys. Rev. Lett., **55**, 2688.

[15] CHEN, J. D. and WILKINSON, D., 1985, Phys. Rev. Lett., **55**, 1985.

[16] STANLEY, H. E., 1977, J. Phys. A, **10**, L211-L220.

[17] CONIGLIO, A., 1982, J. Phys. A, **15**, 3829.

[18] MEAKIN, P., STANLEY, H. E., CONIGLIO, A., and WITTEN, T. A., 1985, Phys. Rev. A, **32**, 2364.

[19] MEAKIN, P., CONIGLIO, A., STANLEY, H. E., and WITTEN, T. A., 1986, Phys. Rev. A, **34**, 3325-3340.

[20] HALSEY, T. C., MEAKIN, P., and PROCACCIA, I., 1986, Phys. Rev. Lett., **56**, 854.

[21] NITTMANN, J., and STANLEY, H. E., 1987a, J. Phys.A, **20**, L1185(1987).

[22] BEN-JACOB, E., GODBEY, R., GOLDENFELD, N. D., KOPLIK, J., LEVINE, H., MUELLER, T., and SANDER, L. M., 1985, Phys. Rev. Lett., **55**, 1315.

[23] NITTMANN, J., and STANLEY, H. E., 1986, Nature **321**, 663-668.

[24] TANG, C., 1985, Phys. Rev. A., **31**, 1977.

[25] KERTÉSZ, J., and VICSEK, T., 1986, J. Phys. A, **19**, L257.

[26] MEAKIN, P. (et al) [unpublished].

[27] BUKA, A., KERTÉSZ, J., and VICSEK, T., 1986, Nature, **323**, 424.

[28] MEAKIN, P., 1986b, Proc. Israel Conference on Fracture.

[29] DOUGHERTY, A., KAPLAN, P. D., and GOLLUB, J. P., 1987, Phys. Rev. Lett. **58**, 1652.

[30] SAITO, Y., GOLDBECK-WOOD, G., and MÜLLER-KRUMBHAAR, H., 1987, Phys. Rev. Lett., **58**, 1541.

[31] BEN-JACOB, E., GOLDENFELD, N., KOTLIER, B. G., and LANGER, J. S., 1984, Phys. Rev. Lett., **53**, 2110.

[32] KESSLER, D., KOPLIK, J., and LEVINE, H., 1984, Phys. Rev. A, **30**, 3161.

[33] NITTMANN, J., and STANLEY, H. E., 1987b J. Phys. A **20**, L981 (1987).

[34] FAMILY, F. and LANDAU, D. P. (eds), 1984, Kinetics of Aggregation and Gelation (Elsevier, Amsterdam).

[35] BOCCARA, N. and DAOUD, M. (eds), 1985, Physics of Finely Divided Matter [Proceedings of the Winter School, Les Houches, 1985] (Springer Verlag, Heidelberg).

[36] PYNN, R. and SKJELTORP, A. (eds.), 1986, Scaling Phenomena in Disordered Systems (Plenum, N.Y.).

[37] STANLEY, H. E., and OSTROWSKY, N. (eds), 1986, On Growth and Form: Fractal and Non-Fractal Patterns in Physics (Martinus Nijhoff, The Hague).

[38] PIETRONERO, L. and TOSATTI, E. (eds), 1986, Fractals in Physics (North Holland, Amsterdam).

[39] STANLEY, H. E., 1989 Introduction to Fractal Phenomena (in press).

[40] AMITRANO, C., CONIGLIO, A., and DI LIBERTO, F., 1986, Phys. Rev. Lett., **57**, 1016.

[41] NITTMANN, J., STANLEY, H. E., TOUBOUL, E., and DACCORD, G., 1987, Phys. Rev. Lett., **58**, 619.

[42] NAKAYA, U., 1954, Snow Crystals (Harvard Univ Press, Cambridge).

[43] LaCHAPELLE, E. R., 1969, Field Guide to Snow Crystals (U. Washington Press, Seattle).

[44] BENTLEY, W. A., and HUMPHREYS, W. J., 1962, Snow Crystals (Dover, NY).

Phase Transition and Fractals:
Fractal Configurations of the Ising Models

N. Ito and M. Suzuki

Department of Physics, Faculty of Science, University of Tokyo,
Bunkyo-ku, Tokyo 113, Japan

Fractal geometric objects behind critical phenomena have been studied in our previous papers[1,2]. In this paper , our results are reviewed and a characterization method of the Ising configurations based on our fractal dimensionality is presented.

What kind of configurations are important to the thermodynamic properties of a statistical mechanical system? To discuss it concretely, we treat the ferromagnetic Ising model in the following. Of course, all configurations contribute to the statistical average in finite systems. But there are some "typical"configurations at every temperature. For example, the ground states are not typical at the disordered phase and the ground states of the antiferro system are not typical at all temperatures. The random spin configurations are typical at the high temperature limit but not at ordered phase.

Our fractal dimensionality defined for Ising configurations can be applied to characterize typical configurations at any temperatures. It is defined like this:

$$D_n(b) = \frac{\log\left(M(T_b^{n-1}\Lambda)/M(T_b^n\Lambda)\right)}{\log b},$$

(1)

where T_b is the majority-rule or box-counting scale transformation with scale b which operates on the Ising configuration, Λ, and M denotes the magnetization of configuration. Our first motivation to investigate the fractal dimensionality of the Ising model is to study the self-similarity of each typical configuration at the critical point[1,3]. The magnetization of such configurations behaves self-similarly with dimensionality D. The value of D is expected to be

$$D = d - \frac{\beta}{\nu'} = \frac{1}{2}(d + \frac{\gamma}{\nu'}),$$

(2)

where ν' denotes the finite-size-scaling exponent which is equivalent to the critical exponent ν for $d \leq 4$ and expected to be $2/d$ from eq.(2) for $d \geq 4$. This eq.(2) is the hyperscaling relations for $d \leq 4$. This fractalness is related with the hyperscaling and finite-size scaling. The discussion on them is given in Ref. [1-3]. For two-, three- and four-dimensional Ising models, this fractalness of the magnetization is studied using Monte Carlo simulations[1,2]. The observed values of D have a small deviation and approach the expected values (2) at the critical point. Therefore our fractal picture at the critical point is confirmed for $d = 2, 3$ and 4.

This is the fractalness of the magnetization which is the difference between the numbers of up sites (N_+) and down sites (N_-). That is

$$N_+ - N_- = a(\frac{L}{b^n})^D \qquad (n \to \infty \text{ at } T_c),$$

(3)

Springer Series in Synergetics, Vol. 43   **Cooperative Dynamics in Complex Physical Systems**
Editor: H. Takayama   © Springer-Verlag Berlin, Heidelberg 1989

where L is the linear dimension of the original system. The total number of sites must be $(L/b^n)^d$. Therefore

$$N_+ + N_- = (\frac{L}{b^n})^d.$$ (4)

From eqs. (3) and (4), the form of $N_\pm$ is given by

$$N_\pm = \frac{1}{2}(\frac{L}{b^n})^d \pm \frac{a}{2}(\frac{L}{b^n})^D.$$ (5)

Therefore the up sites or down sites do not have the dimensionality D, but d.

At temperatures higher than the critical point the values of $D_n(b)$ approach the random percolation value, $d/2$, and at lower temperatures they approach d. At temperature different from T_c, the deviation of D becomes large. This D behaves differently at a different temperature. The behaviors of D are shown in Ref.[1] and [2]. This behavior of D can be used as the index of a typical configuration at a given temperature. Especially, the transient behavior of D before they converge will be suitable for this purpose. If there is a typical configuration of the Ising model at unknown temperature, we will be able to estimate the temperature by measuring the dimensionality and comparing it with the known behavior of D. The simplest method to do so is proposed in the following.

The values of D_n ($n = N, N+1, \ldots, M$ and b is fixed) are regarded as functions of temperature, T, and are rewritten as D_n^T. The values of N and M are some appropriate positive integers. For example $N = 1$ and $M = max\{n : \text{positive}; b^n \le L\}$. The deviations of D_n^T are denoted by σ_n^T. These D_n^T and σ_n^T can depend on the system size. The values of $D_n(b)$ ($n = N, N+1, \ldots, M$) are measured for the relevant configuration and the results are denoted by a_n ($n = N, N+1, \ldots, M$). Then the temperature of the configuration is estimated like this:

$$\frac{\partial F}{\partial T} = 0; \quad F = \sum_{n=N}^{M} \frac{1}{(\sigma_n^T)^2}(a_n - D_n^T)^2.$$ (6)

This is the simplest example. Many other formulations for estimating the temperature are possible based on our fractal dimensionality.

The above mentioned characterization is not the unique method. The spin correlation functions or energy can be also used for this purpose. If we combine these methods, the estimation will be more accurate.

There are some unsolved problems in our fractal picture. The fractal dimensionality of energy is not defined[3]. The relation between cluster distribution theories are not clarified. Although there are such problems, our fractal configuration theory gives a simple geometrical picture of critical phenomena.

Acknowledgment

One of the authors (N.I.) would like to thank T. Ikegami for useful discussion.

References

[1] N.Ito and M.Suzuki,   Prog. Theor. Phys. **77** (1987) 1391.
[2] N.Ito and M.Suzuki: to be published in Proceedings of International Conference on Magnetism (1988, Paris)
[3] M.Suzuki,   Prog. Theor. Phys. **69** (1983) 65.

Aggregation of Particles with Deterministic Trajectories

Y-h. Taguchi

Department of Physics, Tokyo Institute of Technology,
Oh-okayama, Meguro-ku, Tokyo 152, Japan

1. Introduction

The diffusion-limited aggregation (DLA)[1] is currently of major interest in the field of physics of pattern formation, because it is a simple model of several physical phenomena: metal leaves, viscous fingerings, and dielectric breakdown, etc. In all of the phenomena, growth probabilities depend on the gradient of a solution of the Laplace equation. DLA and these physical phenomena are known to show fractal patterns. Hence much effort has been made to obtain the fractal dimension d_f of DLA. Remarkably, in some dimension analysis treatments [2,3], though the Laplace equation is not taken into account at all, the result ($d_f = 5/3$) agrees with that of numerical simulations ($d_f = 1.7$) [1]. The fact they make use of is that only trajectories of particles have fractal dimension d_w of two. Here we notice a possibility that other trajectories with d_w of two can also generate fractal patterns with the same d_f as that of DLA. To clear up this point, we consider trajectories which are not diffusive but have d_w of 2.

2. Model and Result

We employ a sort of Peano curve shown in Fig. 1. It has d_w of two. The growth process is started with a single stationary particle on a square lattice. A particle is launched from a random point on a diamond which encloses the cluster and the direction of an initiator is also randomly chosen from among the four directions. The particle must move along the Peano curve. Reaching a neighboring site of a seed particle, it sticks at that point. Another particle is released from a new random point along a new random direction. The processes are iterated until a large aggregation grows. The model is called 'deterministic DLA' (for short DDLA) [4,5]. The result, shown in Fig.2, looks like DLA and has a diamond shape, i.e. axial anisotropy. In addition, d_f is very close to that of DLA. Hence it may belong to the same universality class as DLA.

The dependence of D_q upon q, where D_q is the q-th general dimension of the growth probability distribution (GPD)[5], is shown in Fig. 3. Contrary to DLA, GPD of DDLA has little multifractality, though DDLA generates fractal patterns. The results of other DDLAs [4] are shown in Fig. 4. These are aggregates of particles moving on deterministic trajectories with d_w of two which are not Peano curves.

3. Conclusions

We propose a new type of aggregation process, DDLA. DDLA may belong to the same universality class as DLA. Because DDLA has no relationship to the Laplace equation, the latter equation does not seem to be a necessary condition for producing DLA-like clusters.

Springer Series in Synergetics, Vol. 43   **Cooperative Dynamics in Complex Physical Systems**
Editor: H. Takayama   © Springer-Verlag Berlin, Heidelberg 1989

In addition to this, DDLA has little multifractality of GPD. The multifractality of GPD is not essential for generating fractal patterns of aggregations.

Fig. 1 An initiator (left) and a generator (right) of the Peano curve used in DDLA simulation.

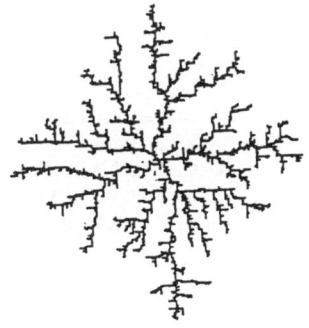

Fig. 2 A cluster of 2964 particles of DDLA [4,5] ($d_f = 1.709 \pm 0.001$).

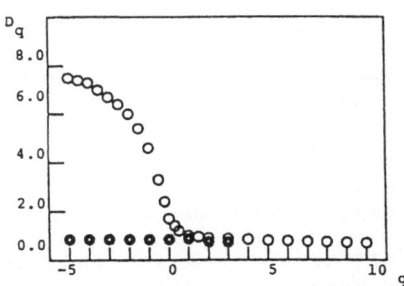

Fig. 3 The dependence of D_q upon q. ⊙are for DDLA, and ○ for DLA.

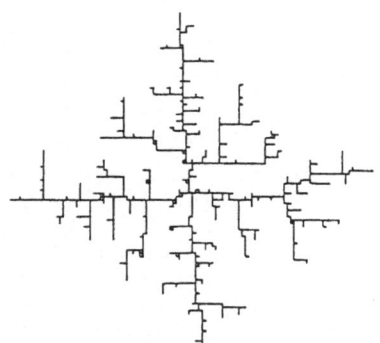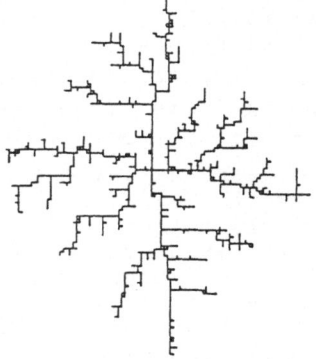

Fig. 4 Other DDLAs [4]. The left pattern has d_f of 1.572 ± 0.002 and the right, 1.614 ± 0.001.

References

[1] P. Meakin, *Phase Transitions and Critical Phenomena* , ed. by C.Domb, J.L.Lebowitz, vol.12 (Academic, New York 1988) p.335

[2] M. Tokuyama, K. Kawasaki: Phys. Lett. **100A**, 337 (1984)

[3] M. Matsushita, K. Honda, H.Toyoki, Y. Hayakawa, H. Kondo: J. Phys. Soc. Jpn. **55**, 2618 (1986)

[4] Y-h. Taguchi J. Phys. A **21**, 4235 (1988)

[5] Y-h. Taguchi J. Phys. A (1988) submitted

Anomalous Diffusion in a Fractal Potential

H. Hayakawa[1], *M. Yamamoto*[2], *and H. Takayasu*[3]

[1]Department of Physics, Kyushu University 33, Fukuoka 812, Japan
[2]Toyo Information Systems Co., Ltd, Suita, Osaka 564, Japan
[3]Applied Physics, Yale University, New Haven, CT06520, USA

Anomalous diffusion on a fractal structure has been investigated by many authors (reviewed by Ref.[1]). Such diffusion is characterized by the the spectral (fracton) dimension[2][3], which is defined through the mean number of distinct sites visited as $S(t) \sim t^{d_s/2}$ (d<2) where d_s and d are the spectral dimension and the spatial dimension respectively.

It is interesting to consider the effect of the temperature in the problem of diffusion on fractal structure. ARGYRAKIS et al./4/,/5/ investigated random walks in a random potential and found that the spectral dimension depends on temperature. In this report we investigate random walks in a fractal potential which is constructed as follows. Let us consider the two dimensional Sierpinski's gasket (SG). We replace the sites cut out by the gasket by hierarchical triangular cones whose heights represent the values of the potential field $E(\vec{r})$ (see Figs.1). Let the hopping rate from site $\vec{r}$ to $\vec{r}'$ be $W(\vec{r},\vec{r}') = z^{-1}\exp(-\Delta E(\vec{r},\vec{r}')/T)$ for $\Delta E(\vec{r},\vec{r}')=E(\vec{r}')-E(\vec{r})>0$ and $W(\vec{r},\vec{r}') =z^{-1}$ for $\Delta E(\vec{r},\vec{r}')<0$ where z is the coordination number (z=6 in our case). We solve the equation of motion for the walker in terms of the Monte Carlo method with 1024x1024 lattice. The Monte Carlo average is taken over 5000 samples during an interval from t=0 to t=50000. The observed time scale is sufficiently smaller than the cross-over time (t~10⁶) in which the walker recovers the usual diffusion. From our simulation the spectral dimension is well-defined at each finite temperature after t≃5x10³. At

(a)   (b)

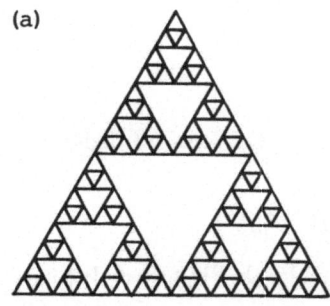 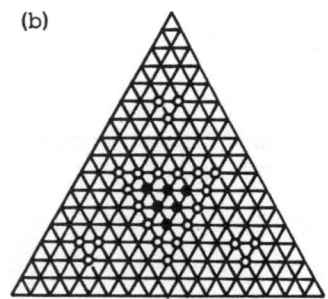

Figs.1: The Sierpinski's gasket (a) and potential field in our simulation with one side having 17 sites (b). In Fig.1(b), vertices, open circles and solid circles are the sites having the values of the potential $E(\vec{r})$=0,1 and 2, respectively.

Springer Series in Synergetics, Vol. 43   **Cooperative Dynamics in Complex Physical Systems**
Editor: H. Takayama   © Springer-Verlag Berlin, Heidelberg 1989

infinite temperature, however, the spectral dimension increases
in proportion to 1/logt, as has been anticipated from the
logarithmic correction for d=2. The spectral dimension increases
with the temperature as is shown in Fig.2. As the temperature
goes to zero, d_s approaches $d_s(T=0)= 2\log3/\log5 \simeq 1.36$ which is
the value on SG[1]. As the temperature goes to infinity, the
spectral dimension shows a rapid change to approach the classical
value $d_s=2$.

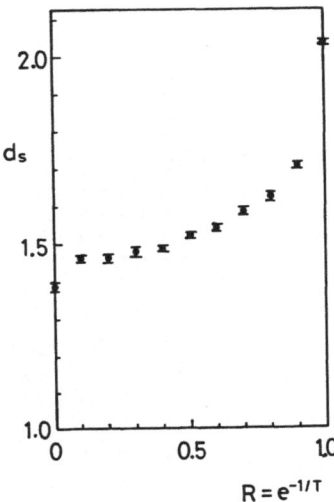

Fig.2: The temperature $R=e^{-1/T}$
dependence of the spectral dimension
d_s.

We conclude our results. We find that the spectral dimension
depends on the temperature. However, it is difficult to
determine the functional form of $\dot{d}_s(T)$. It is the problem to be
studied in the near future.

Main parts of this work have been done in Kobe University.

1. S.Havlin and D.Ben-Avraham: Adv.Phys.**36**,695 (1987)
2. S.Alexander and R.Orbach: J.Physique Lett.**43**,L625 (1982)
3. R.Rammal and G.Toulouse: J.Physique Lett.**44**,L15 (1983)
4. P.Argyrakis, L.W.Anacker, R.Kopelman: J.Stat.Phys.**36**,579 (1984)
5. P.Argyrakis and R.Kopelman: J.Phys.**A21**,2753 (1988)

Dynamical Behavior of Fractal Clusters via Aggregation and Evaporation Processes

K. Honda[1], *Y. Hayakawa*[2], *and M. Matsushita*[3]

[1]Department of Applied Physics, Nagoya University, Nagoya 464-01, Japan
[2]Research Institute of Electrical Communication, Tohoku University, Sendai 980, Japan
[3]Department of Physics, Chuo University, Kasuga, Bunkyo-ku, Tokyo 112, Japan

1. Introduction

Reversible diffusion-limited aggregation (RDLA) model proposed by BOTET et al.[1] is considered to discuss the dynamical properties of the configuration of fractal clusters. RDLA model is defined as follows: We prepare a loopless cluster composed of N particles. One particle linked to the rest of the cluster by only one bond, which will be called a tip hereafter, is chosen randomly and its unique bond is broken. This particle random-walks in a d-dimensional space, following the trajectory with a fractal dimension d_w, until it sticks to the cluster again. This procedure is iterated. The configuration of the cluster is effectively characterized by the radius of gyration, R. The sample average of the radius of gyration approaches asymptotically, with iterative steps, an equilibrium value independent of the initial conditions.

The resulting cluster is fractal in a statistical sense. Recently we obtained the analytic expression of the fractal dimension d_f as [2]

$$d_f = \{d^2 + 2(d_w-1)\}/\{d + 2(d_w-1)\}, \tag{1}$$

which is in good agreement with computer simulation results for the case of $d_w = 2$ (Brownian motion) and $d_w = 1$ (rectilinear motion). However, (1) exhibits a novel nature of cluster dynamics in that the quantity in equilibrium depends on the dynamical one such as d_w. The purpose of this paper is to understand this circumstance and to show the lon-time memory wuth repect to the radius of gyration.

2. Cluster Dynamics

Generally speaking, the probability $P(C,t)$ finding a configuration C at a time t obeys the following master equation: $\partial P(C,t)/\partial t = \Sigma W(C'-C)P(C',t) - \Sigma W(C-C')P(C,t)$, where $W(C-C')$ is the transition rate from a configuration C to another C'. Of course $W(C-C')$ depends on the diffusion processes of the particle and then on d_w. We can assume that the condition of detailed balance, $P_{eq}(C)W(C-C') = P_{eq}(C')W(C'-C)$, $P_{eq}(C)$ being the probability in equilibrium, is satisfied so as to approach an equilibrium state. Therefore $P_{eq}(C)$ is a function of d_w.

To demonstrate this fact, we consider the simplest case of N=4 clusters. See Fig.1, where nine types of N=4 clusters are shown. A pair of these types is connected by a respective line, which represents the changeable path by the diffusion of a single particle. After ascertaining the detailed balance numerically, $P_{eq}(C)$ is calculated. The difference between the cases of $d_w = 2$ and $d_w = 1$ is clearly seen.

The sample-averaged radius of gyration should equal to the ensemble-averaged one, that is, $<R> = \Sigma P_{eq}(C)R(C)$, which leads to the fractal dimension d_f depending on d_w. Then we can conclude that the RDLA cluster has the novel nature mentioned above.

Springer Series in Synergetics, Vol. 43 **Cooperative Dynamics in Complex Physical Systems**
Editor: H. Takayama   © Springer-Verlag Berlin, Heidelberg 1989

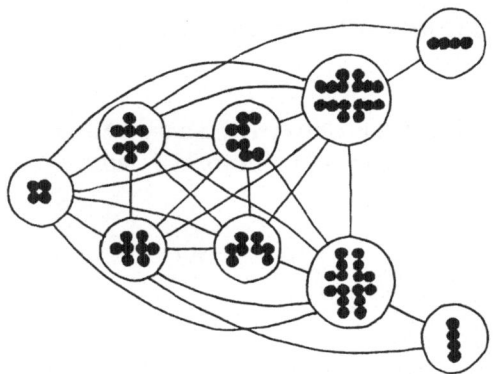

Fig.1 Configuration space of N=4 cluster

3. Long-Time Memory in Cluster Dynamics

We have noted in [2] that the radius of gyration obtained by each trial varies drastically with step and continues to fluctuate highly even after losing the memory of
the initial configuration. To understand this phenomenon further, we calculate numerically the correlation function $C(t)$ of the radius of gyration, defined by $C(t)=$
$(<R(t+t')R(t')>-<R(t')>^2)/(<R(t')^2>-<R(t')>^2)$. From the initial slope of $\ln C(t)$ vs.
t, the relaxation time τ is evaluated for the cases of N=20,30,50,70 and 100 clusters. We also find that the mean number of tips, N_{tip}, is proportional to N. It is
concluded that τ increases with N_{tip} as $\exp(kN_{tip})$ (k is a positive number). The
radius of gyration can vary definitely just after all tips change their positions.
This result suggests that for the infinite cluster the relaxation time diverges and
the correlation function exhibits a very slow decay.

   Lastly it is noted that the configuration space of the RDLA clusters resemble
closely the ultrametric space [3], which often appears in the problems of spin-glasses and neural networks. As seen in Fig.1, the stretched cluster can scarcely reach
the compact one after many kinds of distortion are encountered again and again. This
trend is enhanced in the case of the fractal clusters, for they have hierarchical
branches. Therefore it is reasonable to expect that the radius of gyration obeys
the similar equation of diffusion in the ultrametric space, which leads to the power
law decay [4].

References

1. R.Botet and R.Jullien: Phys. Rev. Lett. 55, 1943 (1985)
2. K.Honda, H.Toyoki and M.Matsushita: J. Phys. Soc. Jpn. 57, 1186 (1988)
3. R.Rammal, G.Toulouse and M.A.Virasoro: Rev. Mod. Phys. 58, 765 (1986)
4. C.P.Bachas and B.A.Huberman: J. Phys. A20, 4995 (1987)

Growth of Clusters Through Evaporation-Condensation Processes

Y. Hayakawa[1] *and M. Matsushita*[2]

[1]Research Institute of Electrical Communication, Tohoku University, Sendai 980, Japan
[2]Department of Physics, Chuo University, Kasuga, Bunkyo-ku, Tokyo 112, Japan

The aggregation and cluster formation of initially dispersed colloidal particles
is often encountered in a variety of natural and industrial processes. The precipitation near the river mouth of tiny mud particles flowing into the sea is a
typical example of such aprocess: The aggregation of mud colloids is initiated by
salt in the sea water (salting-out). This kind of process is now known to be
described by diffusion-limited cluster-cluster aggregation (DLCA) model [1,2].
In DLCA clusters move around via diffusion with appropriate mobility and stick
together irreversibly upon collision to form larger clusters. Quite a few experiments have already been done to elucidate the fractal structure of aggregated
clusters and dynamics of aggregation process for a variety of colloids such as
gold and silica colloids, and found to be explained fairly well by DLCA [3].

However, the precipitation patterns of PbI_2 observed in agar gel with no imposed concentration gradients [4] can hardly be explained by DLCA, in spite of
apparent similarity of the patterns obtained. The reason is simply that macroscopic clusters cannot random-walk in agar gel. Other, more realistic models
should be introduced to explain this kind of pattern formation. One thing we
immediately take notice of is that this phenomenon is closely related to the
Ostwald ripening in spinodal decomposition of alloy systems. There must be cluster formation via evaporation-condensation processes.

From this viewpoint we would like to introduce a simple model to describe this
phenomenon as follows: First of all let $\underline{N}$ particles be dispersed homogeneously in
a $\underline{d}$-dimensional hypercube of size $\underline{L}$. Next let single particles ramdon-walk until
they stick upon collision with each other or clusters to form new clusters.
[Here we take periodic boundary conditions. Then the total number of particles $\underline{N}$
is conserved with the constant density $\rho=N/L^d$.] Resultant clusters never move.
However, every surface particle which is coupling to any cluster with a single
bond can evaporate with equal probability. The particle evaporated starts immediately to random-walk and sticks to other cluster or recombines with the original cluster (but, in general, at a different position).

Repeat these procedures many times, then the clusters grow gradually and exhibit randomly branched, fractal-like structures. In fact, we found the fractal
dimension D≅1.5 for d=2. If there remains only one cluster and yet the above
procedures are repeated, then this model becomes equivalent to Botet-Jullien
model [5], which also yields D≅1.5 for d=2. Moreover, clusters obtained through
various versions of cluster-cluster aggregation model[3] have the fractal dimension D≅1.5 in two dimensions, too. These facts clearly imply that one cannot
identify the underlying mechanism of pattern formation from the fractal dimensions alone. The dynamical properties and cluster statistics must be investigated as well.

It is easy to incorporate physical time of both particle diffusion and evaporation into our model. For simplicity let us here take the time arising from
evaporation events only, by assuming that the particle diffusion is very fast.
Then the time is obtained by adding the increment of $\Delta t=1/N_t$ after each attempt

Springer Series in Synergetics, Vol. 43   **Cooperative Dynamics in Complex Physical Systems**
Editor: H. Takayama   © Springer-Verlag Berlin, Heidelberg 1989

of evaporation–condensation events, where N_t is the number of the surface (or "tip") particles.

At the late stages of the cluster growth average cluster size S was found to be scaled as $S(t) \sim t^z$, where the dynamic exponent $z \cong 1/2$ for d=2. The cluster-size distribution function $n_s(t)$ of cluster-size s is expected to have a scaling form of $n_s(t) \sim s^{-\tau} f(s/t^z)$. Since in the present model the total number of particles N is conserved, the exponent τ is required to satisfy $\tau=2$. The scaling form of $n_s(t)$ with $\tau \cong 2$, $z \cong 1/2$ was in fact confirmed.

These results remind us of Ostwald ripening in the spinodal decomposition. It is well-known that at the late stages of the spinodal decomposition average droplet radius is scaled as $R(t) \sim t^{1/3}$ due to the Ostwald ripening. In the present model also, the average cluster radius is given by $R(t) \sim S(t)^{2/3} \sim t^{1/3}$, because the fractal dimension of the clusters is $D \cong 3/2$. In the case of the Ostwald ripening the droplet growth is driven by the difference of curvature of droplets, while in our model clusters have randomly branched structures and no well-defined curvature at all. It is very interesting that none the less the average radius of both cases is scaled in the same way. This suggests that the effective dimension of cluster surfaces for the evaporation–condensation processes in our model is the same as that of droplets in the spinodal decomposition, in spite of the apparent fractal surfaces.

Since the particle number is conserved, the growth of clusters means nothing but the situation that larger clusters grow more at the sacrifice of smaller clusters which are gradually eaten up and disappear. This is clearly reminiscent of famous "ruin problem of betting" in probability theory [6] as well as the Ostwald ripening. As is expected from this observation, the cluster growth is very slow. In contrast to DLCA clusters which grow completely irreversibly, each cluster in our model becomes sometimes large and sometimes small. The size and form change intermittently. These strong fluctuations seem to come from stubborn memory effect of the size and form of the clusters and their various parts (branches). This may give rise to the power-law distribution of the "relaxation" times. In fact, the power spectrum of the number fluctuation of the surface (or "tip") particles N_t was found to exhibit a prominent 1/f–like spectrum at late stages of the growth.

REFERENCES

1. P. Meakin: Phys. Rev. Lett. 51, 1119 (1983).
2. M. Kolb, R. Botet, and R. Jullien: Phys. Rev. Lett. 51, 1123 (1983).
3. P. Meakin: In Phase Transitions and Critical Phenomena, ed. by C. Domb and J. L. Lebowitz, Vol.12 (Academic Press, New York, 1988); M. Matsushita: In The Fractal Approach to the Chemistry of Disordered Systems: Polymers, Colloids, Surfaces, ed. by D. Avnir (Wiley, New York, 198x).
4. S. Kai, S. C. Muller, and J. Ross: J. Phys. Chem. 87, 806 (1983); S. Kai and S. C. Muller: Science on Form 1, 9 (1985).
5. R. Botet and R. Jullien: Phys. Rev. Lett. 55, 1943 (1985).
6. See e.g., W. Feller, An Introduction to Probability Theory and its Applications Vol.1 (Wiley, New York, 1968).

Part II

Spin Glasses and
Related Random Systems

Slow Relaxations in Spin Glasses and Related Random Systems

J. Souletie

Centre de Recherches sur les Très Basses Températures, C.N.R.S., BP 166 X
F-38042 Grenoble Cedex, France

Spin glasses are magnetic systems whose dynamics is, in many respects, similar to that of usual glasses. This opens the attractive perspective that recent developments of spin glass physics could also be applied to the glass-transition problem.

1. FROZEN DISTRIBUTIONS OF ENERGY BARRIERS /1,2/

The a.c. susceptibility $\chi' = dM/dH$ of a spin glass has a peak at a temperature $T_g(\omega)$. Above $T_g(\omega)$ the susceptibility follows a Curie law: $\chi T = $ const. Well below $T_g(\omega)$ the magnetic response to a step field presents a large elastic (instantaneous) response J_g followed by a slow recoverable non-exponential relaxation $J_d\psi(t)$. One, correspondingly, could describe the strain γ in a rubber, a polymer or an amorphous metal under the effect of a stress applied at fixed temperature. In either case one can define a creep compliance (see e.g. the paper by D.J. Plazek in /2/):

$$J(t) = \gamma(t)/\sigma = M/H = J_g + J_d\psi(t) \tag{1}$$

The recoverable *non-exponential* relaxation $J_d\psi(t)$ is interpreted as evidence for a distribution of relaxation times. With times on the scale of hours or days we are led to postulate a corresponding distribution $P(W)$ of potential wells and hills of sufficient height and depth to inhibit the evolution of the system in phase space. Frustration associated with disorder contains the ingredients which could justify the existence of such a distribution. The time it takes the system to exit from a metastable situation characterized by a particular well is $\tau = \tau_0 \exp(W/kT)$ where W corresponds to the energy difference from the well to the hill. The experimental conditions of time t and temperature T fix a barrier height $W_c = kT\ln(t/\tau_0)$ which separates those states which have relaxed ($t > \tau(T)$) from those which have not and conserve the memory of previous events. For a given distribution $P(W)$ and a given history of the sample, therefore, the compliance $J(t)$ is a function of $W_c = T\ln(t/\tau_0)$ alone. The existence of an annealing in terms of the temperature then implies a relaxation at fixed temperature which is given by the same function of $\ln(t/\tau_0)$ /3/ (Fig. 1). The same model implies that that part of the irreversible response which is annealed by raising the temperature to, say, T_B is then stabilized and found reversible at all $T < T_B$. Annealing proceeds if the temperature is raised over T_B until a maximum annealing temperature T_D is reached where the magnetization, which is now by definition the field-cooled magnetization, is found reversible in this field at all temperatures. T_D which corresponds to the edge W_{max} of the distribution $P(W)$ varies in this model like $W_{max}/\ln(t/\tau_0)$. This type of scaling which is observed in CuMn 5 at% also occurs in many systems where disorder insures a broad distribution of

Springer Series in Synergetics, Vol. 43 **Cooperative Dynamics in Complex Physical Systems**
Editor: H. Takayama © Springer-Verlag Berlin, Heidelberg 1989

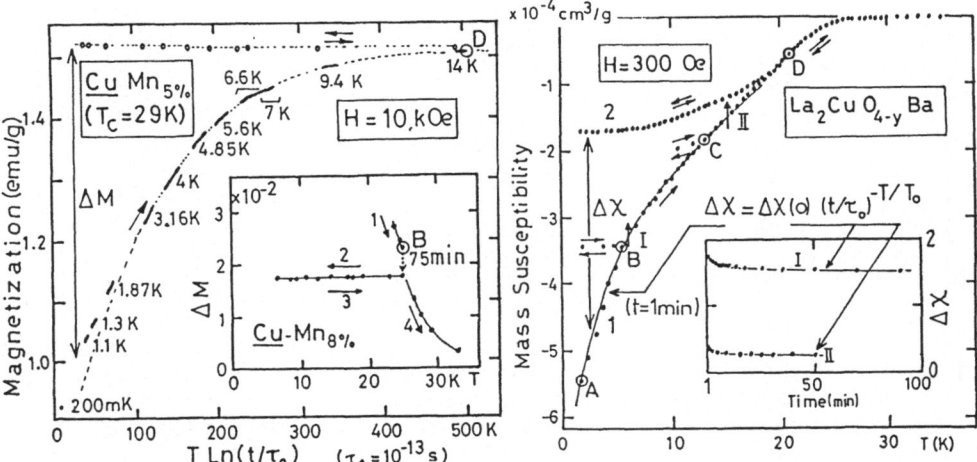

Fig. 1 : The relaxations measured vs time at different constant temperatures $T < T_g$ superimpose upon a unique master curve of $T\ln(t/\tau_o)$ in CuMn 5 at.% /3/. The same correspondence permits one to deduce the time dependence of the relaxation (insert) from the T dependence of M measured in the high Tc superconductor La_2CuO_{4-y}: Ba /4/.

barriers, largely independently of their microscopic origin: superparamagnets, pinning of walls in ferromagnets or of vortices in type II superconductors by the defects of the structure, rubbers, amorphous metals ... . It permits /3/ one to deduce the amplitude and the shape of the pseudologarithmic creep of the magnetization that Müller et al. observed in the high T_c superconductor La_2CuO_{4-y}:Ba /4/ from the temperature dependence of the magnetization in the same system (Fig. 1).

Biology provides similar examples. Myoglobin has the property of fixing CO molecules ; they can be dissociated by a laser flash and will recombine after they have wandered around within a physically restricted area of the huge Mb molecule (the heme pocket for example). Austin et al. /5/ have measured the fraction $N(t)$ of the Mb molecules which have not rebound a CO molecule at time t after photodissociation. The different curves are homothetic in a logt diagram. They collapse onto a unique function of $T\ln(t/\tau_o)$ which reveals the activated nature of this mechanism (Fig. 2). In the three examples considered this curve is not too badly approximated by an exponential of Tlnt, i.e. a power law of time:

$$\psi(T\ln(t/\tau_o)) \sim \exp[(T/T_B)\ln(t/\tau_o)] \sim (t/\tau_o)^{T/T_B}, \qquad (2)$$

which means that we have barriers at all scales in the range explored or conversely the absence of a particular time scale except that which is fixed by the experiment. As the effect of varying the temperature by factors is equivalent to that of varying time by orders of magnitude one explores by a moderate temperature increase regimes of the dispersion which would have required enormous amounts of time to reach at a lower temperature. Often, by varying T, one is able to observe different regimes of the relaxation characterized by a change in the distribution and in the Tlnt response. For example, in myoglobin, just above 220 K the CO molecule has other choices than to stay

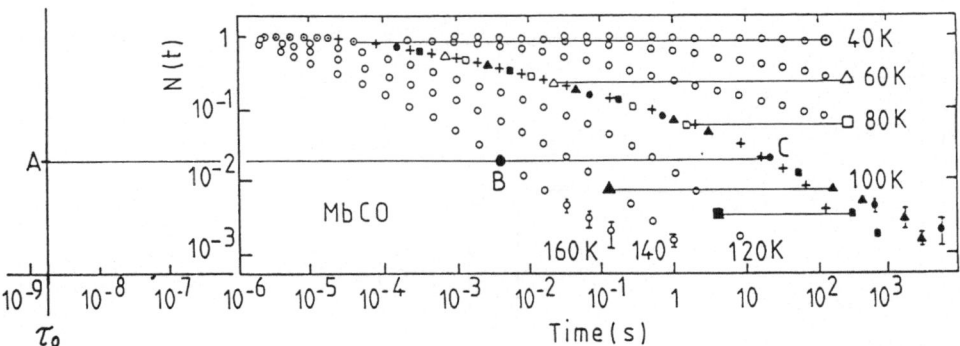

Fig. 2 : N(t) is the fraction of Mb molecules which have not rebound a CO
molecule at time t after photodissociation. The data of /5/ can be
superimposed on the curve at 100 K by taking advantage that
$T\ln(t/\tau_0)$ is the variable ($\tau_0 \sim 10^{-9}$ s). The ratio AB/AC is the inverse
ratio of the temperatures.

in the heme pocket before it rebinds, such as to explore the rest of the protein
molecule or, later on, the solvent. Frauenfelder /5/ reports a hierarchy of 4
such distinct stages, each characteristic of a new mechanism. They correspond to as many time scales and ultimately size scales in the heterogeneous
Mb molecule. Similarly different regimes signal different scales of the interaction of heterogeneous magnetic systems: strong intraparticle vs small
interparticle interactions in fine grains, isolated pairs in dilute insulating
systems near the percolating threshold. These situations are by no means
pathologic although they may be the result of something pathologic having
occurred at higher temperatures and which has been responsible e.g. for the
existence of the P(W) distribution.

2. TEMPERATURE DEPENDENT BARRIERS

Something qualitatively different occurs near the glass transition of say,
rubbers or glass-forming liquids /2/ which is signaled by the occurrence, in
the creep compliance of a non-recoverable deformation proportional to time t

$$J(t) = J_g + J_d \psi(t) + t/\eta. \tag{3}$$

J_g cumulates the effects of previous dispersions J_g+J_d as discussed above.
The presence of a flow t/η defines a liquid. When the viscosity η becomes
infinite, or in practice larger than 10^{14} poises, we have a viscoelastic solid.
In many systems the relaxations measured at different temperatures in this
range, when represented in terms of the logarithm of time, can be
superimposed upon one of them at a reference temperature T_0 by a simple
shift a(T) of the logarithmic scale. This procedure, which was introduced by
Laederman, assumes therefore that even if the relaxation is not an
exponential it is dominated by a unique temperature dependent relaxation
time $\tau(T)$. So that $J[\ln(t/\tau(T))] = J[\ln t - \ln\tau(T)]$ and the shift factor is
$\ln[\tau(T)/\tau(T_0)] = a(T)$. When Tlnt scaling explores a temperature independent
distribution of energy barriers the success of Laederman's procedure would
imply, by contrast, that all barriers have *the same* temperature dependence
whose variation is governed by the relaxation time $\tau(T)$, which can be

deduced from the knowledge of a(T). The study of the viscosity $\eta = G_\infty \tau$ [and also of the shift factor a(T) associated with the transitions $J_{d\psi}(t)$ close to the divergence of η in rubbers and polymers] leads to the same $\tau(T)$ dependence which can be described over many decades by the phenomenological Vogel-Fulcher law (see e.g. the paper by G.P. Johari in /2/):

$$\tau/\tau_0 = \exp\big(B/(T-T_F)\big). \tag{4}$$

The justification of this law remains the primary goal of these theories which try to explain the glass transition: in some respects T_F where τ would diverge is felt as a singularity in the behaviour of the equilibrium liquid, much as the Curie temperature is for a ferromagnet. However, in a ferromagnet critical slowing down would not affect experiments that far from T_c (($T-T_c)/T_c \sim 10^{-4}$ instead of 10^{-1}). The idea was then to construct a theory which could be the dynamic equivalent of the usual theory of phase transitions for the static case. This point of view could change after what we have learnt from the study of the spin glass transition. In spin glasses we still have a distribution of relaxation times in the neighborhood of $T_g(\omega)$. But rather than TInt scaling, the scaling which applies is of the form $f\big[(t/\tau(T))^{\alpha(T)}\big]$: it combines an affinity and a shift in logt diagram /6,7,13/. As for the temperature dependence of the typical relaxation time $\tau(T)$, when it is measured over T_c it can be well approximated by a Fulcher law. However, it results that traditional scaling ideas, as they were established for ordinary continuous transitions, account remarkably well for the static *and* dynamic properties.

3. SPIN GLASSES: A CONTINUOUS TRANSITION WITH LARGER EXPONENTS /1/

The idea of scaling is to describe a transition as the consequence of the divergence of a coherence length ξ which is itself described as a power law of $K-K_c$

$$\xi/\xi_0 = [(K-K_c)/K_c]^{-\nu} = (1-T_c/T)^{-\nu} \tag{5}$$

when the reduced interaction $K = J/T$ reaches a convergence radius $K_c = J/T_c$ which defines T_c. If we develop the magnetization as

$$M = a_1(H/T) + a_3(H/T)^3 + a_5(H/T)^5 + ... \tag{6}$$

the static mark of the spin glass transition is the divergence of the higher order Curie constants $a_3 ... a_{2n+1}$ which behave like a power of ξ hence of $1-T_c/T$. We have /8,9/

$$a_{2n+1} = (1-T_c/T)^{\beta-n(\beta+\gamma)} \text{ for } n > 0. \tag{7}$$

The dynamic scaling hypothesis states that the relaxation time τ is a power of ξ, hence of a_{2n+1},

$$\tau/\tau_0 = (\xi/\xi_0)^z = (1-T_c/T)^{-z\nu}. \tag{8}$$

Equation (8) is a natural substitute for the phenomenological Vogel-Fulcher law. It permits one to obtain better fits to the data provided values $z\nu \sim 8-10$ are accepted. Typical exponents measured in Heisenberg systems at $d = 3$ are $\gamma = 3$, $\beta = 1$ ($\nu \sim 5/3$) with $z\nu \sim 8-10$. This includes

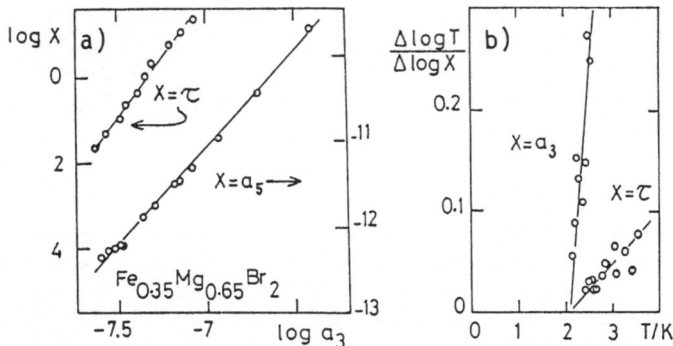

Fig. 3 : a) The logarithm of the relaxation time $\ln\tau$ is proportional to $\ln a_3$ hence to $\ln\xi$, and b) both τ and ξ can be represented as a power of $(1-T_c/T)$ with the same T_c value in the Ising system $Fe_{0.35}Mg_{0.65}Br_2$ /10/.

R.K.K.Y. systems as well as a number of insulators with short range interactions. Figure 3a shows that a_5 and τ are powers of a_3, hence of ξ, in the layered "Ising" system $Fe_{0.35}Mg_{0.65}Br_2$ /10/. Figure 3b is a differential form of (8) (or 5 or 7). We have e.g. that

$$P_\tau(T) = -\partial\ln T/\partial\ln\tau = (T-T_c)/(zvT_c) = (T-T_c)/\theta. \qquad (9)$$

The dynamic *and* the static data aim at the same $T_c = 1.9$ K and exponents can be deduced from the slopes using (9). We found $\gamma = \theta'/T_c \sim 2$, $\beta = 0.5$ (d$v \sim 3$) and $zv \sim 10$. Theory yields comparable values for Ising spins at d = 3. But the prediction is rather that $T_c \sim 0$ for Heisenberg spins at d = 3 and Ising spins at d = 2. These difficulties and the recent observation of a slower dynamics in a number of systems have motivated alternative suggestions. The assumption of critical activated dynamics /11/ states that $\tau/\tau_0 = \exp(W/T)$ where W and *not* τ, is a power of ξ, hence of t:

$$W/W_0 = T\ln(\tau/\tau_0) = (\xi/\xi_0)^\Psi = (1-T_c/T)^{-\Psi v}. \qquad (10)$$

This generalises to finite T_c the suggestion that Binder and Young /1/ first formulated for $T_c = 0$. They reason "rapidly" that $\xi \sim (1-T_c/T)^{-v} \sim (T-T_c)^{-v} \rightarrow T^{-v}$ when $T_c \rightarrow 0$ and they find therefore with the previous analysis that

$$\tau/\tau_0 = \exp(W_0/T^\sigma). \qquad (11)$$

Rammal /12/ by contrast calculates that

$$W = T\ln(\tau/\tau_0) \sim \ln\xi^{\bar d} \qquad (12)$$

on percolating clusters. This would give (8) again, but with a temperature-dependent exponent $zv \sim W_0 v/T$. Note at this point that activated dynamics is also obtained, within the framework of classical scaling (5-8), not as a new hypothesis like in (9) and (10) but as the natural limit of the model in all interesting cases where $T_c \rightarrow 0$ with θ finite (θ is defined in (9)). We have then $\tau/\tau_0 = (\xi/\xi_0)^z = (1-T_c/T)^{\theta/T_c} \xrightarrow[T_c \to 0]{} \exp(-\theta/T)$ /10/.

$Fe_{0.3}Mg_{0.7}Cl_2$, despite its similarities with $Fe_{0.35}Mg_{0.65}Br_2$, is one of those systems which show a slower dynamics $25 < zv < \infty$ with

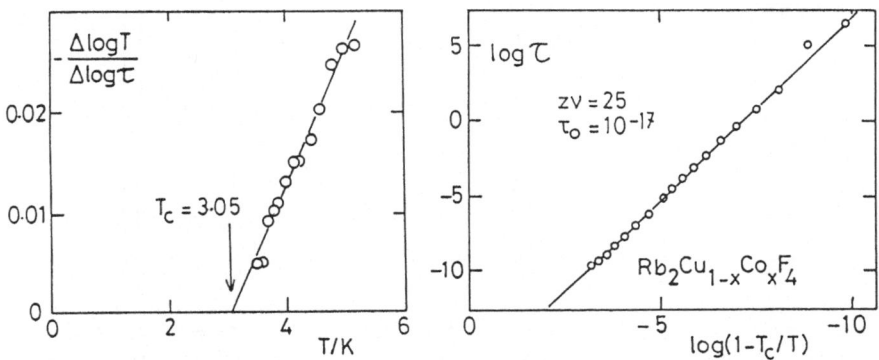

Fig. 4 : $Rb_2Cu_{1-x}Co_xF_4$ is one of a number of systems where an unusually large exponent is measured ($zv \sim 25$ with $T_c \sim 3.1$ K). Alternative expressions give similar fits with a more satisfactory value for τ_o /13/.

$\tau_o \sim 10^{-12}$ s. A comparable behaviour in $Rb_2Cu_{1-x}Co_xF_4$ has been interpreted /13/ as evidence in favour of (10). Fig. 4 shows with these data the difficulty of deciding between two laws. Personally we are not ready to discard a law simply because it would imply a large exponent. First an exponential means an ∞ exponent. Perhaps also our thinking is still biased by the ferromagnetic analogy ! We want for spin glasses a finite number of well-defined universality classes labelled by the spin and the space dimensionality. But what is the space dimensionality of a disordered system ? We know from percolation theory that it differs for the scale of observation being smaller or larger than a structural correlation length ξ_s. In a spin glass, there are simultaneously correlations at many scales, some smaller and some larger than ξ_s. Wouldn't it be more appropriate to introduce some sort of dimension of the disorder which would characterize, say, the way the interaction decreases with distance (if it is long range) or the proximity of a percolation threshold if the interaction is short range ($d' \sim \xi_0/\xi_s$) ? We believe that the question arises naturally within the framework of the "critical fractal model" /7/ which succeeds pretty well in unifying the static and the dynamic scaling properties of spin glasses. Pertinent information could be obtained in this respect, by checking whether exponents in spin glasses do not increase continuously on approaching a percolation threshold.

4. GLASSY GLASSES

Angel in reference /2/ shows a compilation of viscosity data in a number of glass-forming liquids. He classifies them as "strong" or "fragile" depending on whether the temperature dependence of the viscosity is better described by an Arrhenius law or by a Fulcher law. GeO_2 and SiO_2 are typical "strong" systems. 0-terphenyl, and K^+-Ca^{++}-NO_3^- are "fragile". As we have seen in § 2, η is related to a time. The glass transition T_g is defined when this time becomes large as compared with ordinary laboratory times ($\eta \sim 10^{13}$ poises). At these temperatures, which are above the Fulcher temperature where η would diverge, one observes large changes ΔC_p in the specific heat of the fragile systems and smaller changes in that of strong systems. They are the anomalies which lead to the well-known Kautzman's paradox: it seems that the excess entropy of the supercooling liquid over that of the crystal

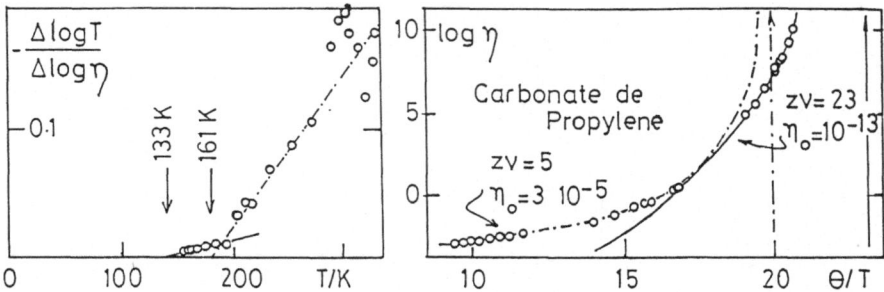

Fig. 5 : The viscosity of many glasses can be described by power laws with large (sometimes infinite) exponents. Note the departure to another regime after many decades are covered with a power law (Data from /14/).

tends to vanish at a temperature T_K which is not very far below T_g (somewhere between T_g and T_f) in a way which suggests that it could be reached at equilibrium ($T_F < T_K < T_g$) (see the paper by G.P. Johari in /2/).

Figure 5 shows that the viscosity of a typical glass-forming system can be analysed as the result of either one or a succession of slowing down regimes characterized by exponents with values similar to those measured in spin glasses. Often a behaviour of the $FeMgCl_2$ type ($25 < zv < \infty$ with a small τ_o) follows a behaviour of the $FeMgBr_2$ type. The critical temperature, in the regime above the glass transition, is very close to T_K. Of course it is expected that if the system is to become a crystal all the processes will be interrupted at the first-order melting transition T_L. On the other hand, if the second-order glass transition T_c were close enough to T_L the nucleation which leads to the solid phase could be blocked by the slowing down associated with T_c. The proximity of T_c to T_L would then decide which liquids easily become a glass and which do not. Even among these liquids which easily become a glass the model would imply tails in the specific heat and in the viscosity which exist and are not easily explained in the framework of the first-order transition. These tails would be stronger in the "fragile" systems where T_c is closer to T_L.

5. CONCLUSIONS AND PERSPECTIVES

Spin glasses have shown us how frozen disorder can produce a new type of transition with larger θ/T_c ratio hence larger exponents. Large exponents mean smooth static properties but non-ergodicity sets in farther from T_c and the dynamical aspect appears dominant. With these large exponents and wide distributions of relaxation times, some of the degrees of freedom are quenched when a typical larger relaxation time freezes: so that we may now be learning from spin glasses how frozen disorder sets in. For example, when a particular species of permutations or rotations of atoms or chains is hampered the result is a modification of the actual Hamiltonian and the occurrence of a new situation which leads, perhaps, to a transition of a different nature. We may thus conceive a hierarchy of aborted transitions before a hierarchy of superparamagnetic states (with $Tlnt/\tau_o$ scaling as described in sect. 1) is ultimately reached. Such changes of regime could account for deviations as observed in Fig. 5 where the data abruptly depart from a power law which has been valid over many decades.

REFERENCES

For general references the reader is referred to two reviews /1,2/ from which further references can be found.

1. K. Binder and A.P. Young, Rev. Mod. Phys. **58**, 801, 1986.
2. Relaxations in complex systems, ed. K.L. Ngai and G.B. Wright (Nat. Techn. Inf. Service of U.S. Dept. of Commerce, 1984).
3. J.J. Préjean and J. Souletie, Phys. Rev. Letters **60**, 1888 (1988) and Phys. Rev. B **37** (1988) and references therein.
4. K.A. Muller, T. Takashige and J.G. Bednorz, Phys. Rev. Letters **58**, 1143 (1987).
5. R.H. Austin, K.W. Beeson, L.E. Eisenstein, H. Frauenfelder and I.C. Gunsalus, Biochemistry **14**, 5355 (1975) and H. Frauenfelder, Helvetica Physica Acta **57**, 165 (1984).
6. P. Granberg, P. Svedlinh, P. Nordblad, L. Lundgren and H.S. Chen, Phys. Rev. B **35**, 2075 (1987) and M. Ocio, J. Hamman, P. Refregier and E. Vincent, Physica B (1988).
7. M. Continentino and A. Malozemoff, Phys. Rev. B **34** (1986) 471 and P. Beauvillain, J.P. Renard, M. Matecki and J.J. Préjean, Europhys. Lett. **2** (1986) 415.
8. M. Suzuki, Progr. Theor. Phys. **58** (1977) 1151 and J. Chalupa, Solid State Commun. **22** (1977) 315.
9. S. Chikazawa, C.J. Sandberg and Y. Miyako, J. Phys. Soc. Jpn. **50** (1981) 2884 ; R. Omari, J.J. Préjean and J. Souletie, J. Physique **44** (1983) 1069.
10. D. Bertrand, J.P. Redoules, J. Ferré, J. Pommier and J. Souletie, Europhys. Lett. **5**, 271 (1988) and D. Bertrand, A.R. Fert, J.P. Redoules, J. Ferre and J. Souletie, Proceedings of ICM Paris 1988, to appear in J. Physique (1988).
11. D.S. Fisher and D.A. Huse, Phys. Rev. Lett. **56**, 1601 (1986).
12. R. Rammal, J. Physique **46** (1987) 1837.
13. C. Dekker, A.F.M. Arts, H.W. de Wijn, A.J. van Duyneveldt and J.A. Mydosh, Proceedings of ICM, Paris 1988, to appear in J. Physique (1988).
14. A. Bondeau and J. Huck, J. Physique **46** (1986) 1717.

Models for Slow Relaxation in Glassy Systems

R.G. Palmer

Department of Physics, Duke University, Durham, NC 27706, USA

1. Slow Relaxation

Many glassy materials display non-exponential relaxation with a non-Arrhenius temperature dependence. We define a *relaxation function* $q(t)$ that decays from 1 to 0 to describe (besides initial transients) the scaled response of some quantity y to a step-function change in another quantity x. For example the control variable x might be temperature, strain, or electric field, and the measured quantity y might be volume, strain, or polarization. The relaxation function $q(t)$ may also be extracted from the AC response or from an autocorrelation function $\langle y(0)y(t)\rangle$. In simple systems controlled by activation over a single free energy barrier ΔF we expect $q(t) = \exp(-t/\tau)$ with an Arrhenius temperature dependence $\tau \propto \exp(\Delta F/k_B T)$ for the relaxation time τ. The non-exponential relaxation seen in glassy systems is slower than exponential, in the sense that $-q(t)/q'(t)$ increases with t. In structural glasses, polymers, and dielectrics it is often well described by a *stretched exponential*, often known as a Kohlrausch [1] law,

$$q(t) = \exp[-(t/\tau)^{\beta}], \tag{1}$$

with $\beta < 1$. This is slower than exponential but faster than power law or logarithmic (though a stretched exponential with small β is indistinguisable from a logarithmic decay). A value of β from 0.5 to 0.7 is common in glasses, while values around 0.3 are often seen in polymers. The law (1) is the best two-parameter fit known across a wide class of glassy materials and properties. Systematic deviations show that it is not, however, the precise functional form of most data. Particular experimental results may be fitted better by other empirical functions.

The non-Arrhenius temperature dependence of τ is often well fitted by a Vogel-Fulcher [2] law

$$\tau = \tau_0 \exp\left(\frac{A}{T - T_0}\right) \tag{2}$$

although crossovers to pure Arrhenius behavior may be seen both at high and at low temperature. Here low temperature means close to the *glass transition* temperature T_f where equilibrium is no longer reached on experimental timescales; this is *kinetic arrest*. The value of T_f depends (slightly) on the experimental timescale adopted and does not reflect a true phase transition. There *may* be an underlying phase transition at T_0, which is typically 10% below T_f, but this is still under debate [3]. In this paper we only consider the viscous liquid regime $T > T_f$, where equilibrium is reached within experimental times.

Springer Series in Synergetics, Vol. 43   **Cooperative Dynamics in Complex Physical Systems**
Editor: H. Takayama   © Springer-Verlag Berlin, Heidelberg 1989

2. Models Involving Constrained Dynamics

Behavior described at least approximately by equations (1) and (2) is ubiquitous in glassy systems and it is natural to ask how it arises. Is there a common explanation, transcending the diverse details? It seems reasonable to suspect so, but such an explanation would have to be at a level of abstraction removed from the microscopic particulars. It might, for instance, be comparable to the explanation of the ubiquitous Gaussian by the central limit theorem.

In the past several years a number of simplified toy models for glassy relaxation have been proposed, and have had some success in reproducing (1) as an outcome of rather general principles. Some of them also reproduce (2) numerically. There are now perhaps *too* many models, based on apparently different principles but rather hard to tell apart in their predictions. In section 3 we will try to see why many of these diverse models all lead to behavior like (1), in terms of even more generic models, but first we review here some of the models. We emphasize those that involve *constrained dynamics*, which will be the central ingredient of the generic models in section 3.

By constrained dynamics is meant that from many configurations most degrees of freedom are "blocked", with only a few at a time free to change without great energetic cost. The free energy landscape must be like a maze, with many dead ends, or blocked passageways, or high walls, and only a few narrow paths leading anywhere. In a structural glass most atoms and groups of atoms are locked in (or *sterically hindered*) by their neighbors and cannot move unless the neighbors move first. Figure 1 (due to Pietronero) is the simplest idealization of constrained dynamics; atoms A and B each have two alternate positions, but when B is in its left-hand position it is in the way of A, which it prevents from moving.

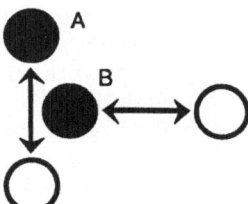

Fig. 1. Example of constrained dynamics

It is worth distinguishing *frustration* and constrained dynamics. Frustration can be expressed in a Hamiltonian formulation, whereas the dynamical constraints considered here are essentially non-holonomic. Dynamical constraints correspond to forbidden (or high energy) *transitions*, while frustration forbids certain *states*; in configuration space the difference is between forbidden paths and forbidden points. The two may however ultimately have similar effects, and indeed frustration can lead to high dynamical barriers [4].

2.1 Constrained Kinetic Ising Models

Kinetic Ising models with constrained dynamics have been studied in one and two dimensions and on a Cayley tree. The Hamiltonian is usually trivial,

$$\mathcal{H} = -h \sum_i S_i, \qquad S_i = \pm 1, \tag{3}$$

where the external field h may even be set to 0. The interest is all in the single spin flip stochastic dynamics, where the flip rate R for S_i takes the form

$$R(S_i \rightarrow -S_i) = e^{-hS_i/k_BT} F_i(\{S_j\}). \qquad (4)$$

Here the first factor takes care of detailed balance, while F_i is a function of the state $\{S_j\}$ of the neighbors j of i. F_i is usually taken to be zero (or small) for many of the possible neighbor states, whereupon spin S_i is unable to flip until released by a suitable change among its neighbors.

In one dimension such models can be treated analytically for certain choices of F_i, and have been studied numerically elsewhere [5,6,7]. Approximately stretched exponential behavior is generally found for correlation functions like $\langle S_i(0)S_i(t)\rangle$. In fact this behavior is very easy to obtain in one dimension, even providing a good fit [6,8,9] to the exactly known correlation function of the Glauber model [10]. One-dimensional defect diffusion, first studied by Glarum [11], is a special case of this behavior. It is worth noting however that defect diffusion in three dimensions gives pure exponential behavior unless modified, perhaps somewhat artificially, by introducing a hierarchical structure either spatially or temporally. In the first case the defects diffuse on a fractal lattice [12,13]; in the second case the defects have a wide distribution of waiting times between jumps [14,15,16].

Fredrickson and Andersen [17] introduced a constrained kinetic Ising model on a two-dimensional square lattice with $F_i \propto m_i(m_i - 1)$, where m_i is the number of nearest neighbors (out of 4) of spin i which are up ($S_j = +1$) in the presence of a field $h < 0$ which biases all spins down. If fewer than two neighbors are up then $F_i = 0$ and spin i is stuck. Up and down spins can be thought of as representing low and high density regions respectively. At high temperature over half the spins are free to flip, but as the temperature T is lowered, more and more spins are frozen. At first it was thought that the model had a purely *dynamic* phase transition ((3) obviously has no static transition), but in fact it does not; there is a glass transition (kinetic arrest) without an underlying static or dynamic singularity. Extensive Monte Carlo simulations [18] have shown that the model exhibits approximately stretched exponential behavior for $\langle S_i(0)S_i(t)\rangle$. The average relaxation time τ has a non-Arrhenius temperature dependence, approximately fitted by a Vogel-Fulcher law. The Adam-Gibbs [19] relation—a common characteristic of glasses—is also satisfied. In all, the model is very glass-like in its characteristics, despite its extreme simplicity.

In the Cayley tree version we allowed a given spin S_i to flip only if all its daughters further out on the tree were up [20]. The spins may be divided into layers or generations, with those outermost on the tree flipping most rapidly and inner layers flipping successively more slowly. The hierarchical organization makes it easy to generate a broad spectrum of relaxation times, but may not in fact be necessary. Depending on the model details we found it easy to obtain *either* power-law *or* stretched-exponential relaxation. Temperature dependence was not fully included, but a simple argument suggested Vogel-Fulcher behavior for the longest (but not for the average) relaxation time τ_∞. In recent work [21] we have obtained appropriate temperature dependence for the average relaxation time too, using "soft" constraints which can be violated at a certain energy price.

Current investigations [22] also include three-dimensional constrained Ising models and models with different constraint rules, including rules that vary randomly from site to site. It will be interesting to classify the dynamic universality classes of all these models, and understand just how wide is the stretched exponential regime.

2.2 Tiling Models

Weber, Fredrickson, and Stillinger [23] have studied two versions of a model in which states of the system correspond to distinct tilings of the plane by integer-sided squares. The square tiles represent well-packed domains, and the domain walls contribute a mismatch energy proportional to their length; the Hamiltonian is just proportional to the total wall length. A possible configuration of a 20 × 20 sample with periodic boundary conditions is shown in Fig. 2.

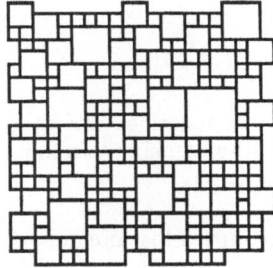

Fig. 2. A configuration of Weber, Fredrickson, and Stillinger's tiling model

The equilibrium behavior can be calculated and exhibits a first order phase transition, below which macroscopic tiles appear. The dynamics is chosen to be highly constrained however, so the system is arrested kinetically and never reaches its underlying transition. The two versions of the model have different allowed dynamical transitions, both of which only just preserve metric transitivity. For instance, one version allows a $pq \times pq$ tile to break up into (or be formed from) p^2 tiles of size $q \times q$ if and only if p is the smallest prime factor of pq. Any configuration can be reached from any other (via a complete 1 × 1 tiling if necessary), but the routes are often quite tortuous. The energy autocorrelation function was studied by Monte Carlo simulation and found to have approximately stretched exponential behavior at not too low a temperature. The temperature dependence of τ was non-Arrhenius, although not quite Vogel-Fulcher. The model also showed rather glassy behavior in heating, cooling, and temperature jump Monte Carlo experiments.

2.3 Sliding Block Models

Inspired by the observation that so many models with constrained dynamics show stretched exponential relaxation, we have performed [24] Monte Carlo simulations on a child's puzzle that has strong constraints. The puzzle consists of $L^2 - 1$ labeled unit squares able to slide vertically and horizontally in a frame of size $L \times L$. A given square can move only if it is adjacent to the single vacancy. A particular state of the puzzle for $L = 4$ might be as shown in Fig. 3.

Rearranging the squares to a desired even permutation can involve a very large number of steps—the system is highly constrained. We let the vacancy move randomly and monitor the configuration space distance from the starting state, defined in the obvious way with a Manhattan metric. The rigid walls are replaced by periodic boundary conditions for convenience. The relaxation from a particular initial state towards the perfectly jumbled "equilibrium" state shows approximately stretched exponential behavior at short times ($t \lesssim L^2 \ln L$) where the constraints are most effective. At longer times regime the relaxation crosses over first to power law decay ($L^2 \ln L \lesssim t \lesssim L^4$)

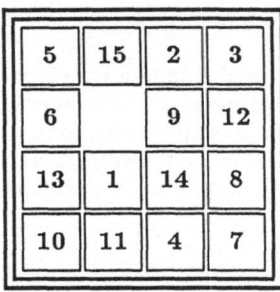

Fig. 3. A configuration of a child's puzzle

and then to pure exponential decay $(t \gtrsim L^4)$. These results for the intermediate and long time regimes were recently derived by Brummelhuis and Hilhorst using random walk theory [25].

Jäckle [26] has examined a related model in which atoms occupy sites and move along bonds on a square lattice, with the constraint that no two adjacent sites be occupied. The relaxational behavior has not yet been studied in detail. Jäckle's model tends to form isolated pockets containing one or more atoms, but not larger scale jams or gridlocks. We find that related sliding block models with longer range constraints can be constructed to overcome this limitation. We [27] are currently exploring a model with $a \times a$ square blocks with centers on a square lattice of lattice constant $a/2$, and a model of hard hexagons on a triangular lattice. All these models, though abstract, embody several glassy ingredients: dynamical constraints; diffusional motion of atoms; defect diffusion; and free volume as a relevant parameter. Some have a crystalline ground state that may be kinetically unreachable. It remains to be seen how glassy their behavior really is, since investigation is only now beginning.

3. Generic Models

In the preceding section we reviewed three classes of models that all involve constrained dynamics and that all lead to glassy behavior. All three classes give slow relaxation, roughly stretched exponential in form. Some of the kinetic Ising models and the tiling model (i.e., all those with a non-trivial energy scale) also show non-Arrhenius temperature dependence, certainly going in the glassy direction (upward curvature on an Arrhenius plot), and sometimes being well described by a Vogel-Fulcher law. Some other glassy properties also emerge.

Nor is this all. There are other models that involve constrained dynamics and exhibit glassy behavior. One can certainly include defect diffusion and trapping [28,29,30], the Gibbs-DiMarzio theory [31], and in some ways the Adam-Gibbs [19] and free volume theories [32]. It is even possible, with a stretch of the imagination, to see ultrametric diffusion models [33,34,25] in this light, though they fall better into the class of configuration space models to be discussed here.

In any case, it is natural to ask whether constrained dynamics might lead generically to slow relaxation and some forms of glassy behavior. If there *were* a generic connection, the ubiquity of what we call glassy behavior would be explained, both for materials and for models.

A good way to answer a generic question is to build a generic model. We [35] have been considering models which embody constrained dynamics and as little as possible else. We discuss two in more detail here. Both take place in configuration space, without a necessary microscopic interpretation. Modelling the structure of configuration space, or a multidimensional energy landscape, has become central to the theory of many complex systems in the past few years [36]. Neither model is fully understood; the following merely sketches some work in progress.

Diluted Hypercubes

We consider a spin model with N Ising spins S_i and single spin flip dynamics. Its configuration space consists of the vertices of an N-dimensional hypercube $\mathcal{Z}_2^N$, with a single spin flip corresponding to motion of the phase point from a vertex to one of its N neighbors, along an edge. We endow the model with the simplest of all possible Hamiltonians, $\mathcal{H} = 0$, but tax it with constrained dynamics. We remove a fraction $1 - p$ of the edges, chosen at random among all $N2^N/2$ edges, and disallow transitions along them. For $p \ll 1$ this captures the essence of constrained dynamics without model-specific details; from most states of the system there are few allowed transitions and so most degrees of freedom are frozen until released by the prior transition of another. All the models of the previous section have a configuration space of this *sparse network* form. Different models have different correlations between which edges are removed, and may also be controlled by energy more than by entropy, but the basic notion remains.

We now consider a random walk—pure diffusion—on the remaining diluted hyper-cube, and ask in particular how *equilibrium* is approached. Equilibrium means that the random walk has lost all memory of its starting point (besides parity) and is equally likely to be found on any accessible vertex of the hypercube. We assume that the probability per unit time of a jump along each non-removed bond is a fixed constant. As a measure of the distance from equilibrium we consider

$$q(t) = \frac{\langle x^2(\infty) \rangle - \langle x^2(t) \rangle}{\langle x^2(\infty) \rangle} \tag{5}$$

where $\vec{x}(t)$ is the vector displacement from the starting point, with components x_i each 0 or 1 and $x^2 \equiv \sum x_i^2$. The Hamming distance $\sum x_i$ is identical to x^2. The averaging is over all walks with a given starting point; we average later over starting points, and over the quenched realizations of the random dilution process. As long as p is not too small, there is (with probability 1) a single "infinite cluster" or "macroscopic component", a connected set of order 2^N vertices that is extended over the whole hypercube, and if the starting point is in this cluster then obviously $\langle x^2(\infty) \rangle = N/2$. If on the other hand we do not start on the large cluster the diffusion is non-ergodic and a smaller $\langle x^2(\infty) \rangle$ is expected.

In fact there are two geometrical transitions [37,38] of a randomly diluted N-cube, at occupation probability $p = 1/2$ and at $p = 1/N$. For $p > 1/2$ there is a single large component and no smaller ones; every site is connected to every other site with probability 1. For $1/N < p < 1/2$ there is still a single large cluster, but there are also smaller ones. For $p < 1/N$ there are no macroscopically large clusters. The transition at $p = 1/N$ is the analog of a percolation transition, though percolation is normally considered for a lattice of infinite extent in a small number of dimensions, instead of finite extent in an infinite number of dimensions (as $N \to \infty$).

We do not yet know for certain how $q(t)$ behaves in each regime of p. We are expecting to have large N simulation results soon. There are however small N simulation results for randomly *site* diluted hypercubes by Campbell [39]. Campbell proposed a site-diluted hypercube as a model for a spin-glass, removing high-energy sites from the configuration space and looking at random diffusion on the remainder. There are good reasons to expect bond and site dilution to lead to similar results in the very dilute regime, especially near the $p = 1/N$ transition. Campbell's results are for a system with $N = 17$, and are somewhat noisy and limited in extent, but there is a strong suggestion of stretched exponential behavior for $q(t)$ in the regime $p < 1/2$. Campbell [39] and colleagues [40] have also advanced heuristic and mathematical arguments for this result, though a complete derivation is not yet at hand.

Campbell's simulation suggests that $q(t)$ is exponential for $p > 1/2$, and there is little doubt about this. Below $p = 1/2$ stretched exponential fits to the simulation show a gradual decrease of β from 1 at $p = 1/2$ to 1/3 at $p = 1/N$. It is intriguing to note that simulated spin glass relaxation [41] when fitted to the function $\exp[-(t/\tau)^\beta]/t^x$ has $\beta \to 1/3$ as $T \to T_c$. Most real glasses and toy models have $\beta \geq 1/3$, though there are a few exceptions.

It is clearly important to investigate this model further, though analytic work is difficult unless drastic further approximations or assumptions are made. We now describe a related model which is easier to treat but perhaps too far from reality.

Diluted Random Graphs

The infinite range version of the preceding model consists of 2^N points connected to each other with probability p, independently for all $2^N(2^N - 1)/2$ pairs. Bray [42] has called this a *diluted hypertetrahedron*; in graph theory it would be a *randomly dilute complete graph*. We expect to need $p \sim N/2^N$ to have the same average connectivity per site as in the hypercube. By throwing out the hypercubic structure we lose the single spin flip character of the original model, but make it more tractable.

Random diffusion on this graph may be described by the Master Equation

$$\frac{d\mathbf{p}}{dt} = \Gamma\mathbf{p}, \tag{6}$$

where $\mathbf{p}(t)$ is a vector with components p_i for the probability of being at each possible vertex (microstate) i. The off-diagonal elements of the transition matrix Γ are randomly 0 or 1 (say), independent besides $\Gamma_{ij} = \Gamma_{ji}$. The diagonal elements must be given by

$$\Gamma_{ii} = -\sum_{j \neq i} \Gamma_{ij} \tag{7}$$

to preserve total probability.

Knowledge of the eigenvalue spectrum $\rho(\lambda)$ of Γ would immediately give a relaxation function $q(t; i_0)$ from an initial state i_0 according to

$$q(t; i_0) = \int_0^\infty \rho(\lambda) e^{\lambda t} \, dt \ . \tag{8}$$

Knowledge of the eigenvectors as well would allow in principle computation of the auto-correlation function $\kappa(t)$ and fluctuation spectrum $S(\omega)$ for any specified state-dependent quantity y_i [43,44]. Moreover, the asymptotic ($t \to \infty$) tail of these functions depends only on the eigenspectrum near $\lambda = 0$ (a proper Γ matrix is negative definite except for one zero eigenvalue), so we only need to be able to calculate the spectrum near 0 to see if asymptotic stretched exponential behavior is implied. The problem is made harder by the diagonal constraint, but is still not far from soluble random matrix problems.

Progress with this problem has been made on three fronts. We [45] have been able to prove that $\rho(\lambda)$ has a mean that scales as $2^N p$ and a variance that scales as $2^N p(1-p)$ as $N \to \infty$, so that it becomes in effect a delta function unless $p \sim 2^{-N}$, which seems too small to be physical (we expect $p \sim N2^{-N}$). In the delta function case (8) would give pure exponential relaxation. There could conceivably be higher order corrections however, and we are striving to understand this in more detail.

A second front is the use of replica theory. Using this to compute the average of the generating function for a Green function (whose imaginary part gives $\rho(\lambda)$), we [46] have found stretched exponential relaxation in a simplified version of the model in which the random graph is replaced by a diluted Cayley tree; this should be a good approximation at sufficiently small p. Bray and Rodgers [42] have gone further and applied a very similar method to the full random graph problem, finding exponential relaxation if p is higher order than 2^{-N}, consistent with our results. In the $p \sim 2^{-N}$ case they have problems extracting the required tail of $\rho(\lambda)$ from the replica theory results.

Finally, Bray and Rodgers [42] have succeeded in obtaining the small λ tail of $\rho(\lambda)$ in the $p \sim 2^{-N}$ case using a Griffiths argument; the small λ values come from rare long sections of the graph which are barely connected at their ends to the rest of the network. The analysis gives $\rho(\lambda) \sim \exp(-A/\sqrt{-\lambda})$, where A is a constant depending on p, leading asymptotically to stretched exponential relaxation of $q(t; i_0)$ with $\beta = 1/3$.

Thus there does seem to be stretched exponential relaxation in the diluted random graph, but only in an over-dilute case, and then only with β strictly $1/3$. But it is not clear yet whether this result carries over to the more realistic hypercube case, except near the percolation point where hypercube, random graph, and Cayley tree all agree on $\beta = 1/3$. Moreover, the quantity $q(t; i_0)$ calculated is *not* really what is wanted; it gives the decay of probability away from an initial point on the graph, and is very rapid indeed. We want the decay of a real-space correlation function, not a configuration space one. It is not clear how to define or calculate the right function for the random graph, while in the hypercube it is readily defined but not easily calculated, except by simulation.

We anticipate that the situation will be greatly clarified in a year or two. Then we will finally be able to tell whether or not slow relaxation of approximately stretched exponential form is a generic consequence of microscopic constraints.

I thank Alan Bray, Ian Campbell, Dan Stein, Y.-Q. Yin, and Stu Kauffman for helpful discussions.

References

1. R. Kohlrausch: *Pogg. Ann. Phys.* **91**, 198 (1854).
2. H. Vogel: *Physik Z.* **22**, 645 (1921); G.S. Fulcher: *J. Am. Ceram. Soc.* **77**, 3701 (1925); G. Tamman, W. Hesse: *Z. Anorg. Allgem. Chem.* **156**, 245 (1926).

3. D.L. Stein, R.G. Palmer: *Phys. Rev. B*, to be published (1988).
4. M. Randeria, J.P. Sethna, R.G. Palmer: *Phys. Rev. Lett.* **54**, 1321 (1985).
5. J.L. Skinner: *J. Chem. Phys.* **79**, 1955 (1983).
6. J. Budimir, J.L. Skinner: *J. Chem. Phys.* **82**, 5232 (1985).
7. J.C. Kimball: *J. Stat. Phys.* **21**, 289 (1979).
8. J.E. Anderson: *J. Chem. Phys.* **52**, 2821 (1970).
9. S. Bozdemir: *Phys. Status Solidi* **B103**, 459 (1981); **104**, 37 (1981).
10. R.J. Glauber: *J. Math. Phys.* **4**, 294 (1963).
11. S.H. Glarum: *J. Chem. Phys.* **33**, 1371 (1960); see also B.I. Hunt, J.G. Powles: *Proc. Phys. Soc.* **88**, 513 (1966); and P. Bordewijk: *Chem. Phys. Letters* **32**, 592 (1975).
12. J. Klafter, A. Blumen: *Chem. Phys. Letters* **119**, 377 (1985).
13. G. Zumofen, A. Blumen, J. Klafter: *J. Chem. Phys.* **82**, 3198 (1985); A. Blumen, G. Zumofen, J. Klafter: *Phys. Rev. B* **30**, 5379 (1984); A. Blumen, J. Klafter, G. Zumofen: *Phys. Rev. B* **28**, 6112 (1983).
14. M.F. Shlesinger, E.W. Montroll: *Proc. Natl. Acad. Sci. USA* **81**, 1280 (1984); M.F. Shlesinger: *J. Stat. Phys.* **36**, 639 (1984).
15. J.T. Bendler: *J. Stat. Phys.* **36**, 625 (1984); E.W. Montroll, J. Bendler: *J. Stat. Phys.* **34**, 129 (1984); J.T. Bendler, M.F. Shlesinger: *Macromol.* **18**, 591 (1985).
16. S. Redner, K. Kang: *J. Phys. A* **17**, L451 (1984).
17. G.H. Fredrickson, H.C. Andersen: *Phys. Rev. Lett.* **53**, 1244 (1984); *J. Chem. Phys.* **83**, 5822 (1985).
18. G.H. Fredrickson, S.A. Brawer: *J. Chem. Phys.* **84**, 3351 (1986).
19. G. Adam, J.H. Gibbs: *J. Chem. Phys.* **43**, 139 (1965). See also H. Tweer, J.H. Simmons, P.B. Macedo: *J. Chem. Phys.* **54**, 1952 (1971).
20. R.G. Palmer, D.L. Stein, E. Abrahams, P.W. Anderson: *Phys. Rev. Lett.* **53**, 958 (1984).
21. R.G. Palmer, D.L. Stein: work in progress (1988).
22. W.J. Motyl, D.L. Stein, R.G. Palmer: work in progress (1988).
23. F.H. Stillinger, T.A. Weber: *Ann. N.Y. Acad. Sci.* **484**, 1 (1986); T.A. Weber, G.H. Fredrickson, F.H. Stillinger: *Phys. Rev. B* **34**, 7641 (1986); T.A. Weber, F.H. Stillinger: *Phys. Rev. B* **36**, 7043 (1987).
24. Ajay, R.G. Palmer: in preparation (1988).
25. M.J.A.M. Brummelhuis, H.J. Hilhorst: Leiden preprint (1988).
26. K. Fröböse, J. Jäckle: *J. Stat. Phys.* **42**, 551 (1986); W. Ertel, K. Frobose, J. Jäckle: *J. Chem. Phys.* **8**, 5027 (1988).
27. W.J. Motyl, Ajay, R.G. Palmer, D.L. Stein: work in progress (1988).
28. J. Klafter, A. Blumen, G. Zumofen: *J. Stat. Phys.* **36**, 561 (1984).
29. G. Zumofen, A. Blumen: *Chem. Phys. Letters* **88**, 63 (1982).
30. S.R. Shenoy: *Phys. Rev. B* **35**, 8652 (1987).
31. J.H. Gibbs, E.A. Dimarzio: *J. Chem. Phys.* **28**, 373 (1958); E.A. Dimarzio, J.H. Gibbs: *J. Chem. Phys.* **28**, 807 (1958).
32. M.H. Cohen, D. Turnbull: *J. Chem. Phys.* **31**, 1164 (1959); D. Turnbull, M.H. Cohen: *J. Chem. Phys.* **34**, 120 (1961); M.H. Cohen, G.S. Grest: *Phys. Rev. B* **20**, 325 (1979); **21**, 4113 (1980); *Solid State Commun.* **39**, 143 (1981); *Phys. Rev. B* **24**, 4091 (1981).
33. A.T. Ogielski, D.L. Stein: *Phys. Rev. Lett.* **55**, 1634 (1985).
34. D. Kumar, S.R. Shenoy: *Phys. Rev. B* **34**, 3547 (1986).
35. R.G. Palmer, D.L. Stein, Y.-Q. Yin, J. Ye: work in progress (1988).
36. D. Stein, ed.: *Lectures in Complexity, Santa Fe Institute Studies in the Sciences of Complexity*, Vol. VIII (Addison-Wesley, Reading MA 1989).
37. P. Erdős, A. Rényi: *Pub. Math. Inst. Hung.Acad. Sci.* **5**, 17 (1960).

38. M. Ajtai, J. Komlós, E. Szemerédi: *Combinatorica* **2**, 1 (1982).
39. I.A. Campbell: *J. Phys. (Paris) Lett.* **46**, L1159 (1985); *Phys. Rev. B* **33**, 3587 (1986); I.A. Campbell, J.M. Flesselles, R. Julien, R. Botet: *J. Phys. C* **20**, L47 (1987).
40. J.M. Flesselles, R. Botet: preprint (1988).
41. A.T. Ogielski: *Phys. Rev. B* **32**, 7384 (1985).
42. A.J. Bray, G.J. Rodgers: Manchester preprint (1988).
43. N.G. van Kampen: *Stochastic Processes in Physics and Chemistry* (North-Holland, Amsterdam 1981).
44. H. Takano, H. Nakanishi, S. Miyashita: *Phys. Rev. B* **37**, 3716 (1988).
45. Y.-Q. Yin, D.L. Stein, R.G. Palmer: work in progress (1988).
46. J. Ye, R.G. Palmer: work in progress (1988).

Experimental Study of the Slow Dynamics in the Spin-Glass Phase

M. Ocio, J. Hammann, and E. Vincent

Service de Physique du Solide et de Résonance Magnétique,
CEN Saclay, F-91191 Gif-sur-Yvette Cedex, France

1. INTRODUCTION

Among the known properties of spin-glasses, one of the most typical is the non-stationary dynamic response of these systems when they are quenched into their low temperature phase. This aging phenomenon has been found in practically all the investigated materials whether metallic or insulating. The non-stationary dynamics is easily evidenced in the relaxation experiments, for instance the time decay of the thermoremanent magnetization (TRM). The decay is found to depend on the time t_w the system has spent at the working temperature after a quench from above the freezing temperature T_g, before the field is switched off. When the observation time t_{obs} ($t_{obs} = 0$ at the field cutoff) becomes larger than t_w, the response function is dominated by the non-stationary contribution.

On the other hand, susceptibility and noise measurements reveal the existence of a stationary contribution to the response. They are indeed made at a constant observation time ($t_{obs} = 1/\nu$ where ν is the frequency), much smaller than the age $t = t_w + t_{obs}$. However, even in these experiments, non-stationary effects can be detected at very low frequencies, where it is possible to fulfill the condition $1/\nu \lesssim t$.

In the following, we mainly describe the results of measurements performed on the chromium thiospinel $CdIn_{0.3}Cr_{1.7}S_4$ ($T_g = 16.6$ K). Similar results have been obtained on several other spin-glasses, metallic (Ag:Mn 2.6%) as well as insulating ($CsNiFeF_6$).

2. SPIN-GLASS SLOW DYNAMICS IN THE LOW TEMPERATURE PHASE

2.1 Stationary response: a.c. susceptibility and magnetic fluctuations

Magnetic noise measurements are a very recent way of exploring the dynamics of magnetic materials without any excitation field [1]. From the noise flux detected by means of a superconducting coil connected to the input of a SQUID, one is able to deduce the time correlation function of the magnetization [2], $C(t)=m(t+t')m(t')$, or its cosine Fourier transform, the magnetization power spectrum $m^2(\omega)$. In an ergodic system, the noise power spectrum can thus be related to the response function through the FDT (Kubo) relation [3] $\sigma(t)=C(t)/k_B T$ where $\sigma(t)$ is the relaxation function and k_B is the Boltzmann constant. One derives the usual frequency dependent relation

Springer Series in Synergetics, Vol. 43   **Cooperative Dynamics in Complex Physical Systems**
Editor: H. Takayama   © Springer-Verlag Berlin, Heidelberg 1989

$$\overline{m^2(\omega)} = \frac{2}{\pi} k_B T \frac{\chi''(\omega)}{\omega} \; , \tag{1}$$

which can be checked by comparing susceptibility and fluctuations data.

After the first work of Ocio et al. [1], several detailed studies were performed on insulating spin-glasses ($CsNiFeF_6$ [1,3], $CdIn_{0.3}Cr_{1.7}S_4$ [3,5], $Eu_{0.4}Sr_{0.6}S$ [6]), as well as on an amorphous metallic system ($(Fe_{0.15}Ni_{0.85})_{75}P_{16}B_6Al_3$ [7]). They all led to the same conclusions:

i) In the low temperature phase, below T_g, the magnetic noise is stationary provided $1/\omega \ll t$ where t is the age. In this stationary range, the power spectrum varies approximately as $1/\omega$ with small deviations depending on the temperature [4,6].

ii) In the stationary range, comparison with the results of susceptibility measurements shows that the FDT (relation (1)) is obeyed in all investigated materials [1,4,6]. This was the first experimental evidence of the applicability of the FDT to spin-glasses.

Figure 1-a displays a log-log plot of $\chi''(\omega)$ in $CdIn_{0.3}Cr_{1.7}S_4$ for several temperatures below T_g, as derived from noise measurements through relation (1). For the lower frequency decade, the age is 24 h. We have proposed in Ref.[4] a description of this result, involving a very smooth crossover between two limiting behaviors, $\chi''(\omega) \propto \omega^{-\alpha'(T)}$ for $\omega \gg \omega_0$ and $\chi''(\omega) \propto \omega^{\alpha(T)}$ for $\omega \ll \omega_0$, where $\alpha(T)$ and $\alpha'(T)$ are slowly increasing functions of temperature, $0 < \alpha(T) \simeq \alpha'(T) \ll 1$ and $\omega_0(T)$ increases very rapidly with temperature. Such a description is in agreement with the prescription that the noise power spectrum must be an integrable function of frequency. It is also reminiscent of the Cole-Cole law already used to analyse susceptibility in Ref.[8]. In $CdIn_{0.3}Cr_{1.7}S_4$, ω_0 is within the experimental frequency range at temperatures below 8 K. We are mainly concerned with the low frequency behavior of the system: thus, we will concentrate only on the results above 10 K.

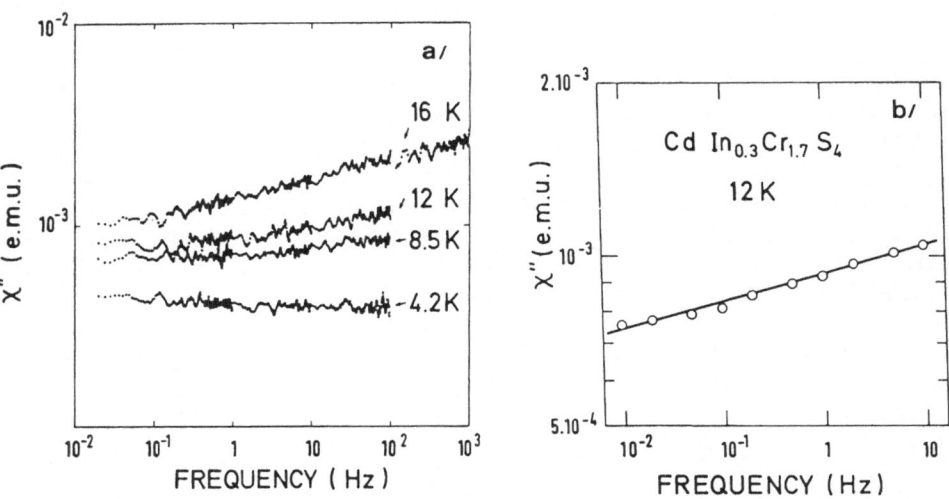

Fig. 1: Out-of-phase susceptibility below T_g in $CdIn_{0.3}Cr_{1.7}S_4$; a) as derived from noise data ; b) as measured after aging for 8 days.

In the lower frequency range in Fig.1-a, there is some upward deviation from the power law behavior of χ''. Since it was suspected that these deviations revealed the effect of a too short aging period, a susceptibility measurement was made after an aging period of 8 days at 12 K. The result, plotted in Fig.1-b, is compatible with a power law down to 10^{-2} Hz and supports the hypothesis that the spin-glass phase remains critical below T_g [9]. Hence we suppose $\chi''(\omega) \propto \omega^\alpha$ for $\omega \to 0$.

A direct consequence of the linear theory is that $\chi''(\omega) \propto \omega^\alpha$ at $\omega \ll \omega_0$ would correspond to a relaxation function $\sigma(t) \propto t^{-\alpha}$ for $t \gg 1/\omega_0$. In that case, the static susceptibility $\chi_0 = \lim_{\omega \to 0} \chi(\omega)$ is given by $\chi_0 = \chi'(\omega) + \cotg(\alpha\pi/2)\chi''(\omega)$.
Figure 3 displays χ_0 derived from the susceptibility measured between 10 K and T_g in $CdIn_{0.3}Cr_{1.7}S_4$, as well as the field-cooled susceptibility $\chi_{FC} = M_{FC}/H$ measured under a field of 0.3 Oe. Since $\chi_{FC} > \chi_0$ it is not possible to describe the relaxation function by a simple power law, but, more generally by

$$\sigma(t) = \chi_0(t/t_0)^{-\alpha} + \Delta\chi , \qquad (2)$$

where $\Delta\chi$ depends on the sample history. For a thermoremanent magnetization (TRM) decay experiment, $\Delta\chi = \chi_{FC} - \chi_0$.

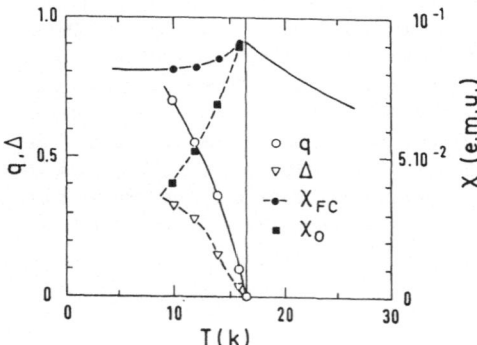

Fig. 2: Susceptibility and order parameters in $CdIn_{0.3}Cr_{1.7}S_4$:
χ_{FC}: field-cooled with a field of 0.3 Oe
χ_0 : extrapolated from a.c. susceptibility

Within the time range of our results (100 s), $\Delta\chi$ is a constant which can be considered as describing the magnetization of the spins frozen during cooling under the field. Nevertheless, a decay of $\Delta\chi$ over periods of time much larger cannot be excluded: relaxation experiments up to more than 10^5 s reveal such a decay. A possible interpretation is that $\Delta\chi$ is dependent on relaxation times which diverge at the thermodynamic limit [10], but depend on size effects in a real material. In the frame of the known mean field theories [11], two order parameters q and Δ are useful. Then: $\chi_0 = \beta(1-q)$ and $\chi_{FC} = \beta(1-q+\Delta)$.
In $CdIn_{0.3}Cr_{1.7}S_4$, $\beta = C/(T-\theta)$ where θ is about -8.8 K [12]. The calculated values of q and Δ are reported in Fig.2. Similar results were obtained from static measurements on a metallic amorphous material [13].

2.2 Non-stationary response

Contrary to the frequency dependent response, the time dependent one is mainly determined by aging, as soon as $t_{obs} \geqslant t_w$. After that point was first stressed by the Uppsala group [14], we were led to propose a time scaling analysis of aging

[15], on the basis of previous works on the strain creep in glassy polymers [16]. TRM decay measurements were made on insulating ($CsNiFeF_6$ [15], $CdIn_{0.3}Cr_{1.7}S_4$ [5]), as well as metallic (Ag:Mn 2.6% [17]) systems, for waiting and observation times up to 10^5 s. In all cases, the results are in good agreement with the time scaling hypothesis.

The basic assumption of time scaling is simply a shift of the distribution of relaxation times such that $a.g(a\tau,t') = g(\tau,t)$ where $g(\tau,t)$ is the amplitude of the distribution for relaxation time τ at age t, and $a=(t'/t)^\mu$, with the scaling exponent $\mu \lesssim 1$.

For instance, suppose a system with only one relaxation time which shifts in time as $\tau = \tau_c^{1-\mu}.t^\mu$. The dynamic equation for the relaxation function $\sigma(t_{obs})$ is thus $d\sigma/dt = -\sigma/(\tau_c^{1-\mu}.t^\mu)$ which is non-stationary. If we use an effective time defined by $\lambda = (t_o^\mu.t^{1-\mu})/(1-\mu)$, or $d\lambda = (t_o/t)^\mu dt$, where t_o is an arbitrary reference time, the equation reads $d\sigma/d\lambda = -\sigma/(\tau_c^{1-\mu}.t_o^\mu)$ and the relaxation function is stationary in λ: $\sigma(\lambda) = \exp-\dfrac{\lambda}{\tau_c^{1-\mu}.t_o^\mu}$. It is simply the relation giving the relaxation function of a stationary system whose relaxation time is held at the value it had at age t_o. For this reason, λ was called _constant age effective time_.

The response to a step function of the field is $\sigma(\lambda_{obs})= \exp-\dfrac{\lambda_{obs}}{\tau_c^{1-\mu}.t_o^\mu}$, with $\lambda_{obs}= (\lambda-\lambda_w)(t_w/t_o)^\mu$. For simplicity, one can choose $t_o= t_w$. Then, $\lambda_{obs}= t_w^\mu((t_{obs}+t_w)^{1-\mu}- t_w^{1-\mu})/(1-\mu)$ and, if $t_{obs}\ll t_w$ or $t_w\to \infty$ then $\lambda_{obs}= t_{obs}$.
The generalization to a distribution of relaxation times is obvious.

In the λ representation, curves corresponding to different waiting times at a given temperature can be superposed by adjusting the exponent μ and a small vertical logarithmic shift which depends on t_w. These are the _master curves_:

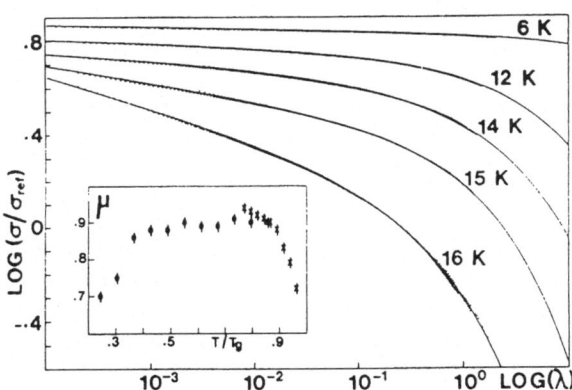

Fig.3: $CdIn_{0.3}Cr_{1.7}S_4$. Master curves from TRM decay data with t_w in the range 10^2-10^3 s. The reference time t_o is 10^4 s. In the inset,corresponding values of the aging exponent μ.

This is the main result of time scaling. At this stage, we can note that the vertical shift reveals the existence of the stationary relaxation. In the aging range ($t_{obs}> t_w$), the latter has a negligible effect on the shape of the relaxation curves. A consequence is that, in the aging range, the response must be invariant by translation in λ scale. This point has been checked. Decay of the magnetization was measured in $CdIn_{0.3}Cr_{1.7}S_4$ after applying a rectangular field pulse, starting from the zero-field-cooled state. Three pulses were tried: they were shifted in age, but calibrated (from the known value of μ) in order to get the same width in λ (see the insets of Fig.4-a & b). When plotted in a real time representation, the curves are rather different, as shown in Fig.4-a. When plotted in the λ representation, they perfectly superpose (Fig.4-b).

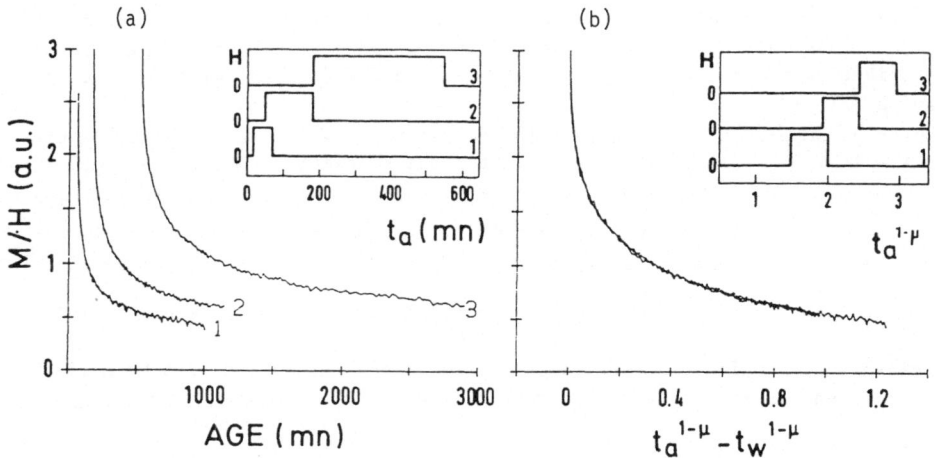

Fig. 4: Response to rectangular field pulses; a) real time plot b) effective time plot. The insets show the pulse shapes.

In the first studies, we tentatively fitted the master curves with the functional form [5] $\sigma/D(t_w) = \lambda_{obs}^{-x} \exp - (\lambda_{obs}/(t_w^{\mu} \cdot \tau_p^{1-\mu}))^{\beta}$, whose stationary part (for $t_{obs} \ll t_w$ or $t_w \to \infty$) was simply t_{obs}^{-x}. Nevertheless, to keep the coherence with the results of 2.1 (relation (2)), we are led to the following choice for the relaxation:

$$\sigma = \chi_o \left(\frac{\lambda_{obs}}{\tau_o}\right)^{-\alpha} . \mathcal{F}\left(\frac{\lambda_{obs}}{t_w^{\mu}}\right) + \Delta\chi \qquad (3)$$

whose small time logarithmic slope has been verified to be in fair agreement with the fitted values of x. The easiest choice for the aging function seems to be $\mathcal{F}(x) = \exp - (x/\tau_p^{1-\mu})^{\beta}$, but since $x \ll \tau_p^{1-\mu}$ even for the longest observation times, this cannot be conclusively supported.

Aging is also detected at small ages in very low frequency a.c. susceptibility and noise measurements [18,19]. It affects mainly the out-of phase susceptibility. It has been shown that the FDT relationship between noise and susceptibility remains valid in the quasi-stationary regime, and that aging of the harmonic response obeys time scaling [19].

3. CONCLUSIONS: TOWARDS A PHENOMENOLOGICAL CLUSTER PICTURE FOR THE SPIN-GLASS PHASE

Below T_g, the power law dependences of the non-linear susceptibility on field [12] as well as of $\chi''(\omega)$ on frequency (see 2.1) suggest that _the collective excitations of the system have the same structure as those at T_g_, provided the system is aged enough to be in equilibrium. Thus, we suggest a picture [20], which is an extension below T_g of the fractal cluster model first applied to the critical behavior above T_g [21]. The clusters are statistical entities representing low energy collective excitations whose power law distribution results from a dynamical equilibrium between aggregation and fragmentation processes. Below T_g, the distribution is normalized to the total number of spins of the system minus the amount of frozen spins which corresponds to the value of the Edwards-Anderson order parameter. A temperature decrease below T_g freezes the larger clusters, introducing a cut-off in the distribution. Aging of the latter corresponds to an increase of the cut-off size with time. This evolution is described by a Smoluchowsky equation for aggregation-fragmentation processes whose long time similarity solution [22] shares the time scaling properties described in 2.2. The proposed picture is coherent with the whole set of experimental results. It yields a power law relaxation at equilibrium and accounts well for the observed effect of temperature changes during aging [23].

REFERENCES

1. M. Ocio, H. Bouchiat and P. Monod: J. Phys. (Paris) Lett. 46, 647 (1985); J. Mag. Mag. Mat 54-57, 11 (1986).
2. P. Refrégier and M. Ocio: Revue Phys. Appl. 22, 367 (1987).
3. R. Kubo: J. Phys. Soc. Japan: 12, 570 (1957).
4. P. Refrégier, M. Ocio and H. Bouchiat: Europhys. Lett. 3, 508 (1987).
5. M. Alba, J. Hammann, M. Ocio and P. Refrégier: J. Appl. Phys. 61, 3683 (1987).
6. W. Reim, R. H. Koch, A. P. Malozemoff, M. B. Ketchen and H. Maletta: Phys. Rev. Lett. 57, 905 (1986).
7. P. Svedlindh, P. Nordblad and L. Lundgren: Preprint.
8. T. Saito, C. J. Sandberg and Y. Miyako: J. Phys. Soc. Japan 52, 3170 (1983); 54, 231 (1985).
9. F. T. Bantilan Jr. and R. G. Palmer: J. Phys. F11, 261 (1981).
10. H. Sompolinsky: Phys. Rev. Lett. 47, 935 (1981).
11. For a review, see K. Binder and A. P. Young: Rev. Mod. Phys. 58, 801 (1986).
12. E. Vincent and J. Hammann: J. Phys. C20, 2659 (1987).
13. Y. Yeshurun and H. Sompolinsky: Phys. Rev. B26, 1487 (1982).
14. L. Lundgren, P. Nordblad, P. Svedlindh and O. Beckman: J. Appl. Phys. 57, 3371 (1985).
15. M. Ocio, M. Alba and J. Hammann: J. Phys. (Paris) Lett. 46, 1101 (1985)
16. L. C. E. Struik: in Physical Ageing in Amorphous Polymers and other Materials (Elsevier, Houston, TX, 1978).
17. M. Alba, M. Ocio and J. Hammann: Europhys. Lett. 2, 48 (1986).
18. L. Lundgren, P. Svedlindh and O. Beckman: J. Mag. Mag. Mat. 31-34, 1349 (1983).
19. P. Refrégier, M. Ocio, J. Hammann and E. Vincent: J. Appl. Phys. 63, 4343 (1988).
20. M. Ocio, J. Hammann, P. Refrégier and E. Vincent: Proc. Int. Conf. Mag. Paris (1988).
21. A. P. Malozemoff and B. Barbara: J. Appl. Phys. 57, 3410 (1985).
22. M. H. Ernst: in Fundamental Problems in Statistical Mechanics VI, 329 (Elsevier Science Publishers B. V., 1985).
23. P. Refrégier, E. Vincent, J. Hammann and M. Ocio: J. Phys. (Paris) 48, 1533 (1987).

Scaling Theory of the Ordered Phase of Real Spin Glasses

M.A. Moore

Department of Theoretical Physics, University of Manchester,
Manchester M13 9PL, UK

1. Introduction

Theoretical approaches to spin glasses can be divided into two main
categories. The first approach envisages constructing the mean-field solution
of (say) the Edwards-Anderson[1](EA) Hamiltonian and then systematically
expanding about it to describe the properties of three-dimensional spin
glasses. Producing a mean-field theory is equivalent to solving the
Sherrington-Kirkpatrick[2] (SK) spin-glass model. This model is now
well understood and the solution reveals a rich structure of many pure states
related by an ultrametric topology[3]. Fig 1(a) shows the expected phase
diagram in a field. The SK model is the infinite dimensional limit of the EA
Hamiltonian. Recent studies by Kondor[4] suggest that the ultrametric
behaviour, de Almeida-Thouless (AT)[5] line etc. will not exist below six
dimensions. Thus, the program of expanding about the customary mean-field
solution to obtain the properties of spin glasses whose dimensionality is less
than six does not look promising.

The alternative approach - called the scaling approach - is of more recent
origin and leads to a quite different physical picture of spin glasses [6,7,8].
From the experimental viewpoint the most striking difference between the two
theories is the absence of a genuine AT line in the scaling approach (see Fig
1(b)) although in Section III we shall show how a pseudo-AT line exists as the
locus of points in the h-T plane at which the system freezes on experimental
time scales. The low-temperature phase is characterised by just two pure
states, i.e. no replica symmetry breaking and only a trivial ultrametric
topology.

The properties of this ordered phase are determined as usual by a
zero-temperature fixed point. The central concept is that of a
scale-dependent coupling J(L). Since the interactions in spin glasses are
random variables we introduce a distribution function $P_L(J)$, whose scale width
is J(L) for the couplings at length scale L, and $J(L) \sim JL^y$. For an Ising

Springer Series in Synergetics, Vol. 43   **Cooperative Dynamics in Complex Physical Systems**
Editor: H. Takayama
© Springer-Verlag Berlin, Heidelberg 1989

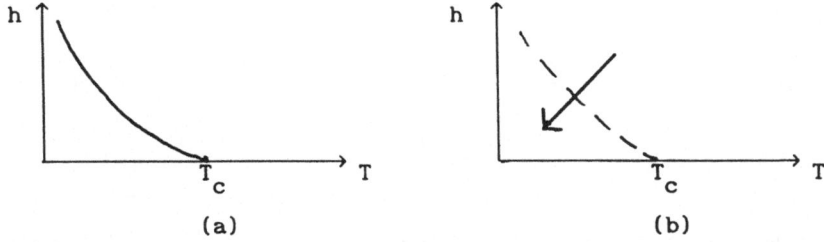

Fig 1 *(a)   AT line as for SK model.   (b)   Pseudo AT line due to freezing on
experimental time scales for d = 3.   Arrow shows how the line moves on
increasing time scales.*

ferromagnet the coupling at length scale L is the domain interface energy
$J(L)=JL^{d-1}$ so for it the zero temperature scaling exponent y has the value
(d-1). For a Heisenberg ferromagnet y=d-2. The value of d for which y(d)=0
determines the lower critical dimension d^* of the system. For $d<d^*$ J(L)
scales to zero as L→∞, since then y<0, but for $d>d^*$, J(L) scales to the strong
coupling (zero-temperature) fixed point for all $T<T_c$. In section II it is
shown how the value of y can be determined numerically for spin glasses.
These investigations imply that for Ising spin glasses there is a phase
transition in three dimensions, but not in two dimensions and that $2<d^*<3$[6].

In Section III we discuss how the dynamics of these systems can be
modelled.   The key concept here is that relaxation takes place via an
activation process over free-energy barriers, which for processes on a length
scale L are typically of magnitude JL^{ψ}, where ψ is another zero-temperature
scaling exponent whose accurate elucidation by numerical methods has so far
proved impossible. However, by studying the dependence of the pseudo-AT line
in the h-T diagram on equilibriation time one can estimate the exponent ψ.

Finally, in Section IV the concept of an overlap length L_p is introduced
[8,9]. This length scale is a measure of the distance within which spins are
correlated after a temperature change δT; $L_p \sim \delta T^{-\zeta}$. We show how the exponent
ζ can be related to the fractal dimension of the spin glass domain wall and
how the existence of an overlap length is revealed by ageing experiments on
spin glasses.

2. The Exponent y

The exponent y for the Ising EA spin glass may be determined numerically by
investigating the sensitivity to boundary conditions of the ground state
energy of finite blocks of linear dimension L. This idea was first proposed

by Banavar and Cieplak[10] and further developed by McMillan[11] and Bray and Moore[12].

We defined a block spin coupling $J' = \frac{1}{2}(E_{\uparrow\uparrow}-E_{\uparrow\downarrow})$ where $E_{\uparrow\uparrow}$ is the ground state energy for random boundary conditions on two opposite edges (d=2) or faces (d=3) of the block and periodic boundary conditions otherwise and $E_{\uparrow\downarrow}$ is the same quantity when the spins on one of the random edges (faces) are reversed holding the spins on the opposite edge (face) fixed. The random boundary conditions are designed to simulate the interaction of the block with neighbouring blocks in a large spin glass system. Exact ground states were obtained by a transfer matrix technique. The exponent y is determined from plots of $\ln(2J(L))$ with $J(L) = [|J'|]_{av}$ versus ℓnL which should have for large L a slope y:

$$y = -0.291 \pm 0.002 \qquad ; \; d = 2$$
$$y = 0.19 \pm 0.01 \qquad ; \; d = 3:$$

av denotes an average over a large number of systems, in which the $\{J_{ij}\}$ exchange interactions are different. The true errors are difficult to estimate as there is always the possibility of systematic error due to a failure to reach asymptopia.

At non-zero temperatures it is useful to define $J' = \frac{1}{2}(F_{\uparrow\uparrow}-F_{\uparrow\downarrow})$ where $F_{\uparrow\uparrow}$ and $F_{\uparrow\downarrow}$ are the free energies associated with the two sets of boundary conditions. One expects[8] $J(L) = [|J'|]_{av} = \Upsilon(T)L^y$ where $\Upsilon(T) \sim \xi^{-y}$ where ξ is a "correlation" length varying as $T \to T_c$ as $\xi_0(1-T/T_c)^{-\nu}$. (NB there is actually only algebraic decay of correlations within the ordered phase[6,8]). $\Upsilon(T)$ is the analog of the interface energy of an Ising ferromagnet or the spin-wave stiffness of a Heisenberg ferromagnet.

In principle the exponent y could be determined experimentally by determining the singular part of the magnetisation of a spin glass in a field for $T<T_c$. For simplicity an Imry-Ma domain argument will be given for this singularity but a semi-quantitative approach is also possible[6]. The magnetisation of a domain of linear size L containing of order L^d spins is $\sim \sqrt{q} \, L^{d/2}$, where q is the EA order parameter $N^{-1}\sum_i <S_i>^2$. q vanishes as $T \to T_c$ as $(1-T/T_c)^\beta$. A domain of size L will flip to take advantage of an applied field h if

$$\Upsilon(T)L^y \approx h\sqrt{q} \, L^{d/2} . \tag{1}$$

The left-hand side is the free-energy cost of flipping the domain; the right-hand side is the field energy gained. Solving eq.(1) for $L = L^*$,

$$L^* \sim \xi(T,h) \sim [\Upsilon(T)/\sqrt{q} \; h]^{2/(d-2y)} . \tag{2}$$

136

In a field spin correlations do decay exponentially with a correlation length $\xi(T,h)$. The magnetisation M per spin $\sim (L^{*d/2}\sqrt{q})/L^{*d}$, so

$$M = M_0\sqrt{q}[\sqrt{q}h/\Upsilon(T)]^{d/(d-2y)} + \chi h \tag{3}$$

where χh is the non-singular term. Measurements of M for small fields could in principle serve to determine y. However, it should be stressed that eq.(3) is an equilibrium result and thus only measurements on very long time scales are likely to see evidence of the non-singular behaviour.

3. Dynamics and the Pseudo AT Line

McMillan[7] argued that barriers against relaxation at length scale L were typically of magnitude B(L) where

$$B(L) = b(T)L^{\psi} \sim J(L/\xi)^{\psi} \tag{4}$$

so[8] $b(T) \sim J(1-T/T_c)^{\nu\psi}$. In a magnetic field the barrier heights will not increase on all length scales, but only up to a maximum set by the correlation length $\xi(T,h)$, ie the maximum barrier height B_{max} will be of order $b(T)\xi^{\psi}(T,h)$, and hence there will be a longest relaxation time of order τ_R, where

$$\tau_R = t_o \exp(B_{max}/T) \tag{5}$$

and t_o is some microscopic attempt time. We expect (5) to hold even in the vicinity of T_c for values of h and T such that $B_{max} \gg T_c$ (where $B_{max} = T_c$, (5) should be replaced by the dynamic scaling expression $\tau_R \sim \xi^z$). Now associated with every experiment there is a time τ_e such that spin re-orientation process which take place on time scales longer than τ_e are not equilibrated or cannot be followed. The usual hall-mark of the AT line is the line of onset of non-equilibrium effects (such as irreversibility) and will occur at fields and temperatures such that $\tau_e = \tau_R$. Using (5) for τ_R gives near T_c the equation

$$\left[\frac{h}{T_c}\right]^2 \sim \frac{1}{\left[\ln\frac{\tau_e}{t_o}\right]^{\frac{d-2y}{\psi}}} \left[1 - \frac{T}{T_c}\right]^{\Delta} \tag{6}$$

where $\Delta = \gamma+\beta = \nu(d+2-\eta)/2$. Notice that the relation between $(h/T_c)^2$ and $(1-T/T_c)^{\Delta}$ is just what has been obtained from scaling apart from a prefactor

137

which depends on the experimental time scale τ_e. Note that the condition $B_{max} \gg T_c$ requires $\tau_e \gg t_o$. Measurements of the time-dependence of the pre-factor may provide an experimental method of determining the exponent ψ.

4. Nature of the Ordered Phase

According to the scaling picture the ordered phase of a spin glass has some remarkable "chaotic" features[9]. For an Ising ferromagnet at T=0 all the spins are (say) up, $S_i^0 = +1$ and at finite temperatures the magnetisation projected on to its ground state $\Phi = \sum_i \langle S_i \rangle S_i^0 / N$ is non-zero for all $T < T_c$. However, in a spin glass the analogous quantity is zero in the thermodynamic limit. Even for an infinitesimally small temperature Φ is zero implying that any change in temperature can produce large changes in site magnetisations $\langle S_i \rangle = m_i$.

To quantify this effect it is useful to introduce the concept of the overlap length $L_p(\delta T)$. This is the size of a typical domain for which a temperature change δT causes a change of sign of the m_i for spins within the domain. We shall argue that[9, 14]

$$L_p = \xi \left[\frac{T_c - T}{\delta T} \right]^{1/(d_s/2 - y)} , \tag{7}$$

where d_s is the exponent which relates the area of a domain interface A to its linear dimension L, $A = L^{d_s}$. The value of d_s has been studied using transfer matrix techniques for d=2 and is of order 1.26[9]. Its value for d=3 is at present unknown.

Eq.(7) can be derived heuristically as follows. Let $F_{int} = -\frac{1}{2}(F_{\uparrow\uparrow} - F_{\uparrow\downarrow})$ be the interface free energy. $F_{int} = E_{int} - TS_{int} \sim \Upsilon(T)L^y$. Now the interface entropy S_{int} and the interface energy E_{int} will each be of order $L^{d_s/2}$ as they are the sum of L^{d_s} essentially random terms. The average value of F_{int}; $[|F_{int}|]_{av} \sim f(L/\xi) \rightarrow (L/\xi)^y$ for $L \gg \xi$ so

$$[|S_{int}|]_{av} = \frac{\partial \, [|F_{int}|]_{av}}{\partial T} \sim (L/\xi)^{d_s/2} / (1 - T/T_c) . \tag{8}$$

Notice that since $d_s/2 > y$, F_{int} for a given sample must be a very rapidly varying function of temperature. In particular if the temperature is changed from T to $T + \delta T$, F_{int} will change sign if

$$F_{int} \sim S_{int} \, \delta T \sim \Upsilon(T) \, L^y . \tag{9}$$

Solving for $L(=L_p)$ gives equation (7). Within a region of size $L<L_p$ the change in temperature by δT causes few sign change of the m_i. However at length scales $L>L_p$ there will take place many domain reorientations to take advantage of the sign changes in F_{int}.

After a time t regions of size R(t) are equilibrated, where

$$t = t_o \exp\left[(R(t)/\xi)^{\psi}/T\right] . \tag{10}$$

If the temperature is now altered by δT, there will be spin reorientations on length scales $L>L_p$. If $L_p>R(t)$ there will be no consequences for the time dependence of, say, the growth of the magnetisation after the application of a small field. However, if $L_p<R(t)$ new domains are formed and produce a change in the growth of the magnetisation. Let ΔT be the value of δT for which $L_p=R(t)$. ΔT is the smallest δT for which there is no consequence for ageing. Then

$$\Delta T \sim (T_c-T)\ (T_c/T)^{(d_s/2-y)/\psi} . \tag{11}$$

The linear dependence of ΔT as $T\rightarrow T_c$ is consistent with experimental data[16].

5. Conclusions

The scaling theory is able to provide a semi-qualitative description of both the static and dynamic properties of spin glasses. However many basic problems remain. In particular, systematic field theoretic calculations of the zero-temperature scaling exponents y, d_s and ψ are not as yet possible.

References

1. S.F. Edwards and P.W. Anderson: J.Phys F5, 965 (1975)

2. D. Sherrington and S. Kirkpatrick: Phys.Rev.Lett. 35, 1972 (1975)

3. M Mezard, G. Parisi and M.A. Virasoro: Spin Glass Theory and Beyond, (World Scientific, Singapore 1987)

4. I. Kondor: to be published

5. J.R.T de Almeida and D.J. Thouless: J.Phys. A11, 983 (1978)

6. For a review of the scaling theory see A.J. Bray and M.A. Moore: in Heidelberg Colloquium on Glassy Dynamics", Lecture Notes in Physics, 275, (Springer Verlag 1987)

7. W.L. McMillan: J.Phys C17, 3179 (1984)

8. D.S. Fisher and D.A. Huse: Phys.Rev.Lett. 56, 1601 (1986) and to be published

9. A.J.Bray and M.A. Moore: Phys.Rev.Lett. 58, 57 (1987)

10. J.R. Banavar and M. Cieplak: Phys.Rev.Lett. 48, 832 (1982)

11. W.L. McMillan: Phys.Rev. B31, 340 (1985), B29, 4026 (1984)

12. A.J.Bray and M.A. Moore: J.Phys. C17, L469 (1984)

13. Y. Imry and S.K. Ma, Phys.Rev.Lett. 35, 1399 (1975)

14. A.J. Bray and J.R. Banavar, Phys.Rev. B35, 8888 (1987)

15. G.J.M. Koper and H.J. Hilhorst, J.de Physique, 49, 429 (1988)

16. L. Sandlund, P. Svedlindh, P. Granbery, P. Nordblad and L. Lundgren, to be published

Phase Diagram and Critical Properties of the Finite-Dimensional Spin Glasses

Hidetoshi Nishimori

Department of Physics, Tokyo Institute of Technology,
Meguro-ku, Oh-okayama, Tokyo 152, Japan

I explain a variety of techniques to investigate the properties of spin glasses with emphasis on the multicritical region and the Griffiths singularity.

1. INTRODUCTION

Randomness in spin interactions sometimes gives rise to a spatially disordered low-temperature phase called a spin glass (SG) state. The central issues in the research of spin glasses have been to determine the condition for the existence of a thermodynamically stable SG phase and to characterize this unusual state [1]. A consistent picture has gradually emerged at the level of the mean field theory, and many experimental observations appear to be qualitatively understood within the framework of the mean field theory.

However, it is important to know how the mean field predictions are modified in the real world of finite dimensions, as they are in the case of conventional critical phenomena. This general idea has prompted recent theoretical studies, mostly numerical, of finite-dimensional spin glasses. It has been established fairly convincingly that the three-dimensional random Ising model has a finite SG transition temperature, whereas the two-dimensional counterpart seems to be lacking one[1]. The vector spin glasses (XY, Heisenberg) are unlikely to have a thermodynamically stable SG phase in three dimensions unless the range of interactions is sufficiently long[2].

The problem of phase diagram — concentration vs. temperature — is also of continuing interest: how does the SG phase (if any) compete the ferromagnetic (FM) (or antiferromagnetic) phase? Where is the multicritical point? Is there really a reentrant transition (from an FM phase to an SG phase as the temperature decreases)? Although the mean field theory provides answers to these questions, one should again realize the limitations of the artificial infinite-range model (for which the mean field theory is exact). I am going to discuss here the problem of phase diagram of finite-dimensional spin glasses. Both analytical and numerical approaches will be employed.

2. GAUGE TRANSFORMATION

The model system of a spin glass

$$H = -\sum J_{ij} S_i S_j \tag{1}$$

has a special symmetry, gauge symmetry, which leads to quite useful information on the possible structure of the phase diagram[3]. For simplicity, I restrict the argument to the $\pm J$ Ising model ($J_{ij}=\pm J$, $S_i =\pm 1$), although generalization is possible. The Hamiltonian (1) is invariant under the gauge transformation

$$J_{ij} \rightarrow J_{ij}\sigma_i\sigma_j \ , \ S_i \rightarrow S_i\sigma_i, \tag{2}$$

where σ_i is an Ising-like gauge variable independent of S_i or J_{ij}. Correspondingly, the partition function is also gauge invariant.

The gauge invariance of the partition function is used in the following to rewrite the average free energy

$$-\beta F = \sum_{\{\tau_{ij}\}} p^M(1-p)^L \ln Z(\{\tau_{ij}\},K), \tag{3}$$

where p is the probability that τ_{ij}(the sign of J_{ij}) is positive, M (or L) denotes the total number of positive (or negative) bonds, and $K=\beta J$. It is easy to verify that the probability part of (3) is also expressed as[3]

$$p^M(1-p)^L = (2\cosh K_p)^{-M-L}\exp(K_p\sum_{\langle ij\rangle}\tau_{ij}) \ ,$$

where $\exp(-2K_p)\equiv(1-p)/p$. Since the partition function Z in (3) is gauge invariant and so is the sum over $\{\tau_{ij}\}$, the free energy (3) after the transformation (2) reads

$$-\beta F \equiv (2\cosh K_p)^{-M-L}\sum_{\{\tau_{ij}\}} \exp(K_p\sum\tau_{ij}\sigma_i\sigma_j)\ln Z(\{\tau_{ij}\},K). \tag{4}$$

The fact that both sides of (4) are independent of $\{\sigma_i\}$ allows us to sum (4) over all possible values of $\{\sigma_i\}$ and divide by 2^{-N} (N is the total number of spins), yielding

$$-\beta F \equiv 2^{-N}(2\cosh K_p)^{-M-L}\sum_{\{\tau_{ij}\}} Z(\{\tau_{ij}\},K_p)\ln Z(\{\tau_{ij}\},K). \tag{5}$$

It is remarkable that the probability part in (3) has been expressed in (5) in terms of the partition function of the same system with the effective coupling K_p.

Using (5), we can calculate the internal energy exactly on a special line in the phase diagram. The internal energy is essentially the temperature derivative of the free energy:

$$E = -J2^{-N}(2\cosh K_p)^{-M-L}\sum_{\{\tau_{ij}\}} Z(\{\tau_{ij}\},K_p)\frac{\partial Z(\{\tau_{ij}\},K)/\partial K}{Z(\{\tau_{ij}\},K)} \ . \tag{6}$$

If K is equal to K_p, the Z in the denominator of (6) cancels out the Z in the numerator. Then the problem reduces to the evaluation of the sum of the derivative of the partition function over $\{\tau_{ij}\}$. This

summation is easily found to be proportional to tanh$K$.   In this way
we have proved that the internal energy is exactly $-J(M+L)$tanh$K$ provided $K=K_p$.

The condition $K=K_p$ corresponds to a line (the Nishimori line,
hereafter called the N-line for short) in the phase diagram (Fig. 1).

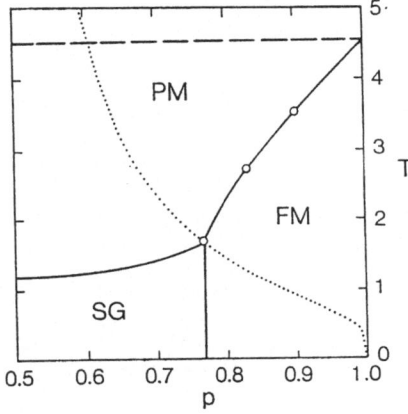

Fig. 1. Phase diagram of
the three-dimensional $\pm J$
Ising model. The dotted
line is the N-line. The
PM phase below the dashed
line is the Griffiths
phase.

It should be remembered that the above exact solution of the internal
energy has been obtained for any lattice in any dimension.   The N-
line apparently intersects the phase boundary between an ordered (FM)
and a disordered (paramagnetic; PM) phases. Nevertheless the exact
energy $-J(M+L)$tanh$K$ shows no singular behavior. This is not a contradiction, because the amplitude of the singular part of the internal
energy could accidentally vanish on the line.

It is not possible to carry through the calculation of the specific heat on the same line; the partition function in the denominator
of (6) is squared this time, which prevents one from making use of
lucky(!) cancellation encountered above. However, an upper bound of
the specific heat is calculable with the aid of the Schwartz inequality.

The result for the specific heat proves that this quantity is
always finite on the N-line, in spite of the line's crossing the
phase boundary. Since the FM Ising model (corresponding to $p=1$) has
a divergent specific heat at the critical point in two and higher
dimensions, finiteness of the same quantity on the N-line implies
strong suppression of critical divergence by randomness. Let me
stress again the enormous generality of the result (any lattice, any
dimension, and no approximations involved!).

The spin-spin correlation function is important in characterizing
the ordered phase. I have applied the present technique to this
gauge covariant quantity to find

$$[\langle S_i S_j \rangle_K] = [\langle S_i S_j \rangle_K \langle S_i S_j \rangle_{K_p}], \tag{7}$$

where the inner brackets $\langle \cdots \rangle_K$ denote the thermal average for a
fixed distribution of $\{\tau_{ij}\}$ with coupling K, and the outer ones $[\cdots]$

represent the average over configurations of $\{\tau_{ij}\}$. By setting $K=K_p$ and $|i-j| \to \infty$ in (7), one readily obtains the relation

$$m(\text{FM order parameter}) = Q(\text{SG order parameter}). \tag{8}$$

This equation implies that the N-line does not enter the SG phase ($Q > 0$, $m=0$). If the SG phase exists, it should be located below the N-line (Fig. 1). Equation (7) leads also to an inequality

$$|[\langle S_i S_j \rangle_K]| \leq [|\langle S_i S_j \rangle_{K_p}|]. \tag{9}$$

By taking the limit $|i-j| \to \infty$ in (9), I conclude that, if the right hand side of (9) (which is an order parameter on the N-line) vanishes, then the FM order parameter on the left hand side (evaluated at the same p as in the right hand side but with an arbitrary K) is zero. This implies that the FM - non FM boundary below the N-line should be either reentrant or vertical (see Fig. 1).

As is now apparent, the N-line plays a quite important role in determining the phase diagram of spin glasses. The very special character of the N-line strongly suggests that the multicritical point (where the FM, PM and SG phases merge) is located on the line. This conjecture is verified in some special cases[3,4].

3. NUMERICAL APPROACHES

The method of gauge transformation explained in the previous section gives generic structure of the phase diagram common to all lattices. Special features of each lattice system, such as the precise location of the multicritical point, must be investigated by other techniques. This section is devoted to a short summary of our numerical studies on the phase diagram of model spin glasses.

As mentioned in the previous section, the N-line plays the key part in drawing the phase diagram. In particular, the multicritical point is very likely to be just on the line. For this reason, we have carried out Monte Carlo renormalization group calculations along the N-line on the simple cubic lattice.[5].

We followed KIKUCHI and OKABE[6] and calculated numerically the renormalized susceptibility

$$\chi_n(u_T) = \chi(b^{ny_T} u_T) = b^{n(d-2y_H)} \chi(b^{2ny_T} u_T), \tag{10}$$

where n is the number of renormalization processes, b is the scaling factor, and u_T denotes the deviation from the critical point T_c. Since u_T should vanish at the critical point by definition, the ratio

$$\Delta_{nm}(T) \equiv \frac{\log[\chi_n(u_T)/\chi_m(u_T)]}{\log[b^n/b^m]} \tag{11}$$

is independent of the set $[n,m]$ at $T=T_c$:

$$\Delta_{nm}(T_c) = d-2y_H = -\frac{\gamma}{\nu} = \eta-2.$$

The data in Fig. 2 show that the critical point is at T_c=1.68±0.025 (p_c=0.767±0.004) and γ/ν=1.97±0.1 on the N-line. This latter value is close to 1.968 obtained for the non-random system [5].

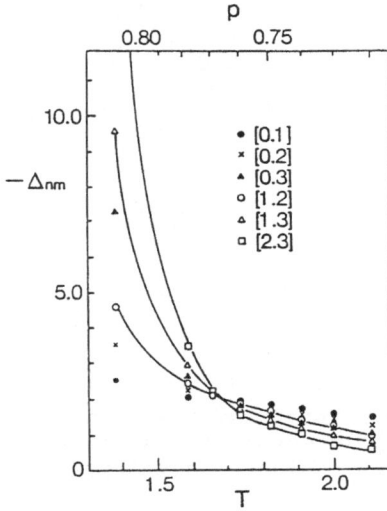

Fig. 2. Scaling analysis of χ of the three-dimensional $\pm J$ Ising model

Another critical exponent $\nu(=1/y_T)$ can be calculated by expanding (10) and (11) in the neighborhood of the critical point. The result is ν=0.51±0.06. This value is somewhat puzzling because it violates the inequality

$$\nu \geq \frac{2}{d} , \qquad (12)$$

which is derived from finiteness of the specific heat ($\alpha \leq 0$), the Rushbrooke inequality ($\alpha+2\beta+\gamma \geq 2$) and the hyperscaling relation ($2\beta+\gamma =d\nu$). CHAYES et al have also obtained (12) from general scaling arguments[7]. A possible reason for this difficulty is breakdown of conventional scaling properties in the vicinity of the multicritical point. I believe that the resolution of this problem would greatly enhance our understanding of critical properties of the $\pm J$ spin glass near the multicritical point.

The two-dimensional system is more efficiently investigated by the numerical transfer matrix method[8]. We have calculated the susceptibility χ and the correlation length ξ defined by

$$|\langle S_0 S_R \rangle| \sim \exp(-\xi/R)$$

for long strips up to the size 14×10⁵. The scaling analyses show that χ and ξ diverge simultaneously along the FM-PM phase boundary above the N-line (Fig. 3). On the other hand, below the N-line, ξ behaves in a singular manner at definitely higher temperatures than χ. This observation suggests the existence of a non-FM ordered phase (the random antiphase state, RAS) adjacent to the FM region, as first proposed by MAYNARD and RAMMAL[9]. There is a numerical study by MORGENSTERN [10] who reaches a negative conclusion on the existence of RAS. However, we believe that ours is more reliable because our system size is by far larger (14 × 10⁵ vs 12 × 20) and our analysis is more

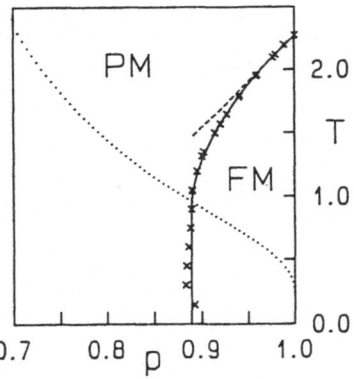

Fig. 3a. Critical points
determined from scaling analysis
of χ in two dimensions

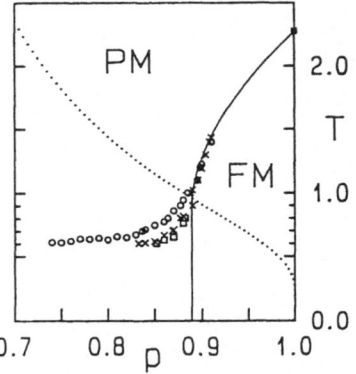

Fig. 3b. Same as in Fig. 3a
obtained from ξ

systematic (scaling argument vs simple extrapolation). In any
event, numerical approaches never give final answers to this sort of
subtle problem. A new analytical approach should be developed to
settle the controversy on RAS.

4. GRIFFITHS SINGULARITY

It was pointed out by RANDERIA et al.[11] that there may exist a
novel phase, a "Griffiths phase", in the temperature range $T_g < T < T_c$
(Fig. 1). Their idea comes from a theorem proved by GRIFFITHS [12]
for a dilute ferromagnet. Therefore let me explain this case first.

In a randomly diluted ferromagnet, there exist arbitrarily large
connected clusters. If one applies an external magnetic field h at
$T < T_c$, these large clusters show strong response because they "almost"
have spontaneous magnetization. GRIFFITHS rigorously investigated
the response of isolated clusters, taking careful account of the
probability of the existence of large clusters, to conclude that the
free energy, as a function of h, is singular at $h=0$ if $T < T_c$ irrespective of the probability of a single bond's being missing. His proof
explicitly relies upon the theory of LEE and YANG [13] who pointed
out importance of the distribution of zeros of the partition function
in the complex h-plane.

Now back to the SG problem; RANDERIA et al. suggested that something similar to that above could happen in the SG system, say the $\pm J$
Ising model. If the dynamics of the system is dominated by
unfrustrated clusters of spins moving coherently within each cluster,
the whole system may behave quite differently above and below the FM
critical point; an unfrustrated cluster, if isolated, is essentially
an FM system. Although OGIELSKI[14] disputed their claim on the
basis of his extensive numerical simulations, OGIELSKI's analysis
does not appear to give a clear-cut answer[15]. In consideration of
this situation, I try to put the analysis of the Griffiths
singularity on a more sound footing than before. A price to be paid
is the restriction to the two-dimensional $\pm J$ Ising model with p
(probability that a bond is ferromagnetic)=1/2.

146

The average free energy of the $\pm J$ Ising model with $p=1/2$ is, according to (3),

$$-2^{M+L}\beta F = \sum_{\{\tau\}} \ln Z(\{\tau_{ij}\},K) = \ln \prod_{\{\tau\}} Z(\{\tau_{ij}\},K). \qquad (13)$$

This expression indicates that the singularity of the free energy is determined by the superposition of the zeros of each $Z(\{\tau_{ij}\},K)$ in the complex temperature plane. So it makes sense to investigate the distribution of Lee-Yang zeros of the partition function for a fixed $\{\tau_{ij}\}$.

If the system is on the two-dimensional square lattice with free boundary conditions, the partition function of the $\pm J$ Ising model is equal to the following expectation value up to a trivial factor [16],

$$G(A) \equiv \sum_{\{S\}} (\prod_{j\in A} S_j)\exp(K^*\sum_i S_i S_j) \qquad (14)$$

on the dual square lattice. Here K^* is the dual coupling defined by $\exp(-2K^*)=\tanh K$, and A is the set of sites on the dual lattice on which the original $\pm J$ model has frustration. (Remember that a plaquette in the original lattice corresponds to a site in the dual lattice.) Thus one should look for the zeros of $G(A)$ of a **regular** (non-random) system.

If the system size is finite, all zeros of $G(A)$ are off the real temperature axis. As the system size grows, some of the zeros would approach the FM critical point on the real axis because, otherwise, the regular system is non-singular at the critical point. It looks apparent now that the distribution of zeros of the average free energy (13) has an accumulation point at $T=T_C$ in the thermodynamic limit, which implies the existence of a Griffiths singularity in the average free energy F (per spin) as a function of the temperature.

A clear advantage of this formulation is that one can avoid a hand-waving argument concerning the role of "clusters" in the $\pm J$ Ising model. One should realize, however, that the present argument is still far from a proof of the existence of a Griffiths singularity. Possible pitfalls lie in the following points[17]:

1) The thermodynamic limit may obscure the singular behavior even if zeros of finite-size systems approach T_C. For instance, the residue of a pole of $\partial \log Z/\partial T$ may become infinitesimally small as the system size N tends to infinity. What is more, we do not have a circle theorem, which played a decisive role in Griffiths' proof, in the case of the complex temperature plane.

2) The dilute system has finite clusters even in the thermodynamic limit. This fact greatly facilitated the original argument of Griffiths. In the $\pm J$ model, one has to consider the full-size system, not a finite cluster, in each term (factor) appearing in (13).

3) In relation to the point 2), I note that the size $|A|$ of the set A in (14) is unbounded: Arbitrarily large numbers of frustrated plaquettes can show up in (13) as the system size grows. It is not necessarily a trivial matter whether $G(A)$ has an accumulation point of zeros at T_C in the limit $N\to\infty$ with $|A|/N$ fixed.

In spite of those difficulties, I feel that we are now at a much closer position to the final resolution of the problem than before.

REFERENCES

1. K. Binder, A.P. Young: Rev. Mod. Phys. $\underline{58}$, 801 (1986)
2. A.J. Bray, M.A. Moore, A.P. Young: Phys. Rev. Lett. $\underline{56}$, 2641 (1986)
   Y. Ozeki, H. Nishimori: J. Phys. Soc. Jpn. (submitted)
3. H. Nishimori: Prog. Theor. Phys. $\underline{66}$, 1169 (1981); J. Phys. C$\underline{13}$, 4071 (1980); J. Phys. Soc. Jpn. $\underline{55}$, 3305 (1986); Prog. Theor. Phys. $\underline{76}$, 305 (1986)
4. P. Le Doussal, A. George: preprint
   P. Le Doussal, A.B. Harris: Phys. Rev. Lett. $\underline{61}$, 625 (1988)
5. Y. Ozeki, H. Nishimori: J. Phys. Soc. Jpn. $\underline{56}$, 1568 (1987); $\underline{56}$, 2992 (1987)
6. M. Kikuchi and Y. Okabe: Prog. Theor. Phys. $\underline{75}$, 192 (1986)
7. J.T. Chayes, L. Chayes, D.S. Fisher, T. Spencer: Phys. Rev. Lett. $\underline{57}$, 2999 (1986)
8. Y. Ozeki, H. Nishimori: J. Phys. Soc. Jpn. $\underline{56}$, 3265 (1987)
9. R. Maynard, R. Rammal: J. de Phys. Lett. $\underline{43}$, L347 (1982)
10. I. Morgenstern: Phys. Rev. B$\underline{25}$, 6071 (1982)
11. M. Randeria, J.P. Sethna, R.G. Palmer: Phys. Rev. Lett. $\underline{54}$, 1321 (1985)
12. R.B. Griffiths: Phys. Rev. Lett. $\underline{23}$, 17 (1969)
13. C.N. Yang, T.D. Lee: Phys. Rev. $\underline{87}$, 404 (1952)
    T.D. Lee, C.N. Yang: Phys. Rev. $\underline{87}$, 410 (1952)
14. A.T. Ogielski: Phys. Rev. B$\underline{32}$, 7384 (1985)
15. H. Takano: private communication
16. F.J. Wegner: J. Math. Phys. $\underline{12}$, 2259 (1971)
    E. Fradkin, B.A. Huberman, S.H. Shenker: Phys. Rev. B$\underline{18}$, 4789 (1978)
17. R.B. Griffiths: private communication

Griffiths Singularities and the Dynamics of Random Systems

A.J. Bray

Department of Theoretical Physics, University of Manchester,
Manchester M13 9PL, UK

I. INTRODUCTION

Two decades ago Griffiths [1] showed that the free energy F of a diluted ferromagnet is non-analytic as a function of the applied magnetic field h, at $h = 0$, at all temperatures T below the critical point $T_C(1)$ of the undiluted system. This singularity in $F(h)$ at $h = 0$ has since been termed a 'Griffiths singularity'. Its physical origin is the presence in the diluted system of arbitrarily large regions which locally resemble a system below its ordering temperature. Such regions occur due to random statistical fluctuations in the quenched disorder. Subsequently, the concept of Griffiths singularities has been extended [2,3] to more general kinds of quenched disorder than simple dilution.

In 1980, Dhar [4] pointed out that the same statistical fluctuations which lead to singularities in $F(h)$ will also lead to anomalously slow dynamics below $T_C(1)$. The reason is that a large region below its local ordering temperature will be quasi-ordered, the temporal persistence of the order being limited only by finite size effects. For an Ising system, relaxation of a quasi-ordered region of linear dimension L requires passing a domain wall through it, and takes a time of order $\exp\{\sigma L^{d-1}/T\}$ where σ is the surface tension and d the spatial dimension. By weighting the relaxation function for such a region by its probability of occurrence, and summing over L, Dhar obtained

$$C(t) \sim \exp\{-A(\ln t)^{d/(d-1)}\} \tag{1}$$

for the spin autocorrelation function $C(t) = [\langle S_i(t)S_i(0)\rangle]$, where $\langle \ldots \rangle$ and $[\ldots]$ indicate thermal and disorder averages respectively.

Similar arguments have been applied [5] to the Ising spin glass, where the asymptotic dynamics is dominated by regions free of frustration. Again weighting the contribution to $C(t)$ from such regions by their probability of occurrence, the authors of [5] recovered the form (1). These authors also introduced the term 'Griffiths phase' [6] to describe the temperature regime between the critical point of the random system and the highest critical temperature achieved by any member of the ensemble of systems generated by the disorder distribution. We call this latter temperature the 'Griffiths temperature' T_G. For a diluted ferromagnet $T_G = T_C(1)$, the critical temperature of the undiluted system. For the $\pm J$ spin-glass model T_G is the critical temperature of the unfrustrated system.

For the $\pm J$ spin-glass, extensive numerical simulations [7] fail to reveal the predicted form (1), but instead are better fitted by the 'stretched-exponential' or 'Kohlrausch' form $C(t) \sim \exp\{-(t/\tau)^\beta\}$ with a temperature-dependent β. On the other hand, there are convincing arguments [8] that (1) provides an exact lower bound on $C(t)$, so that the Kohlrausch form cannot be asymptotically correct. We will return to this point later.

In this article we will discuss the dynamics of random Ising and Heisenberg (or more generally $O(m)$ with $m \geq 2$) systems in the Griffiths phase. For simplicity we will restrict ourselves to ferromagnets (or two-sublattice antiferromagnets) with simple dilution (more complicated kinds of disorder can be handled with similar techniques) and to simple relaxational dynamics with no conservation laws (i.e. 'model A' dynamics [9]). In the vicinity of T_G we derive [10] a novel form of dynamic scaling, involving the scaling variable t/ξ^{d+z}, where ξ is the correlation length of the pure system at the given temperature, and z the pure system dynamical exponent. Below T_G, the form (1) is recovered for Ising systems. For Heisenberg systems we obtain

$$C(t) \sim \exp\{-(Bt)^{1/2}\} \quad . \tag{2}$$

The faster relaxation of (2) is a consequence of the continuous symmetry of Heisenberg systems: there is no activation barrier to relaxation, which proceeds instead by 'diffusion of the magnetisation' [10]. These methods can be extended to obtain results for the non-local correlation functions $C(r,t) = [\langle \mathbf{S}_i.\mathbf{S}_j \rangle]$, where $r = |r_i - r_j|$ is the spin separation, for both Ising and Heisenberg systems [11].

The statics of the Griffiths phase is also non-trivial [12]. The same 'clustering' phenomena which lead to anomalously slow dynamics also yield interesting behaviour in the eigenvalue spectrum of the equal-time correlation matrix $\chi_{ij} \equiv \langle \mathbf{S}_i.\mathbf{S}_j \rangle$ (N.B.: there is no disorder average here!). Using arguments analogous to those used for the dynamics yields, for the disorder-averaged eigenvalue density of the inverse matrix χ^{-1}, the result

$$\rho(\lambda) \sim \exp(-A/\lambda) \tag{3}$$

for $\lambda \to 0$. This should be contrasted with the eigenvalue spectrum for an undiluted ferromagnet above its critical point, which has a gap equal to the inverse of the susceptibility. For a random system, this gap disappears at T_G. As the temperature is lowered further, the amplitude A in (3) decreases, vanishing at the onset of long-range order [3,12].

Finally, we will consider a suggestion of Campbell and coworkers [13] that glassy relaxation in general can be understood in terms of the diffusion of the phase point in a sparsely (and randomly) connected configuration space. A simple model in which the elements M_{ij} of the Markov transition matrix are independent random variables (subject to the symmetry requirement $M_{ij} = M_{ji}$), with a finite mean number of non-zero elements per row, is proposed and solved. The solution yields asymptotic Kohlrausch relaxation with $\beta = 1/3$.

II. DYNAMICS OF THE GRIFFITHS PHASE

We consider a site-dilute ferromagnet (bond dilution requires a trivial modification) with nearest-neighbour interactions. We define $T_C(p)$ to be the transition temperature of the dilute system, where p is the fraction of sites occupied. The inverse function $p_C(T)$ gives the critical occupation for fixed T: $p_C(0)$ is the percolation threshold, and $p_C(T_G) = 1$. The basic idea is that the asymptotic dynamics is dominated by quasi-ordered regions (henceforth called 'clusters') whose average concentration p' satisfies $p' > p_C(T)$. These clusters should not be confused with percolation clusters: in particular, they need not be isolated from the rest of the system. The autocorrelation function is then given by a sum over clusters:

$$C(t) = \sum_{cl} \mathrm{Pr}_{cl} \exp\{-t/\tau_{cl}\} \tag{4}$$

where Pr_{cl} is the probability for a given site to belong to the cluster, and τ_{cl} is the cluster relaxation time. In writing the cluster relaxation function as a single exponential in (4) we are assuming that t is large enough that all 'internal' relaxation modes are fully equilibrated: τ_{cl} is then the 'ergodic time' for relaxation of the total cluster magnetisation.

Loosely speaking, clusters can be characterised by three concepts: size, shape and 'composition' (i.e. distribution of occupied sites). The key assumption underlying the philosophy of our approach is that for large t the sum in (4) can be evaluated by the method of steepest descents: for any large, fixed t the sum will be dominated by clusters of a particular size, shape and composition. Since we cannot evaluate τ_{cl} for a cluster of arbitrary shape, we will *assume a priori* that the steepest descent calculation will pick out clusters with spherical symmetry. This seems plausible on symmetry grounds, but requires further justification (see later). For similar reasons we will characterise the cluster composition by the single parameter p' that gives the mean site occupancy of the cluster. A true steepest descent calculation might give (after suitable coarse-graining) a non-trivial concentration profile $p'(r)$, where r is the distance from the cluster centre. For these reasons our final result will be technically a *lower bound* on the true $C(t)$, though it should have the correct functional form.

With the above caveats, (4) can be rewritten

$$C(t) \geq \max_{L,p'} \mathrm{Pr}(L,p') \exp\{-t/\tau(L,p')\} \tag{5}$$

where L is the cluster 'size'(i.e. linear dimension) and [10]

$$\ln \Pr(L, p') = -L^d \{ p' \ln(p'/p) + (1 - p') \ln[(1 - p')/(1 - p)] \} \equiv -L^d f(p') . \quad (6)$$

The relaxation time $\tau(L, p')$ has to be evaluated separately for Ising and Heisenberg systems.

1. Ising systems

The cluster relaxation time can be estimated as

$$\tau(L, p') \sim \tau(p') \exp\{ \sigma(p') L^{d-1}/T \} \quad (7)$$

where $\tau(p')$ and $\sigma(p')$ are the characteristic relaxation time and surface tension respectively for a bulk system with occupation probability p'. Use of (7) requires $L \gg \xi(p')$, the correlation length. Inserting (6) and (7) in (5), and extremising with respect to L, yields [10,14] a result of the form (1):

$$C(t) \geq \exp\{ -A[\ln(t\{\sigma(p^*)/T\}^{d/(d-1)})/\tau(p^*)]^{d/(d-1)} \} \quad (8)$$

where, ignoring numerical factors,

$$A = \min_{p'} f(p') \{ T/\sigma(p') \}^{d/(d-1)} \quad (9)$$

and p^* is the minimising value of p'. For $T \to T_G$ one must have $p^* \to 1$ to ensure $p^* > p_C(T)$. In this limit $\tau \sim \xi^z$ and $\sigma \sim \xi^{-(d-1)}$, where ξ is the correlation length of the bulk pure system, giving, up to numerical factors,

$$C(t) \geq \exp\{ -f(1) \xi^d [\ln(t/\xi^{d+z})]^{d/(d-1)} \} \quad . \quad (10)$$

The condition $L \gg \xi$ requires $t \gg \xi^{d+z}$. More generally one obtains near T_G the scaling form [10]

$$C(t) \geq \exp\{ -\xi^d g(t/\xi^{d+z}) \} \quad (11)$$

valid for both Ising and Heisenberg systems, above, at and below T_G.

2. Heisenberg Systems

Relaxation in this case is by 'diffusion of the magnetisation' [10]: driven by the thermal noise provided by the heat bath, the cluster magnetisation vector undergoes a random walk on the surface of a sphere. For a cluster of N spins, the integrated thermal noise acting in time t is (ignoring temperature-dependent factors) of order $(Nt)^{1/2}$. This needs to be of order N for complete decorrelation of the magnetisation vector with its original direction, giving $\tau(L, p') \propto N \sim L^d$. A more precise expression is obtained if one expresses $\tau(L, p')$ in units of the characteristic relaxation time $\tau(p')$ of the bulk system, and L in units of the correlation length $\xi(p')$:

$$\tau(L, p') \sim \tau(p') \{L/\xi(p')\}^d \quad . \tag{12}$$

Using (12) and (6) in (5), and extremising with respect to L, yields [10] a result of the form (2), where, ignoring numerical factors,

$$B = \min_{p'} f(p') \xi(p')^d / \tau(p'). \tag{13}$$

For $T \to T_G$ one has $p^* \to 1$, and $\tau \sim \xi^z$, giving $B \sim \xi^{d-z}$ consistent with the scaling form (11). Once again, $t \gg \xi^{d+z}$ (equivalent to $L \gg \xi$) is necessary for (2) to hold.

For Heisenberg systems, a non-trivial result is obtained also for $T \to T_C(p)$, when (13) yields a p^* close to $p_C(T)$ [10]:

$$C(t) \sim \exp\{-[(T - T_C)^{\alpha_r} (t/\xi_r^{z_r})]^{1/2}\} \tag{14}$$

where subscripts r indicate that the relevant quantities pertain to the *random* system (i.e. to a bulk system with $p = p^*$). Although (14) is not part of the scaling function near $T_C(p)$, it dominates at sufficiently long times [10].

The weakest link in the above chain of argument is the assumption that spherical clusters dominate. For Ising systems this is plausible since, for clusters of a given volume, a spherical cluster has the largest relaxation time. For Heisenberg systems it is less obvious, since, provided the spins are strongly correlated, the relaxation time depends only on the cluster volume and not on its shape. In fact, the assumption may not be strictly necessary. As an extreme example, consider $p < p_C(0)$ and T infinitesimal. Then the clusters really are percolation clusters, and every cluster of N sites relaxes as $\exp(-tT/N)$ irrespective of its shape. ($T \to 0$ is needed to treat the spins on the (ramified) percolation clusters as strongly correlated. At any non-zero T, large clusters break up into almost independently relaxing subunits). Weighting by the number of N-clusters per site [15], $n_N(p) \sim \exp(-a(p)N)$, and summing over N using steepest descents, yields once more a result of the form (2) which does, therefore, seem to have some general validity. In fact this argument can be extended, for $p < p_C(0)$, to general temperatures $T < T_G$, to derive upper and lower bounds for $C(t)$ which have the same asymptotic forms, given by (1) for Ising systems and (2) for Heisenberg systems [16]. This establishes that the asymptotic forms (1) and (2) are correct in at least part of the Griffiths phase.

An obvious final question concerns how large t has to be for (1) and (2), which are only claimed to be valid 'asymptotically', to hold. Although Ogielski's spin-glass results [7] are not encouraging, recent Monte Carlo simulations of $d = 2$ bond-dilute

Ising ferromagnets at the percolation threshold [17] seem to approach the form (1) for $C(t) < 10^{-2}$, although better statistics for small $C(t)$ will be required to establish this convincingly. It is interesting that the 'pre-asymptotic' decay is well fitted [17] by the ubiquitous Kohlrausch law, whose origin remains a challenge to the theorist [13].

III. STATICS OF THE GRIFFITHS PHASE

We concentrate on one aspect of the statics – the eigenvalue spectrum $\rho(\lambda)$ of the inverse χ^{-1} of the 'correlation' (or 'susceptibility') matrix $\chi_{ij} = \langle S_i . S_j \rangle$. Whereas non-exponential relaxation is the 'dynamical signature' of the Griffiths phase, the 'static signature' [12] is the existence of arbitrarily small eigenvalues of χ^{-1}. Using similar cluster arguments to those used for the dynamics, the smallest eigenvalue (higher eigenvalues are irrelevant for $\lambda \to 0$) associated with a (spherical) cluster of size L and 'composition' p' may be estimated using a variational approach with a trial eigenfunction which is constant inside the cluster and vanishes outside. This gives $\lambda(L, p') \sim 1/m(p')^2 L^d$, where $m(p')$ is the magnetisation per site for a bulk system with mean site occupancy p'. Summing over clusters with weight $\Pr(L, p')$ yields

$$\rho(\lambda) \geq \sum_{L, p'} \exp\{-L^d f(p')\} \, \delta(\lambda - 1/m(p')^2 L^d) \sim \exp\{-A/\lambda\} \tag{15}$$

$$A = \min_{p'} f(p')/m(p')^2 \quad . \tag{16}$$

One expects [3] the amplitude A to *diverge* for $T \to T_G$, and to *vanish* for $T \to T_C(p)$. Equation (16) fulfills these expectations. For $T \to T_G$, one has $p^* \to 1$ (p^* is the minimising value of p'), so A diverges as $m(1)^{-2}$, i.e. as $(T_G - T)^{-2\beta}$. For $T \to T_C(p)$, (16) yields a non-trivial p^* close to $p_C(T)$, and A vanishes as $[T - T_C(p)]^{2(1-\beta_r)}$ [12]. Here the exponents β, β_r are the magnetisation exponents for the pure and random systems respectively.

We conclude that statistical fluctuations in a random magnet lead to a non-vanishing density of 'quasi-soft' modes in the Griffiths phase, associated with 'quasi-ordered' regions similar to those responsible for the anomalous dynamics.

IV. DIFFUSION ON A RANDOM SPARSELY CONNECTED SPACE

In this final section the (at first sight) unrelated problem of diffusion on a random, sparsely connected space is briefly considered. This study was motivated by a suggestion of Campbell and coworkers [13] that glassy relaxation can be understood in terms of a phase point restricted to a decreasing subset of configuration space as T is decreased. An analogy with percolation was invoked, according to which the available configuration space becomes increasingly ramified as the glass (or spin-glass) transition is approached, and finally breaks into finite disconnected pieces below T_C.

As the simplest model of such a process, we consider a random walk among a set of N sites $\{i\}$ (representing states of the system), with the allowed steps defined by a Markov transition matrix (or 'connectivity matrix') whose elements M_{ij} are independent random variables (subject to $M_{ij} = M_{ji}$) taking the values $1/p$, with probability p/N, and 0, with probability $1 - p/N$. Thus the average connectivity p remains fixed as $N \to \infty$. We calculate the probability $f(t)$, averaged over sites, for a walker to be at the same site at time t as at time zero. It is easily shown [18] that the matrix M describes a set of randomly branching Cayley trees, with a percolation threshold $p_C = 1$. Consider a walk on the infinite cluster. Slow relaxation occurs if the particle finds itself many steps from the nearest 'node' (defined as a site from which there are at least three routes to infinity), since such nodes effectively act as 'traps' [18]. Weighting the result for a node-free section of a particular length by its probability of occurrence yields the asymptotic result $f(t) \sim \exp\{-(t/\tau)^{1/3}\}$, i.e. a Kohlrausch relaxation with $\beta = 1/3$. We refer the reader to [18] for details. Although the $\beta = 1/3$ Kohlrausch behaviour is asymptotically valid for all p, its domain of applicability is very small except near $p = 1$ where τ diverges ($\tau \sim |p - 1|^{-3}$ [18]). Away from $p = 1$ the pre-asymptotic decay is faster, which might be interpreted in simulations or experiments as a Kohlrausch law with a larger β. It is interesting that Ogielski's data [7] is well fitted by values of β which decrease from 1 at T_G to about $1/3$ at T_C.

The relevance (if any) of such phenomenological models for the dynamics of random magnets needs to be clarified. The qualitative agreement with the simulation data may be coincidental, or may be very significant. Since the Kohlrausch law must ultimately fail for Ising systems ((1) is a lower bound), it can at best describe the pre-asymptotic regime. Since most of the decay of $C(t)$ seems to belong to this regime [7,17], an understanding of the pre-asymptotic behaviour is a very worthwhile goal.

Note Added

Very recent Monte Carlo simulations (S. G. W. Colborne and A. J. Bray, to be published), on bond-dilute Ising and Heisenberg ferromagnets with relaxational dynamics, indicate that (1) and (2), with only minor modifications, can describe the decay of $C(t)$ quite accurately for all but the shortest times.

REFERENCES

1. R. B. Griffiths: Phys. Rev. Lett. **23**, 17 (1969)

2. R. Fisch: Phys. Rev. B **24**, 5410 (1981)

3. A. J. Bray and M. A. Moore: J. Phys. C **15**, L765 (1982)

4. D. Dhar: In *Stochastic Processes: Formalism and Applications*, ed. by G. S. Agarwal and D. Dattagupta (Springer, Berlin, 1983)

5. M. Randeria, J. P. Sethna and R. G. Palmer: Phys. Rev. Lett. **54**, 1321 (1985)

6. The term 'Griffiths phase' was suggested by M. E. Fisher [5]

7. A. T. Ogielski: Phys. Rev. B **32**, 7384 (1986)

8. D. Dhar, M. Randeria and J. P. Sethna: Europhys. Lett. **5**, 485 (1988)

9. P. C. Hohenberg and B. I. Halperin: Rev. Mod. Phys. **49**, 435 (1977)

10. A. J. Bray: Phys. Rev. Lett. **60**, 720 (1988)

11. A. J. Bray and G. J. Rodgers: J. Phys. C **21**, L243 (1988)

12. A. J. Bray: Phys. Rev. Lett. **59**, 586 (1987)

13. I. A. Campbell: J. Phys. (Paris) Lett. **46**, L1159 (1985); Phys. Rev. B **33**, 3587 (1986); I. A. Campbell, J.-M. Flesselles, R. Jullien and B. Botet: preprint; see also R. G. Palmer: In *Heidelberg Colloquium on Glassy Dynamics*, ed. by L. van Hemmen and I. Morgenstern (Springer, Heidelberg,1986)

14. A. J. Bray and G. J. Rodgers: Phys. Rev. B, in press

15. H. Kunz and B. Souillard: J. Stat. Phys. **19**, 77 (1978)

16. A. J. Bray, submitted to J. Phys. A

17. S. Jain: preprint

18. A. J. Bray and G. J. Rodgers: Phys. Rev. B, in press

Moment Instabilities and Reentrant Spin Glass Behavior in 3d Transition Metal Alloys

E.F. Wassermann

Tieftemperaturphysik, Universität-GH-Duisburg, Lotharplatz 1,
D-4100 Duisburg 1, Fed. Rep. of Germany

1. INTRODUCTION

It is well known that spin glass (SG)- and reentrant spin glass (RSG)- behavior can be observed not only in dilute but also in concentrated 3d-alloy systems. As examples we show in Fig. 1 the magnetic phase diagrams of the ternary systems in the fcc γ-range FeNiCr /1,2/, FeNiMn /3-6/ and FeCoMn /7,8/ as compiled from data in the literature and our own experimental investigations. As one can see all three systems show ferromagnetic (FM) and antiferromagnetic (AF) order in certain composition regions. The concentration dependence of the respective Curie temperatures T_C and Néel temperatures T_N projected into the basal plane are given by the full lines in Fig. 1. In between the two types of ordered phases one finds pure SG-regions (dotted in Fig. 1) and RSG-regions (hatched in Fig. 1), with transition temperatures $T_f < T_C$, T_N. Similar ternary magnetic phase diagrams can be established for other combinations of 3d-elements.

With respect to the scope of the present paper the following observation on the FM- and AF sides of the diagrams in Fig. 1 is of importance. In both ranges (even when reentrant) a broad spectrum of magnetovolume effects -when large called "Invar-effect"- is observed. The origin of these anomalies has long been debated /9,10/, but it is certain to date that they are neither related to the occurrence of the mixed magnetic phases nor due to any chemical inhomogeneity.

In a series of recent investigations, summarized in two review articles on the Invar problem /11,12/ we have demonstrated that a general picture for the understanding of the Invar-effect can be developed to date on the basis of modern band calculations /13/. These calculations give clear evidence for the instability of the magnetic moments of fcc Fe, Cr and Mn with respect to changes in the lattice constant. Though finite temperature calculations are not yet available, we have shown experimentally /12,14/ that the moment instabilities exist up to temperatures $T > T_C$, T_N and manifest themselves even in the liquid state /15/.

In the present paper we put forward the following question: if the magnetoelastic and magnetic properties of FM and AF long range ordered alloys in concentrated 3d-systems can be understood within band calculations showing moment-volume instabilities of their constituents, what physical picture does evolve, if the band description is used to understand the SG- and RSG-alloys of these systems?

2. MOMENT-VOLUME INSTABILITIES of 3d-ELEMENTS

Figure 2 shows (for T=0) the moment μ versus radius of the Wigner Seitz cell r_{WS} for fcc Fe and Co /13/ and fcc Mn /16/. It is clearly seen that γ-Fe can exist in three different states, a FM high spin (HS)-state, an AF low spin (LS)-state and a no moment (NM)-state. The three states overlap in the instability range as indicated in Fig. 2. Fcc Co shows a FM HS- to NM-transition, fcc Mn a transition from an AF HS-state to an AF LS-state at $r_{WS} = 2.84$. In this context it is most important to note that a computed moment of zero indicates only that the ground state is not

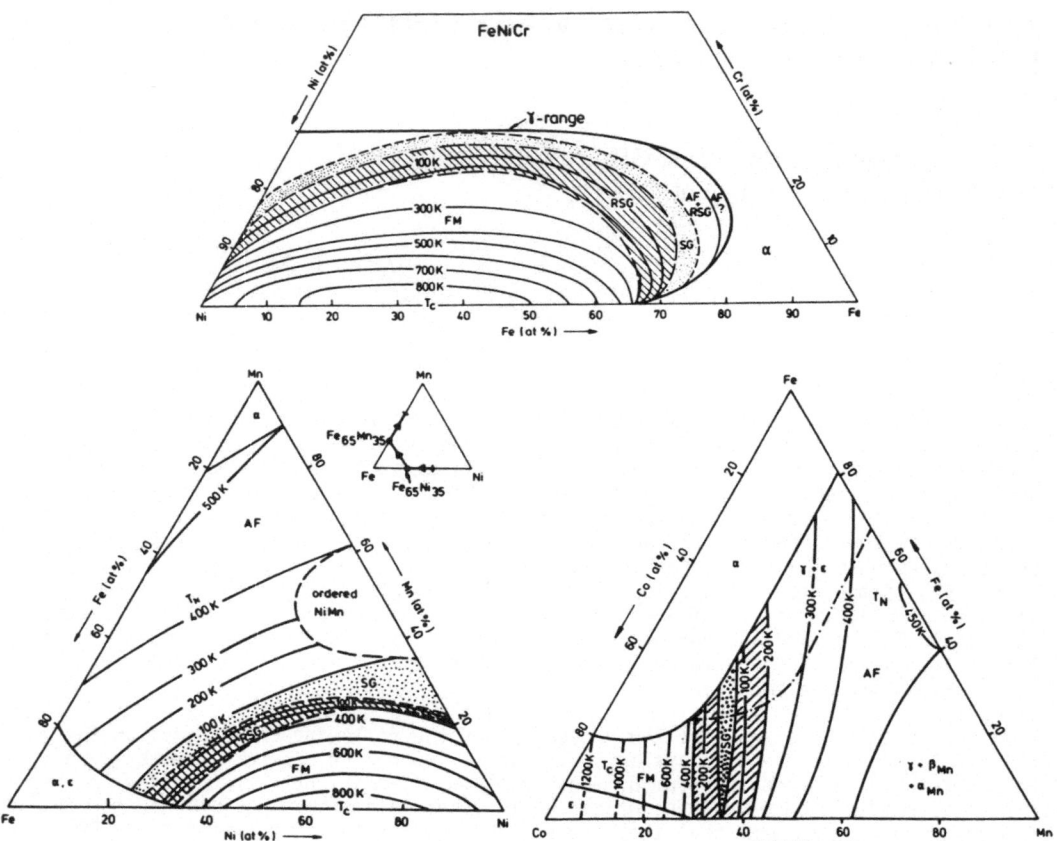

Fig. 1: Ternary magnetic phase diagrams of FeNiCr /1,2/ FeNiMn /3-6/ and FeCoMn
/7,8/ in the fcc γ-range

ferromagnetic. The calculations cannot determine whether the ground state is non-
magnetic or AF! For fcc Mn it has been shown by additional wave vector dependent
susceptibility calculations /17/ that the ground state has AF order. From the experimental side /18,19/ it is also evident to date that at r_{WS}=2.65 a.u. the (NM) ground
state of γ-Fe is antiferromagnetic. Finally we note that fcc Ni shows a continuous
transition from $0.5\mu_B$ (FM-state) to zero moment (NM-state).

The existence of HS-, LS- and NM-states for fcc Fe is best revealed in an energy
versus moment diagram (with r_{WS} as a parameter) which is shown in Fig. 3a (after
ref. /13/). The states are characterized by energy minima in the curves and their
energy difference as well as their relative position to each other depends on the
lattice constant. This is better seen in Fig. 3b (from ref. /13/), where within the
instability range the total energy versus r_{WS} for γ-Fe is plotted. Note that the
LS-state which has AF order as well as the NM-state (which almost definitely has AF
order) can energetically lie below as well as above the FM HS-state. The energy
difference between the states in this range of lattice constant is maximal of the
order of 2 mRy (or ~ 300K) so that the LS- or NM-state is easily accessible with
temperature. As we have shown /11,12,14/ the HS-LS or HS-NM transition is fundamental
to understand the Invar-effect. More important with respect to the present discussion
is the observation that in the vicinity of r_{WS}=2.673 a.u. (c.f. Fig. 3b) all three
states are energetically equal. This means that FM- and AF-order are equally allowed.
If the system fluctuates thermally between the two states, we infer from a band
type magnetic picture that we are dealing with an "itinerant" spin glass.

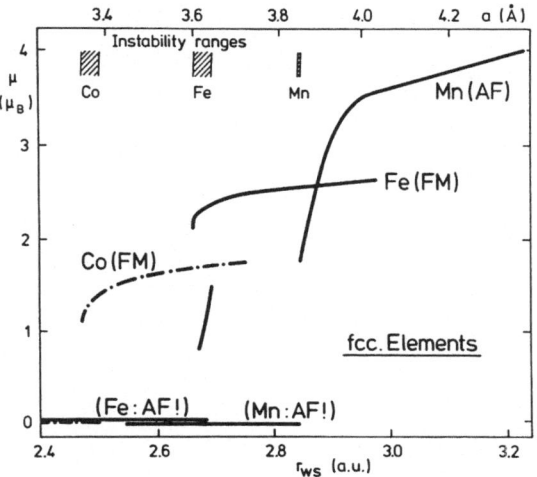

Fig. 2: Magnetic moment versus radius of the Wigner-Seitz cell r_{WS} for Fe, Co (after ref. /13/) and Mn (after ref. /16/) in fcc lattice structure. Note the instability ranges as indicated by the hatched regions.

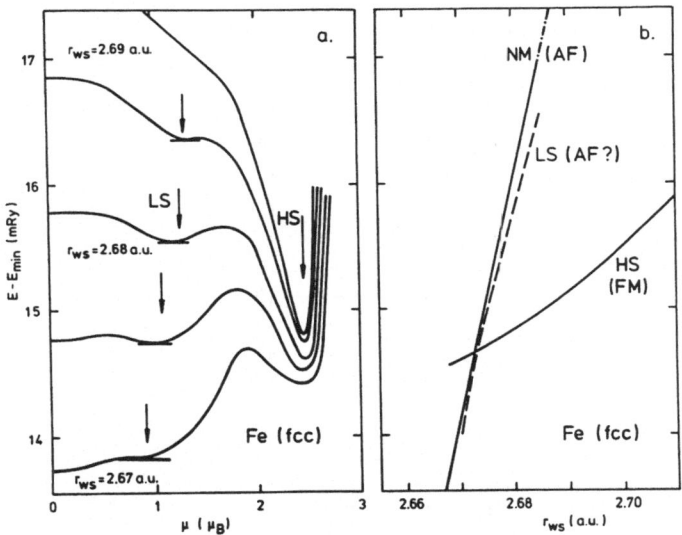

Fig. 3a: Total energy E (relative to a minimum energy E_{min}) versus magnetic moment for fcc Fe at different radii of the Wigner-Seitz cell r_{WS} within the instability region. High spin (HS) and low spin (LS) states are indicated by arrows.
Fig. 3b: $E-E_{min}$ versus r_{WS} for fcc Fe (both Figs. after ref. /13/)

3. MOMENT-VOLUME INSTABILITIES and SPIN GLASS BEHAVIOR in FeNiMn-ALLOYS

"Unfolding" the ternary FeNiMn-diagram along the lines shown in the inset of Fig.1, results in the series of binary magnetic phase diagrams as shown in Fig. 4 (top row). In this way one achieves a continuous phase diagram, revealing the transition from FM to AF long range ordered phases through a SG (hatched in Fig. 4) and a RSG-phase in the range ~ 25 at % Ni to 33 at % Ni. The AF side also might be reentrant into a SG-phase above x=20 at % Ni. The existence of the SG- and RSG-phases and their freezing temperatures have been proven by us through measurements of the a.c. susceptibility (χ' and χ''). Further support comes from our own specific heat measurements (see below) and data for the values of the electronic γ-term in $c_p(T)$, which

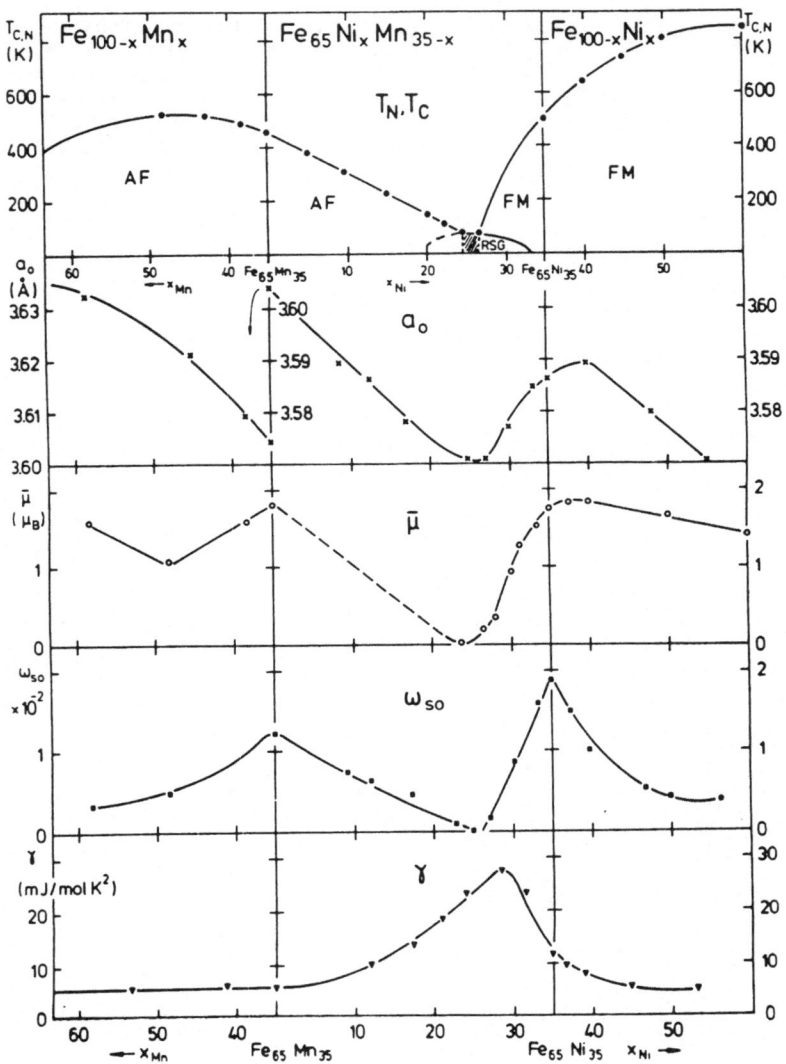

Fig. 4: Top row: magnetic phase diagram along the "cut" through the fcc-range of the ternary system FeNiMn as shown in the inset in Fig. 1. The rows below show the concentration dependence of the lattice constant at zero temperature a_o /21/, the average magnetic moment $\bar{\mu}$ /12/, the spontaneous volume magnetostriction ω_{so} /21/ and the γ-term of the electronic specific heat /20/.

have been determined by DERYABIN et. al. /20/. These data are given in the bottom row, Fig. 4, and one can see that due to the magnetic contribution from the SG-ordering, which is linear in T like the γ-term, unusually high γ-values result in the range where the system shows SG- and RSG-phases. On the AF and FM-side "normal" electronic γ-terms of the order of 5 mJ/mol K^2 are observed /21/.

An interpretation of the experimental data in Fig. 4 within the results of the LSDA-calculations presented in Figs. 2 and 3 can be done in the following way. We start in the FeNi-phase diagram (on the right side in Fig. 4) and observe that the magnetovolume effect, i.e. ω_{so} /21/ increases with increasing lattice constant a_o

/21/, until maximum values are reached at around $Fe_{65}Ni_{35}$, the archtypical Invar composition. In Fig. 3b we are then in the range of lattice constant above the tricritical point, where the FM HS-state is lower in energy than the LS- and NM-states, and thus forms the ground state of the system. Going beyond $Fe_{65}Ni_{35}$ into the $Fe_{65}Ni_xMn_{35-x}$ diagram, we observe that ω_{so} and $\bar{\mu}$ decrease with decreasing lattice constant a_o. In Fig. 3b this means we are approaching the tricritical point. At the same time the energy difference between the HS- and the LS-state is decreasing and finally at around $r_{ws}=2.672$ the FM HS-state and AF LS- and NM-states are energetically equal. Then we are in the SG-state, where ω_{so} and $\bar{\mu}$ vanish, when the lattice constant a_o reaches a minimum value (c.f. Fig. 4). The equality of the energy levels on reaching the SG-transition point can also make plausible the absence of a peak in the specific heat of a SG at T_f. It "costs" no energy to install the SG-phase.

Going beyond the SG-phase into the AF region, the lattice constant increases again due to the large atomic volume of Mn. The magnetovolume effects reappear and should be governed by transitions from an AF ground state to an AF exited state. Though energy versus r_{ws} diagrams like the one in Fig. 3b are still lacking for Mn, we see from Fig. 2 that indeed the HS- and the NM-state of Mn are both antiferromagnetic. Thus the instability of Mn is now governing the magnetovolume effects of the system, though in FeNiMn and FeMn alloys we could deal with the instability of two 3d-elements, so that the HS-LS-transition behavior will be quite complex.

A long time ago it was proposed to describe the low temperature behavior of spin glasses in terms of magnetic two-level systems (MTLS) /22/, analogous to the structural two-level systems introduced for the understanding of amorphous, nonmagnetic materials. Experimentally, the presence of MTLS in spin glasses has been shown by us /23/. Yet, a picture for the physical nature of the MTLS in spin glasses could not directly be given. This is, however, possible within the present analysis, where the existence of HS, LS and NM-states with well defined energies and possibly very small energy differences in the vicinity of the multicritical point is demonstrated.

4. MOMENT-VOLUME INSTABILITIES and RSG-BEHAVIOR

In this chapter we ask the question if the occurrence of RSG-behavior can be made plausible within the instability picture. For the discussion we use again the ternary system FeNiMn, but an alloy series along the composition line $Fe_{50}Ni_xMn_{50-x}$. In Fig. 5 we show the low temperature behavior of the thermal expansion coefficient α versus temperature as determined by us /24/ for several alloys of this system. One can see that within accuracy limits (size of the data symbols) a minimum is absent for the SG-alloy (x=32 at % Ni), a very weak minimum in the range $\alpha>0$ is present in the AF sample (x=20), but a pronounced minimum in the range $\alpha<0$ is seen with increasing Ni concentration (x=35,38,40) in the FM-RSG alloys. That means a positive magnetovolume effect (with respect to a non-magnetic Grüneisen behavior) develops at low T, when the alloys are reentrant. The depth of the minimum reduces when the alloys order purely ferromagnetic. This is seen in Fig. 6, where the magnetic phase diagram (fig. 6a) as determined by us /24/, the γ-term of the electronic specific heat (from ref. /20/ (triangles) and own data (open circles) in zero field and B=6T /25/), and the minimum temperatures T_{min} (Fig. 6c) as well as the values of the expansion coefficient at T_{min} (α_{min}) both taken from Fig. 5 versus concentration are shown. Similar behavior is found on alloys in the system $Fe_{80-x}Ni_xCr_{20-x}$ (c.f. Fig. 6e-h). One can see that the low temperature minimum in $\alpha(T)$ is most prominent in the FM - RSG - range in this system too.

An anomalous increase of the hyperfine field B_{HF} has been observed in Mössbauer investigations on the RSG alloys of $Fe_{50}Ni_xMn_{50-x}$ /4/ with x=36.5, 35 and 33.5 % Ni. It is absent in the SG (x=32) and the AF-alloys. The enhancement of B_{HF} is known as "canting" /26/ and found in many RSG alloys. It is explained by an increase of the value of the local moment, but so far only detected within the time window of the Mössbauer effect, i.e. around 10^{-8} sec. Our results show that "canting" in RSG alloys

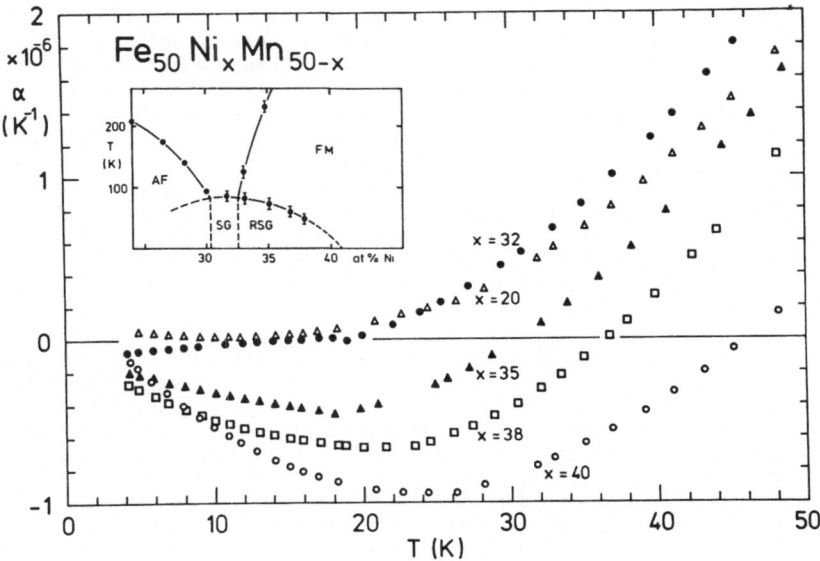

Fig. 5: Thermal expansion coefficient α versus temperature as measured /24/ on Fe$_{50}$Ni$_x$Mn$_{50-x}$ alloys with different Ni-concentrations x (in at %). The inset shows part of the magnetic phase diagram of the alloy system.

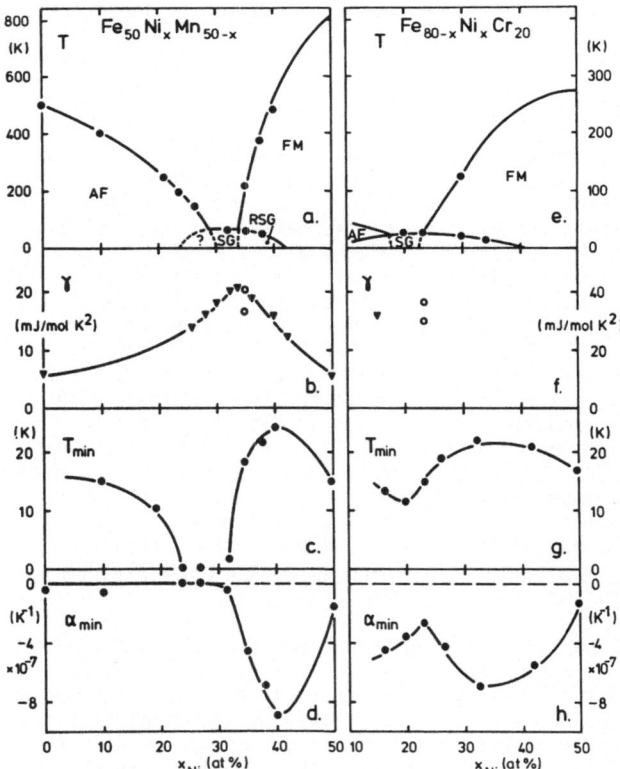

Fig. 6: a) Magnetic phase diagram of Fe$_{50}$Ni$_x$Mn$_{50-x}$ /24/; b) γ-term of the low temperature electronic specific heat (triangles from ref. /20/; circles from ref. /25/); c) minimum temperatures T_{min} in the α(T) curves and d) minimum values of the thermal expansion, both taken from Fig. 5. Figs. 6 (e-h) show analogous data for Fe$_{80-x}$Ni$_x$Cr$_{20}$.

is obviously related to a magnetovolume effect at low temperatures. This means it can be observed on an almost infinite time scale as compared to the Mössbauer effect.

5. MOMENT-VOLUME INSTABILITIES at HIGH TEMPERATURES

Figure 7 shows the temperature dependence of the specific heat c_p as a function of temperature as measured on the alloy $Fe_{50}Ni_{35}Mn_{15}$. The data between 4.2 and 120K (full dots), revealing a SG-like maximum in $c_p(T)$ (see inset of Fig. 7) have been determined by us /25/. The data between 120K and 320K (open triangles) are taken from ref. /27/, the high temperature values in the range 325 to 1150K (dots) from ref. /28/. The full curve is calculated from $c_p(T) = \gamma T + \beta T^3$, with $\gamma = 4.0$ mJ/molK^2 and a Debye temperature $\Theta_D = 360K$. The most surprising result is that at the Curie temperature T_c there is obviously no anomaly due to the magnetic ordering, but there occurs an excess specific heat for $T > T_c$ around 500K. Simultaneously, we have plotted in Fig. 7 the temperature dependence of the thermal expansion coefficient $\alpha(T)$. Data are a combination of own results and from ref. /29/. The dashed line is a Grüneisen curve, calculated with $\Theta_D = 360K$ and fitted to the experimental results in the low temperature range. If one tries to fit the $\alpha(T)$ data to a Grüneisen curve at high temperatures, meaningful results cannot be obtained.

One can see from Fig. 7 that the behavior of $c_p(T)$ and $\alpha(T)$ is analogous. There is no effect at T_c, i.e. neither an excess specific heat due to the onset of FM-order nor a magnetovolume effect. On the other hand, around 500K, where we observe an excess contribution to $c_p(T)$, there is -with respect to the Grüneisen curve- a negative magnetovolume effect. Similar results as in Fig. 7 are found on other alloys of FeNiMn and FeNiCr /27-29/. Even if the order is AF no anomaly in $c_p(T)$ at T_N, but maxima in $c_p(T)$ and $\alpha(T)$ around 500K are observed. Though it is too early to understand these findings in detail within the instability picture, we note k the following. A negative magnetovolume effect with respect to a non-

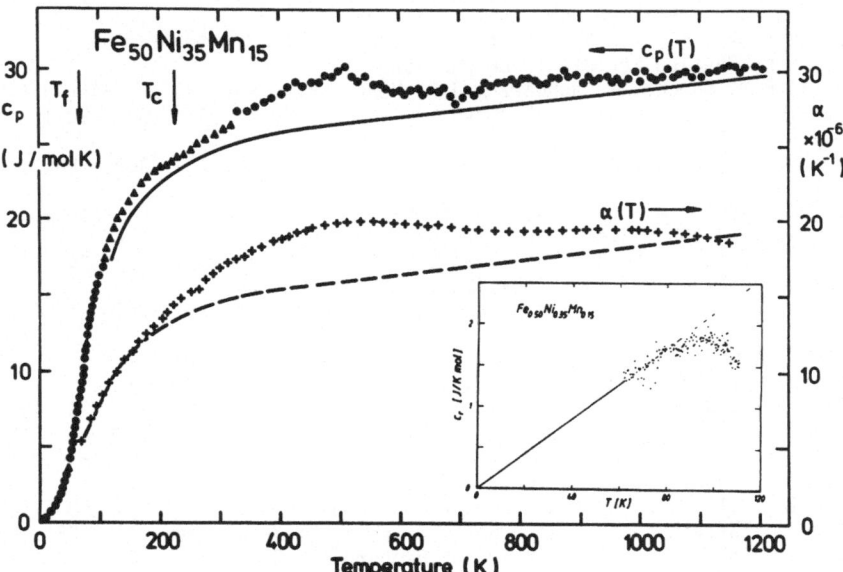

Fig. 7: Total specific heat c_p versus temperature for $Fe_{50}Ni_{35}Mn_{15}$; low temperature data (see also inset) are own results from ref. /25/; triangles from ref. /27/ and high temperature data (dots) from ref. /28/. The full curve is calculated from $c_p(T) = \gamma T + \beta T^3$ with $\gamma = 4.0$ mJ/mol K^2 and $\Theta_D = 360K$. The crosses show the temperature dependence of the thermal expansion coefficient $\alpha(T)$ (from ref. /29/). The dashed curve is a calculated Grüneisen curve with $\Theta_D = 360K$.

magnetic reference calls for a LS-HS-state transition, i.e. a reversal of the order of the levels at high temperatures ($T>T_C$, T_N) as compared to low temperatures ($T<T_C$, T_N). Secondly, if no magnetovolume effect occurs at T_C or T_N and no excess heat necessary to install long range magnetic order, we speculate that in the vicinity of the ordering temperatures HS, LS and NM-states are again energetically equal, but here cross as a function of temperature. Consequently it is not surprising that the $c_p(T)$ behavior of these concentrated alloys reminds us of that of a pure spin glass: no effect in $c_p(T)$ at the ordering temperature, but a pronounced maximum above it. We will analyze this further in a future publication.

ACKNOWLEDGEMENTS

The author is very much indebted to W. Pepperhoff for numerous valuable discussions. I also thank M. Acet, G.V. Lecomte, N. Schubert and W. Stamm for their continuous support. Work was supported by Deutsche Forschungsgemeinschaft within Sonderforschungsbereich 166.

LITERATURE

1. A.K. Majumdar and P.v. Blanckenhagen: Phys. Rev. B29, 4079 (1984)
2. A.Z. Menshikov, G.A. Takzey and A. Ye. Teplykh: Phys. Met. Metall. 54, 41 (1982)
3. M. Hayase, M. Shiga and Y. Nakamura: J. Phys. Soc. Japan 30, 729 (1971)
4. H.H. Ettwig and W. Pepperhoff: phys. stat. sol. (a) 23, 105 (1974)
5. A.Z. Menshikov, V.A. Kazantzev and N.N. Kuzmin: Sov. Phys. JETP Lett. 23, 5 (1976)
6. M. Acet, H. Zähres, W. Stamm and E.F. Wassermann: J. Appl. Phys., 63, 3921 (1988)
7. M. Matsui, K. Sato and K. Adachi: J. Phys. Soc. Japan 35, 419 (1973)
8. Y.A. Dorofeyev, V.A. Kazantzev, A.Z. Menshikov and A.E. Teplykh, preprint
9. Honda Memorial Series on Material Science, No. 3: The Physics and Application of Invar Alloys, ed. by H. Saito (Marunzen Comp. Ltd.) Amsterdam, 1979
10. Proc. Int. Symp. on the Invar Problem, Nagova, Japan, 1978, ed. by. A.J. Freeman and M. Shimizu (North Holland Pub. Comp., Amsterdam, 1979)
11. E.F. Wassermann: Advances in Sol. State Phys. 27, 85 (1987)
12. E.F. Wassermann: Physica Scripta to be published
13. V.L. Moruzzi, P.M. Marcus, K. Schwarz and P. Mohn: Phys. Rev. B34, 1784 (1986)
14. E. Kisker, E.F. Wassermann and C. Carbone: Phys. Rev. Lett. 58, 1784 (1987)
15. I. Renz and S. Methfessel: (ICM 88) to be published
16. N.E. Brener, G. Fuster, A.J. Callaway, J.L. Fry and Y.Z. Zhao: J. Appl. Phys. 63, 4057 (1988)
17. J.L. Fry, Y.Z. Zhao, P.C. Pattniak, V.L. Moruzzi and D.A. Papaconstantopoulos: J. Appl. Phys. 63, 4060 (1988)
18. Y. Tsunoda, N. Kunitomi and R.M. Nicklow: J. Phys. F: Metal Phys. 17, 2447 (1987)
19. W.A.A. Macedo and W. Keune: Phys. Rev. Lett. (1988) in press
20. A.V. Deryabin, V.I. Rimlyand and A.P. Larionov: Sov. Phys. Solid State 25, 1109 (1983)
21. M. Hayase, M. Shiga and Y. Nakamura: J. Phys. Soc. Japan 30, 729 (1971)
22. P.W. Anderson, B.I. Halperin and V.C. Varma: Phil. Mag. 25, 1 (1972)
23. D.M. Herlach, E.F. Wassermann and R. Willnecker: Phys. Rev. Lett. 50, 529 (1983)
24. E.F. Wassermann, H. Zähres, M. Acet and W. Stamm: J. Mag.Magn. Mat. 70, 434 (1987)
25. G.V. Lecomte, N. Schubert and E.F. Wassermann: J.Mag. Magn. Mat. 71, 318 (1988)
26. J. Lauer and W. Keune: Phys. Rev. Lett. 48, 1850 (1982)
27. A.V. Deryabin, V.I. Rimlyand and A.P. Larionov: Sov. Phys. JETP 57, 1298 (1983)
28. W. Bendick, H.H. Ettwig and W. Pepperhoff: J. Phys. F: Metal Phys. 8, 2525 (1978)
29. W. Bendick, H.H. Ettwig, F. Richter and W. Pepperhoff: Z. Metallk. 68, 103 (1977)

Ultradiffusion in the SK Spin Glass

K. Nemoto

Department of Physics, Faculty of Science, University of Tokyo,
Hongo 7–3–1, Bunkyo-ku, Tokyo 113, Japan

1. Introduction

Various complex systems are known to have a number of metastable states which are separated from each other by large barriers. It is obvious that the multi-valleyed landscape makes the system very glassy because there exist wide-spread time scales. Therefore the structural property of the landscape is important for an understanding of the long-time behavior of these systems.

Recently it has been realized that the multi-valleyed landscape has an exact or approximate hierarchical structure. An example of a such system is the SK spin glass model (for a recent review see [1]). Indeed we can construct a hierarchical tree indexed by energy level and barrier height at zero temperature for the model [2]. The tree characterizes the property of glassy dynamics at low temperature. In this work we consider a diffusive process in the hierarchical tree (ultradiffusion [3]) and try to examine numerically the relation between the structure of the tree and glassy relaxation.

2. Ultradiffusion

The diffusive process considered here is described by the equation

$$\frac{dP_i}{dt} = \sum_j w_{ij} P_j, \tag{2.1}$$

where i labels metastable states, $P_i(t)$ is the probability of finding the system in state i. The transition rate from j to i, w_{ij}, is defined by

$$w_{ij} = \exp(-D_{ij}/T) \qquad (i \neq j)$$
$$w_{jj} = -\sum_{i \neq j} w_{ij} \tag{2.2}$$
$$D_{ij} = B_{ij} - E_j,$$

where E_j is the energy of state j, and B_{ij} is the lowest barrier level between states i and j, satisfying hierarchical condition $B_{ij} \leq \max(B_{ik}, B_{kj})$ for all i, j and k. Here we assume that temperature T is effectively introduced in w_{ij} whereas E_i and B_{ij} are evaluated at zero temperature ($T = 0$).

There is a static solution P_i^{eq} proportional to the Boltzmann weighting $\exp(-E_i/T)$. By introducing Green's function $G_{ij}(t)$ for (2.1), the general solution can be written as $P_i = \sum_j G_{ij} P_j(t = 0)$. Then using G_{ij} we can define autocorrelation function $C(t)$:

$$C(t) = \sum_i \{G_{ii}(t) - G_{ii}(t = \infty)\} P_i^{\text{eq}}. \tag{2.3}$$

It is our target to examine long time glassy dynamics.

Springer Series in Synergetics, Vol. 43   **Cooperative Dynamics in Complex Physical Systems**
Editor: H. Takayama   © Springer-Verlag Berlin, Heidelberg 1989

3. Numerical result

We have obtained numerically E_i and B_{ij} of many samples for the SK spin glass [2]. Therefore $C(t)$ can be evaluated directly from these data. The resulting function seems to have a stretched exponential form:

$$< C(t) >_J \approx \exp(-At^\alpha), \tag{3.1}$$

where $< ... >_J$ denotes sample average. The T-dependence of the exponent is estimated from the function. As shown in Fig.1, the exponent is almost a linear function of T and no systematic size dependence is found.

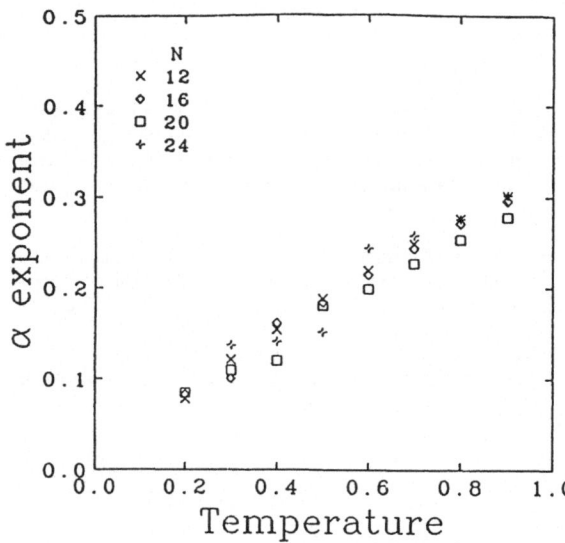

Figure 1 Temperature dependence of the exponent of (3.1). N denotes the system size.

4. Discussion

BACHAS and HUBERMAN have examined a problem of ultradiffusion in the system where all states have the same energy, and concluded that the autocorrelation function obeys a power law under proper ratio of tree branching [3]. On the other hand, KOPER and HILHORST have found that ultradiffusion in a random energy model having only one branching point exhibits a power law relaxation [4]. In the present case the tree has both many branching points and distributed energy levels, so that the situation is quite different. As shown in Fig. 5 of [2], the hierarchical tree of the SK model is highly unbalanced. Therefore the middle part of the tree, where the branching ratio is highest, contributes dominantly to the autocorrelation function, and makes relaxation faster than a power law. Another possibility is that we observe the transient part of the autocorrelation function because of the rather small size of the system.

In conclusion, we have numerically demonstrated that ultradiffusion in the SK spin glass exhibits a stretched exponential relaxation, whose exponent is a linear function of temperature.

This work was financially supported by a Grant-in-Aid for Scientific Research from the Ministry of Education, Science and Culture of Japan. The author is indebted to the Japan Society for the Promotion of Science for financial support.

1. K. Binder, A.P. Young: Rev. Mod. Phys. **58**, 801 (1986)
2. K. Nemoto: J. Phys. **A21**, L287 (1988)
3. P. Bachas, B.A. Huberman: J. Phys. **A20**, 4995 (1987)
4. G.J.M. Koper, H.J. Hilhorst: Europhys. Lett. **3**, 1213 (1987)

Critical Behavior of the Uniaxially Anisotropic Spin Glass ZnMn

S. Murayama[1], Y. Miyako[1], and E.F. Wassermann[2]

[1]Department of Physics, Faculty of Science, Hokkaido University,
Sapporo 060, Japan
[2]Tieftemperaturphysik, Universität Duisburg,
D-4100 Duisburg, Fed.Rep. of Germany

Critical behaviors of the nonlinear susceptibility in spin glasses have been extensively studied by various experiments and numerical calculations. It has been suggested from both experimental [1,2] and theoretical [3] points of view that the anisotropy in real systems plays an important role for the presence of spin-glass transition and the scaling behavior in the nonlinear susceptibility. Recently, we have studied the uniaxially anisotropic spin-glass ZnMn [4,5] over a wide range of the Mn impurity concentration. Since in ZnMn it is possible to control the ratio of the anisotropy to the variance of the random exchange interaction (D/J) by changing the Mn concentration, systematic study for the effect of the uniaxial anisotropy D on the spin-glass critical behavior could be done in this system.

Therefore, we have performed detailed measurements of the longitudinal ($\parallel$c) and transverse ($\perp$c) ac susceptibilities $\chi\parallel$ and $\chi\perp$ of ZnMn single crystal at various fields parallel to the ac field. As we previously reported, ZnMn with 270, 390 and 600 ppm Mn show the two successive paramagnetic(P)-longitudinal(L)- mixed longitudinal and transverse(LT) transitions [5]. The nonlinear part of the susceptibility is extracted from $\chi\parallel_{NL} = \chi\parallel(0) - \chi\parallel(H\parallel)$ in the longitudinal direction and $\chi\perp_{NL} = \chi\perp(0) - \chi\perp(H\perp)$ in the transverse direction. The obtained temperature and field dependence of the nonlinear susceptibility for the two directions is independently analyzed in the framework of the scaling hypothesis: $\chi_{NL}/\tau^{\beta} = f(H^{2}/\tau^{\beta+\gamma})$, where τ is a reduced temperature $(T-T_g)/T_g$, β and γ are a critical exponent. Here, we determined $T_g\parallel$ and $T_g\perp$, respectively, from the maximum of $\chi\parallel$ and $\chi\perp$. Figure 1 is the double logarithmic scaling plot of for ZnMn with 600 ppm Mn in the longitudinal and transverse directions respectively, by using the best fitted value of β and γ. We have done the same scaling analysis for ZnMn with 150, 270 and 390 ppm Mn. We found that the best fitted value of β is in the range of $\beta = 1.0 \pm 0.2$ for all the present samples in the longitudinal and transverse directions. The γ value

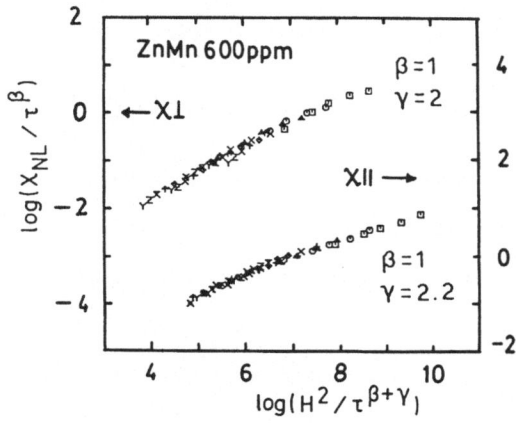

Fig. 1: Best fitted logarithmic scaling plot of χ_{NL}/τ^{β} vs $H^{2}/\tau^{\beta+\gamma}$ for ZnMn (600 ppm), where $\beta = 1.0 \pm 0.2$ and $\gamma = 2.2 \pm 0.3$ in the longitudinal direction, $\beta = 1.0 \pm 0.2$ and $\gamma = 2.0 \pm 0.3$ in the transverse direction.

is dependent on the Mn concentration. The obtained critical exponent $\phi = \beta + \gamma$ is shown as a function of D/J in Fig. 2. Here, we use the D/J value estimated previously in Ref. 5.

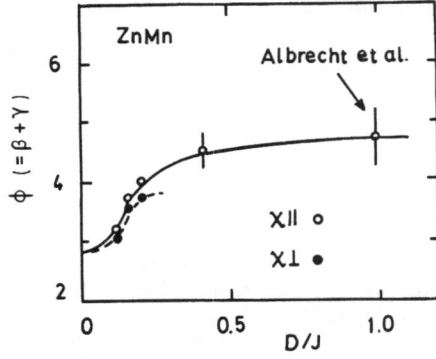

Fig. 2: Critical exponent ϕ vs relative anisotropy D/J for $\chi_{\parallel NL}$ and $\chi_{\perp NL}$ of ZnMn including the results of Albrecht et al.(4). The solid line and dashed line serve only to guide the eyes.

We can roughly estimate the value in the "Heisenberg" limit from the extrapolation of D/J $\rightarrow$ 0 as $\beta_H = 1.0 \pm 0.2$, $\gamma_H = 1.8 \pm 0.3$, and therefore $\phi_H = 2.8 \pm 0.4$ for the two directions on the assumption of $\phi_\parallel = \phi_\perp$ in this limit as shown in Fig. 2. These "Heisenberg" values should be compared with the results of typical metallic spin glasses such as CuMn AgMn and AuFe. The recent precise nonlinear magnetization measurements(2) for AgMn and CuMn show the critical exponents $\gamma = 2.2$ and $\beta = 1$ ($\therefore \phi = 3.2$) in the range of temperature near T_g and small magnetic field. Similar values of β and γ are reported in recent nonlinear ac susceptibility measurements(6) on AuFe and AgMn in the same range of temperature and field. The present extrapolated values β_H and γ_H for ZnMn are determined also in the comparable range. It is possible that the present β_H and γ_H already include the effects of the DM or dipole anisotropy and that they are not pure Heisenberg exponents. Thus, we can conclude that the critical behavior of ZnMn in the limit of D/J $\rightarrow$ 0 (high Mn concentration limit) is the same as that of CuMn, AgMn and AuFe in the range of small field and temperature near T_g. Increasing the relative anisotropy D/J, we find that ϕ for the longitudinal direction increases and seems to approach $\phi_\parallel \sim 4.6$ for D/J > 1, where the Ising behavior is well observed. We estimate the "Ising" limit values as $\beta_I = 0.9 \pm 0.2$, $\gamma_I = 3.7 \pm 0.5$, and therefore $\phi_I = 4.6 \pm 0.7$. It should be noted that these values are slightly larger than the exponents calculated in 3D short-range Ising spin glass, but seem consistent within the error bars(7). The intermediate value of the exponents between the two limits arises from the crossover effect due to the uniaxial anisotropy.

References

1. Y. Yeshurun, H. Sompolinsky: Phys. Rev. Lett. 56 984 (1986)
2. N. de Courtenay, H. Bouchiat, H. Hurdequint, A. Fert: J. Physique 47 1507 (1986)
3. R. E. Walstedt: Physica 109&110 B+C 1924 (1982); A. J. Bray, M. A. Moore: Phys. Rev. B 34 6561 (1986); A. Chakrabarti, C. Dasgupta: Phys. Rev. B 36 793 (1987)
4. H. Albrecht, E. F. Wassermann, F. T. Hedgcock, P. Monod: Phys. Rev. Lett. 48 819 (1982)
5. S. Murayama, K. Yokosawa, Y. Miyako, E. F. Wassermann: Phys. Rev. Lett. 57 1785 (1986)
6. T. Taniguchi, Y. Miyako: J. Phys. Soc. Jpn., to be published.
7. R. N. Bhatt, A. P. Young: Phys. Rev. Lett. 54 924 (1985); A. T. Ogielski: Phys. Rev. B 32 7384 (1985); R. R. P. Singh, S. Chakravarty: Phys. Rev. B 36 546 (1987), 567 (1987)

Mixed Phase in the Reentrant Ising System $Fe_xMn_{1-x}TiO_3$

H. Yoshizawa[1], S. Mitsuda[1], H. Aruga[2], and A. Ito[2]

[1]The Institute for Solid State Physics, The University of Tokyo,
Roppongi, Minato-ku, Tokyo 106, Japan
[2]Department of Physics, Faculty of Science, Ochanomizu University,
Bunkyo-ku, Tokyo 112, Japan

Despite a number of experimental reports on the magnetic ion concentration x versus temperature phase diagram on spin glass systems,[1] little attention has been paid to the difference between the pure spin glass phase and the reentrant spin glass phase. This is partly because the spin glass phase appears in a very narrow region on the concentration axis, and partly because almost all reentrant spin glass materials belong to the class of the ferromagnetic interaction dominated system, in other words, they reenter to the spin glass phase after establishing the ferromagnetic long range order. The former reason means that the concentration gradient in the sample is crucial in order to perform a reliable study of spin glass behavior as a function of the concentration x. The latter causes the technical difficulty in neutron scattering experiments.[2] Since a ferromagnetic Bragg reflection always overlaps with a structural Bragg reflection, it is difficult to obtain definite information on the magnetic long range order. If the system exhibits the antiferromagnetic order, such a difficulty is obviously resolved. Recently A. Ito and coworkers discovered an antiferromagnetism-based spin glass system $Fe_{0.5}Mn_{0.5}TiO_3$, and carried out intensive studies on relaxational phenomena.[3]

In this short note we report systematic studies of the reentrant Ising spin glass system $Fe_xMn_{1-x}TiO_3$. Using neutron scattering technique, we have demonstrated that the x-T phase diagram of the system exhibits a beautiful qualitative correspondence with the phase diagram of the SK model.[4] The SK phase diagram consists of four phases: a paramagnetic phase, an (anti)ferromagnetic phase, a spin glass phase, and a mixed phase. In the mixed phase the spin glass ordering coexists with the (anti)ferromagnetic long range order. Between the mixed phase and the spin glass phase, a few theories predict the second order phase boundary.[5,6] By observing the temperature dependence of the magnetic Bragg reflection as well as the

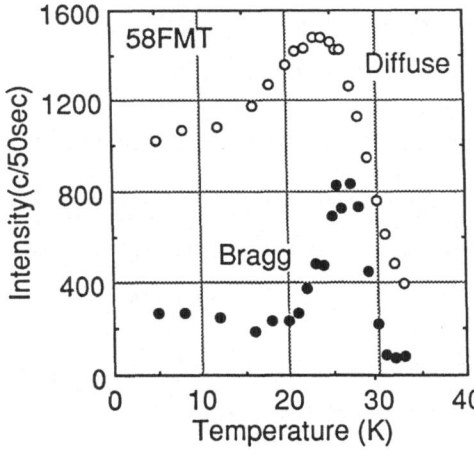

Fig. 1. Bragg and diffuse intensity versus temperature observed at Q= (1,1,1.5) in the $Fe_xMn_{1-x}TiO_3$ sample with x=0.58. By fitting the scan profiles to Lorentzian, the peak diffuse intensity at Q = (1,1,1.5) is determined. After subtracting the diffuse intensity from the observed peak intensity, the remaining intensity is plotted as the Bragg intensity.

Springer Series in Synergetics, Vol. 43   **Cooperative Dynamics in Complex Physical Systems**   169
Editor: H. Takayama   © Springer-Verlag Berlin, Heidelberg 1989

diffuse scattering, one can identify two critical temperatures: the (anti)ferromagnetic phase transition temperature and the reentrant transition temperature. Typical data observed in the $Fe_xMn_{1-x}TiO_3$ sample with x=0.58 is shown in Fig. 1. Among the samples we studied, the x=0.58 sample is the closest to the multicritical point (MCP) between the paramagnetic, the antiferromagnetic, and the spin glass phases. The Bragg intensity in this particular sample is weaker than the diffuse scattering because the static moment which contributes to the long range order vanishes at the MCP. The Neel temperature is ~30.5K. The Bragg intensity decreases below 27K due to the reentrant spin glass transition. When the Bragg intensity decreases most steeply, the diffuse scattering also shows the broad maximum. Below 20K, the Bragg intensity levels off. The Bragg intensity below 20K is, however, significantly larger than the paramagnetic value, indicating that the antiferromagnetic long range order survives below the reentrance and it coexists with the spin glass ordering. Four other samples with x = 0.33, 0.38, 0.60, and 0.65 in the reentrant regime of x showed the qualitatively identical behavior. Based on these observations, we constructed a schematic phase diagram of the $Fe_xMn_{1-x}TiO_3$ system as shown in Fig. 2. In each side of the figure, it has the antiferromagnetic phase of either $MnTiO_3$ type structure or the $FeTiO_3$ type structure. We tentatively assign the reentrant transition to the AT line.[7] In order to emphasize the mixed phase and to separate it from the spin glass phase in the middle, we drew thick solid line as the phase boundaries. An additional support to the existence of such boundaries is found in the concentration dependence of the diffuse scattering at the low temperature limit. If they are the second order transition lines as suggested by theories,[5,6] there should be critical behavior. By plotting intensity of the diffuse scattering as well as its full width at half maximum as a function of the concentration x, we have, indeed, demonstrated the divergence of both quantities near the boundaries as reported in ref. 4.

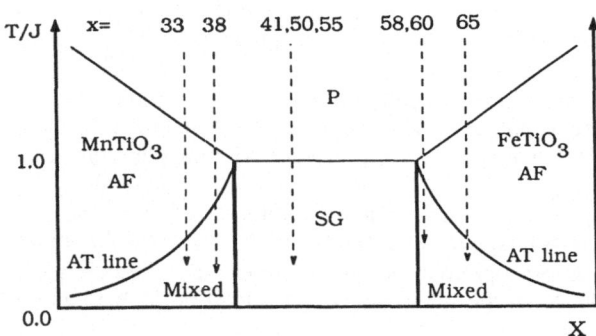

Fig. 2 Schematic phase diagram of the $Fe_xMn_{1-x}TiO_3$ system. The horizontal line corresponds to the Fe ion concentration x in units of percentage.

1. C.Y. Huang, J. Magn. Magn. Mater.**51**,1(1985).
2. For example, see H. Maletta, G. Aeppli, and S.M. Shapiro, Phys. Rev. Lett.**48**,1490 (1982).
3. A. Ito, H. Aruga, E. Torikai, M. Kikuchi, Y. Shono, and H. Takei, Phys. Rev. Lett. **57**,483(1986); K. Gunnarsson, P. Svedlindh, P. Nordblad, L. Lundgren, H. Aruga, and A.Ito, ibid**61**,754(1988); A. Ito, H. Aruga, M. Kikuchi, Y. Shono, and H. Takei, Solid State Commun.**66**,475(1988).
4. H. Yoshizawa, S. Mitsuda, H. Aruga, and A. Ito, Phys. Rev. Lett.**59**,2364(1987).
5. G. Toulouse, J. Phys.(paris) Lett.**41**,L447(1980).
6. H. Nishimori, Prog. Theor. Phys.**66**,1169(1981);H. Nishimori, Y.Taguchi, and T. Oguchi, J. Phys. Soc. Jpn.**55**,656(1986);H. Nishimori,J. Phys.Soc.Jpn. **55**,3305 (1986);Y. Ozeki and H. Nishimori, J. Phys.Soc. Jpn.**56**,1568(1987).
7. J.R.L. de Almeida and D.J. Thouless, J. Phys.A**11**,983(1978).

Dynamic Properties
of a Reentrant Spin Glass $Pd_{1-x-y}Fe_xMn_y$

H. Takano [1] *and Y. Miyako* [2]

[1]Department of Dental Radiology,
 Higashi Nippon Gakuen University School of Dentistry,
 Ishikari-Tobetsu, Hokkaido 061-02, Japan
[2]Department of Physics, Faculty of Science, Hokkaido University, Sapporo 060, Japan

We have previously studied the reentrant behavior of $Pd_{1-x-y}Fe_xMn_y$ alloy and obtained a phase diagram for the alloy[1]. A Gabay-Toulouse transition [$T_{GT}(\omega)$] was found in $Pd_{1-x-y}Fe_xMn_y$ by Mössbauer spectroscopy[2] and magnetization measurement[3]. Recently, the frequency dependence of $T_{GT}(\omega)$, the peak temperature of the imaginary part of the ac-susceptibility of $Pd_{0.920}Fe_{0.015}Mn_{0.065}$, has been well analyzed using a dynamic scaling relation $\omega/\omega_0 = [\{T_{GT}(\omega) - T_{GT}(0)\}/T_{GT}(\omega)]^{z\nu}$ [4] proposed by SOULETIE and THOLENCE[5] and SOULETIE[6]. From the analysis, the finite temperature Gabay-Toulouse transition was suggested. We study here the time decay of a thermoremanent magnetization (TRM) $M_{TRM}(t)$ for $Pd_{1-x-y}Fe_xMn_y$. In the ferromagnetic phase of two samples, $Pd_{0.920}Fe_{0.015}Mn_{0.065}$ ($T_C=26.5$ K, $T_{GT}=3.6$ K) and $Pd_{0.881}Fe_{0.027}Mn_{0.092}$ ($T_C=34$ K, $T_{GT}=13.2$ K), short-time relaxation ($t < \sim 30$ sec) curves at the cooling field $H=500$ Oe are represented by a single exponential form. Its characteristic time τ is about 4 sec for $Pd_{0.920}Fe_{0.015}Mn_{0.065}$. On the other hand, the long-time tail of the log-log plot of the TRM is nearly represented by a straight line for these samples. This fact shows that the time behavior of the TRM in this region follows a power law. Consequently, the entire decay form of the TRM in the ferromagnetic region for $Pd_{1-x-y}Fe_xMn_y$ is described as

$$M_{TRM} = M_{0S}\exp(-t/\tau) + M_{0L}t^{-\alpha}. \tag{1}$$

Here M_{0S} and M_{0L} is a fitting parameter for short and long relaxation, respectively. This feature in the ferromagnetic region did not change even if the applied field was lowered (H>8 Oe).

The relaxation of the TRM in the spin glass phase for $Pd_{0.881}Fe_{0.027}Mn_{0.092}$ ($T_C/T_{GT} \sim 2.6$) is different from that for $Pd_{0.920}Fe_{0.015}Mn_{0.065}$ ($T_C/T_{GT} \sim 7.4$). This behavior is also distinguished from that in the ferromagnetic region, and the whole relaxation in the spin glass phase is slower than that in the ferromagnetic phase. For $Pd_{0.881}Fe_{0.027}Mn_{0.092}$, an algebraic decay is also found, i.e. a power law decay at H=500 Oe as well as in the ferromagnetic region:

$$M_{TRM}/M_0 \propto t^{-\alpha}. \tag{2}$$

While it is impossible to fit the time decay of the TRM for the same sample at lower cooling fields (<30 Oe at T=4.2 K) by a simple power law (Fig. 1), excellent fits, as represented by solid lines in Fig. 1, of the stretched exponential

$$M_{TRM}/M_0 \propto \exp[-(t/\tau)^{1-n}] \tag{3}$$

to the experimental data can be made in the time window of log(t)≳1. But the decay at T=5.1 K is different from the one at T=4.2 K and is represented by a straight line in log-log plots of TRM and time. It can therefore be described not by a stretched exponential form but by a power law. From these results, the cross-over from power law decay to stretched exponential decay seems to occur in the spin glass phase.

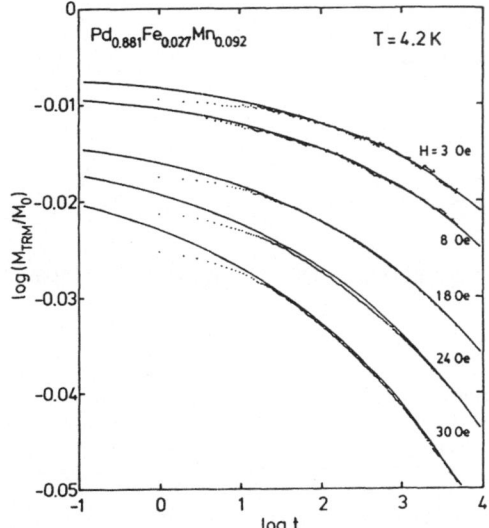

Fig. 1: Logarithmic plots of M_{TRM}/M_0 for $Pd_{0.881}Fe_{0.027}Mn_{0.092}$ as a function of time t for various fields at 4.2 K.

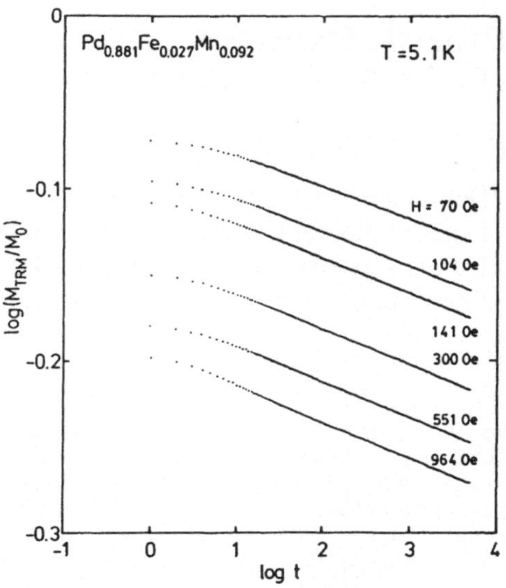

Fig. 2: Logarithmic plots of M_{TRM}/M_0 for $Pd_{0.881}Fe_{0.027}Mn_{0.092}$ as a function of time t for various fields at 5.1 K.

In the spin glass region of $Pd_{0.920}Fe_{0.015}Mn_{0.065}$, which has a large ratio of T_C/T_{GT}, the time decay of the TRM for H=500 Oe cannot be analyzed in terms of either a simple power law of time or a simple stretched exponential, and it has to be described by a modified stretched exponential function,

$$M_{TRM}/M_0 \propto t^{-\alpha} \exp[-(t/\tau)^{1-n}]. \tag{4}$$

GRANBERG et al.[7] and OCIO et al.[8] found the same function.

As shown above, the characteristic time in the spin glass region of the $Pd_{1-x-y}Fe_xMn_y$ reentrant spin glass system depends on the applied cooling field and the sample. Moreover we find that the relaxation in the spin glass phase is different from that in the inhomogeneous ferromagnetic phase. From these experiments and an interpolation of a power law and a stretched exponential, we finally obtain a modified stretched exponential form as the general relaxation function in the spin glass region. Power law or stretched exponential decay becomes the leading term in the observed time window depending on sample, applied field strength and the temperature measured.

References

1. Y.Miyako, T.Nishioka, T.Sato, S.Morimoto and A.Ito: J. Mag. Mag. Mater. 54-57, 149 (1986)
2. Y.Takeda, S.Morimoto, A.Ito, T.Satō and Y.Miyako: J. Phys. Soc. Jpn. 54, 2000 (1985)
3. T.Satō, T.Nishioka, Y.Miyako, Y.Takeda, S.Morimoto and A.Ito: J. Phys. Soc. Jpn. 54, 1989 (1985)
4. H.Takano, Y.Miyako, M.Godinho and J.L.Tholence: J. Phys. Soc. Jpn. 57, 3514 (1987)
5. J.Souletie and J.L.Tholence: Phys. Rev. B32, 516 (1985)
6. J.Souletie: J. Phys. (Paris) 41, 1211 (1988)
7. P.Granberg, P.Svedlindh, P.Nordblad, L.Lundgren and H.S.Chen: Phys. Rev. B35, 2075 (1987)
8. M.Ocio, M.Alba and J.Hammann: J. Phys. (Paris) 46, L1101 (1986)

Time Dependent Magnetic Phenomena in Dilute Ising Systems $Fe_{1-x}Mg_xCl_2$

T. Kamai, K. Iio, H. Tanaka, and K. Nagata

Department of Physics, Faculty of Science, Tokyo Institute of Technology,
Oh-okayama, Meguroku, Tokyo 152, Japan

Dilute Ising systems $Fe_{1-x}Mg_xCl_2$ are a random mixtures of $FeCl_2$ and $MgCl_2$ both crystallizing in the $CdCl_2$ structure with a hexagonal layered lattice. A pure $FeCl_2$ has ferromagnetic c planes stacked antiferromagnetically along the c axis. There exists fairly strong uniaxial anisotropy parallel to the c axis.[1] Recently these dilute systems have attracted considerable attention because of their interesting magnetic properties inherent in a typical Ising-like random magnet with short range interaction. In the region where x is beyond the site percoration threshold($x_c=0.5$) of the triangular net for spin sites, the system exhibits Ising spin glass behavior at low temperatures.[2] For x<0.5, the coexistence of spin glass and antiferromagnetic ordering has been observed.[3] Then, focusing on these characteristic concentration ranges below and above x_c, we will show the results of the dynamical properties of the remanent magnetization of this Ising system through the measurement of the Faraday effect.

Faraday rotation for the present uniaxial magnetic mixture has been revealed to be proportional to the total magnetization of the system.[4] Thus, the specific rotation angle θ(rad/mm) and that divided by a field strength θ/H shown below are regarded as the magnetization and the parallel susceptibility, respectively. Figure 1 shows the temperature dependence of θ/H of specimens with two different concentrations under various applied fields. With decreasing temperature, the field cooled(FC) curves deviate from the zero-field cooled(ZFC) curves. We designate the temperatures at which the irreversibility between θ/H(ZFC) and θ/H(FC) appears as $T_g(H)$ for x>0.5 and $T_R(H)$ for x<0.5, respectively. For x=0.52(the spin glass regime), the cusp of θ(T) is broadened in association

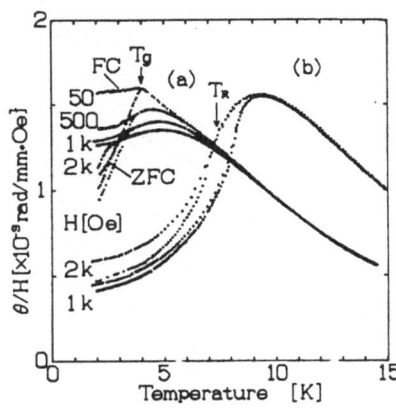

Fig. 1. Temperature dependence of ZFC and FC susceptibilities (θ) for various magnetic fields in (a) $Fe_{0.48}Mg_{0.52}Cl_2$ (the spin glass regime) and (b) $Fe_{0.58}Mg_{0.42}Cl_2$ (the dilute antiferromagnetic regime).

Springer Series in Synergetics, Vol. 43   **Cooperative Dynamics in Complex Physical Systems**
Editor: H. Takayama   © Springer-Verlag Berlin, Heidelberg 1989

with the increase in the strength of H and, in addition, the temperature $T_g(H)$ is shifted to lower temperature. For x=0.42(the dilute antiferromagnetic regime), the temperature $T_R(H)$ is reduced in the same way as $T_g(H)$ of the spin glass system, as the field intensity is increased. However, a shape of θ/H as a function of temperature around $T_R(H)$ resembles that of uniaxial antiferromagnets around the Néel temperature.[5]

Next, we show the time dependence of thermoremanent magnetization (TRM) measured in a zero field after cooling the samples in a field from $T > T_g$(for $x > x_c$) and T_R(for $x < x_c$). Below T_g(and T_R), the long time-persistence in TRM was observed, where the relaxation rate becomes higher in the vicinity of T_g(and T_R) and under higher intensity of annealing field as well. It can be realized from Fig. 2 that the functional form of the relaxation in the system with x of 0.52 cannot be fit either by a log t(Fig. 2a) or a power-law(Fig. 2b). A computer simulation has predicted that TRM for short range Ising spin glass like this system obeys a power-law decay below T_g.[6] With regard to the aging process, our measurement performed at various waiting times($10^0 < t_w < 10^4$) shows that the time dependence of the long time-relaxation in $Fe_{1-x}Mg_xCl_2$ was not modified meaningfully by t_w. This fact indicates that the system on the FC-process is probably in a state of thermal equilibrium. The behavior of TRM similar to that mentioned above was also observed in the system x<0.5. Therefore, the time decay of TRM can be governed essentially by the same mechanism over both concentration ranges.

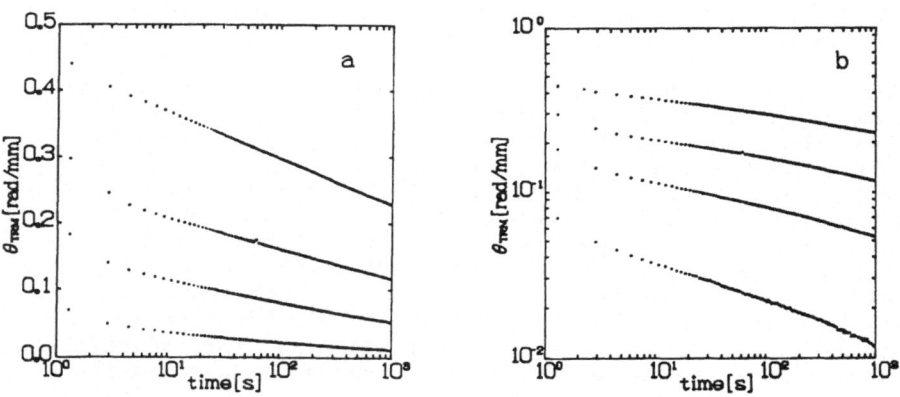

Fig. 2. Semilog (a) and log-log (b) plots of TRM vs t in $Fe_{0.48}Mg_{0.52}Cl_2$. From top to bottom, T/T_g=0.492, 0.525, 0.703 and 0.838, respectively. The measuring field is 2 kOe.

1. I. S. Jacobs and P. E. Lawrence: Phys. Rev. 164(1967)866.
2. D. Bertrand, A. R. Fert, M. C. Schmidt, F. Bensamka and S. Legrand: J. Phys. C15(1982)L883.
3. Po-zen Wong, S. von Molnar, T. T. M. Palstra, J. A. Mydosh, H. Yoshizawa, S. M. Shapiro, and A. Ito: Phys. Rev. Lett. 55(1985)2043.
4. H. Yamasita, K. Iio, M. Sano, H. Masuda, H. Tanaka and K. Nagata: J. Magn. Soc. Jpn. Vol. 11, Supplement, S1(1987)87.
5. Po-zen Wong, S. von Molnar and P. Dimon: J. Appl. Phys. 53(1982)7954.
6. A. T. Ogielski : Phys. Rev. B32(1985)7384.

Nonlinear Susceptibility and Magnetic Ordering of MCl$_2$-GICs

M. Hagiwara, T. Kawaguchi, and M. Matsuura

Department of Material Physics, Faculty of Engineering Science,
Osaka University, Toyonaka, Osaka 560, Japan

Graphite intercalation compounds (GIC) are complex systems in which each intercalant layer is separated by a certain number (stage num.) of carbon layers. Besides such a staged structure, GICs have a characteristic island structure, in which each intercalant layer is divided into "mesoscopic" clusters (islands) [1]. Among these, NiCl$_2$- and CoCl$_2$-GICs have been studied as interesting model systems of a two-dimensional (2D) XY-type ferromagnet [1]. An observed two-step magnetic phase transition at T_{cu} and T_{cl} ($< T_{cu}$) with a purely 2D long-range order in the intermediate temperature range can be understood by a combined effect of good 2D nature and mesoscopic cluster size as a successive ordering such as disorder $\rightleftarrows$ intraisland (2D) order $\rightleftarrows$ interisland (3D) order [2,3]. The detail of the ordered structure below T_{cu}, however, is not yet clear although a neutron scattering experiment [4] and a series of magnetic measurements [2,3] have shown very interesting phenomena suggesting a XY type of intraisland interaction and a glass-like nature of complicated interisland interaction.

For further information, we investigate here the nonlinear susceptibility of 2nd-stage NiCl$_2$- and CoCl$_2$-GICs based on single crystal Kish graphites. Linear and nonlinear responses of magnetization M for weak sinusoidal field (h sin ωt) are detected by SQUID magnetometer. Fourier analysis of M with a micro-computer gives us any harmonic components $M'_{n\omega}$, where the in-phase components of 1st and 3rd harmonics M'_{ω} and $M'_{3\omega}$ are coupled with linear and nonlinear susceptibility χ_0 and χ_2 by relations as $M'_{\omega} \doteqdot \chi_0 h$ and $M'_{3\omega} \doteqdot -(1/4)\chi_2 h^3$, respectively.

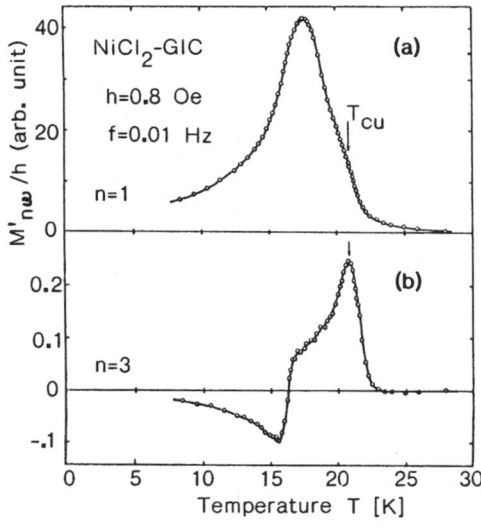

Fig.1. Harmonic components $M'_{n\omega}$ of AC-magnetization vs T for NiCl$_2$-GIC

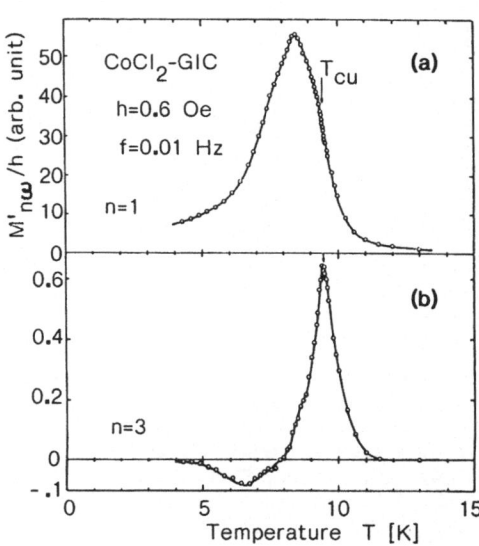

Fig.2. Harmonic components $M'_{n\omega}$ of AC-magnetization vs T for CoCl$_2$-GIC

Springer Series in Synergetics, Vol. 43   **Cooperative Dynamics in Complex Physical Systems**
Editor: H. Takayama   © Springer-Verlag Berlin, Heidelberg 1989

Figures 1 and 2 are the experimental results of $M'_{\omega}(T)$ (a) and $M'_{3\omega}(T)$ (b) for $NiCl_2$- and $CoCl_2$-GICs, respectively. $M'_{\omega}(T)$ shows only a very weak anomaly at T_{cu},[2] but $M'_{3\omega}(T)$[2] exhibits a distinguishable peak indicating a divergent feature and therefore a negative divergence of χ_2 at T_{cu}. While around T_{c1}, $M'_{3\omega}(T)$ shows another different anomaly in which the sign of $M'_{3\omega}$ changes from plus to minus as temperature decreases across T_{c1}, such a temperature dependence of $M'_{3\omega}$ around T_{c1} is similar to theoretically predicted one for antiferromagnet at T_N [5] and seems consistent with an observed antiferromagnetic correlation below T_{c1} through neutron diffraction [4].

The divergent nature of $M'_{3\omega}$ is certainly attributed to the singularity of the transition at T_{cu} which is brought about two-dimensionally by intraisland ferromagnetic interaction. While, the singularlity of χ_2 for ferromagnet at T_c is theoretically predicted to be of negative divergence above Tc but positive one below T_c [5,6]. Thus one may think that the present anomaly of $M'_{3\omega}$ at T_{cu} is rather inconsistent with intraisland (2D) ferromagnetic order. Such an inconsistency, however, can be eliminated, if we take the interisland magnetic fluctuations into account below T_{cu}, referring to the finite size of each island and to the interisland disorder in the intermediate temperature range ($T_{c1}<T<T_{cu}$).

If it is the case, we may expect the following effects. Firstly, if the frequency of applied ac-field increases, the contribution of interisland fluctuatons will decrease because of the slower correlation time. Secondly, if the applied field intensity is decreased, such an additional contribution will decrease, too. As a result, a change of shape as well as a decrease of amplitude of $M'_{3\omega}(T)$ is expected from symmetric into antisymmetric divergence. As an example, $M'_{3\omega}(T)$ at various frequencies for $NiCl_2$-GIC is shown in Fig.3. The amplitude of $M'_{3\omega}(T)$ decreases with increasing frequency, and then its shape appears to change as expected above. Similar change is also found when h is decreased in a preliminary experiment. These supplemental data may apparently confirm the above speculation.

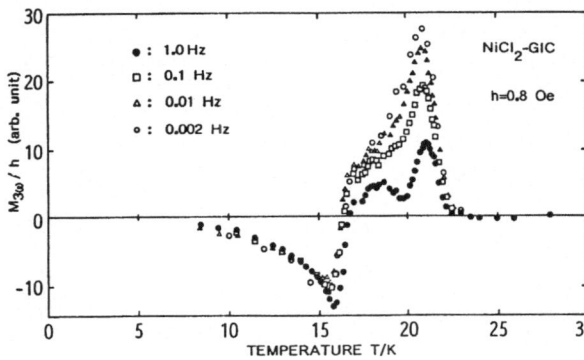

Fig.3. Temperature dependences of $M'_{3\omega}$ with various frequencies for 2nd-stage $NiCl_2$-GIC

References

1. M.Matsuura, Y.Murakami, K.Takeda, H.Ikeda and M.Suzuki: Synth. Met. 12, 427 (1985)
2. M.Matsuura: Ann. Phys. Suppl. (Paris) 11, 117 (1986)
3. Y.Murakami and M.Mastuura: J. Phys. Soc. Jpn. 57, 1056 (1988)
4. D.G.Wiesler, M.Suzuki and H.Zabel: Phys. Rev. B36, 7051 (1987)
5. S.Fujiki and S.Katsura: Prog. Theor. Phys. 65, 1130 (1981)
6. K.Wada and H.Takayama: Prog. Theor. Phys. 64, 327 (1980)

On the Three-Dimensional $\pm J$ Ising Model by the Transfer Matrix Method

H. Kitatani and T. Oguchi

Department of Sciences and Mathematics, Nagaoka University of Technology,
Kami-Tomioka, Nagaoka 940-21, Japan

1. The Value of Critical Exponent η at p=0.5

Recently, the three-dimensional ±J Ising model has been recognized to undergo a phase transition from a paramagnetic phase to a spin glass phase mainly by the study of Monte Carlo simulations[1] and high temperature series expansion methods[2]. In this paper, we investigate the above model with binary distribution $[P(J_{ij})=p\delta(J_{ij}-1)+(1-p)\delta(J_{ij}+1)]$ $(0<p\le 1)$ by the screw transfer matrix method and a finite size scaling for $L\times L\times N$ lattices(L=3,4,5).

In this section, we calculate averaged correlation functions at $T=T_{sg}$ for the symmetric distribution case (p=0.5). Extrapolating to $L\to\infty$, we estimate the value of a critical exponent η. First we investigate the pure ferromagnetic case(p=1) to present the utility of our method. We consider $L\times L\times\infty$ lattices for L=3, 4 and 5, and calculate the correlation functions $\langle S_0 S_r\rangle_L$ assuming $T_c=4.5115$, which is the most reliable value, where the distance of two spins, r, takes $1,2,\cdots,L$. Then we assume the following form for the correlation function

$$\langle S_0 S_r\rangle_L = A\exp(-r/\xi_L)/r^{d-2+\eta}L. \tag{1.1}$$

We estimate η_L and ξ_L for each L by the least squares method. Figures 1 show the plots of η_L versus 1/L for both two- and three-dimensional cases. (For two-dimensional case we consider $L\times\infty$ lattices and use $T_c=2.2692$.) As is easily seen from the figures, the points are fairly well arranged in a straight line. The extrapolated value of η is 0.029 in three-dimensional case, while the most reliable one is 0.031. (In two-dimensional case, the extrapolated value is 0.2496, while the exact one is 0.25.) Now we apply the above method to the random(p=0.5) case. We assume that $T_{sg}=1.2$. We consider $L\times L\times(n+N+n)$ lattices for L=3 and 4, where we cannot calculate for L=5 due to the limited memory space of computer. $L\times L\times n$ spins on both sides remove the boundary effects. We calculate $f(L,r)$ defined as

$$f(L,r)=\sum_{i\in L^2 N}\langle S_i S_{i+r}\rangle^2/L^2 N, \tag{1.2}$$

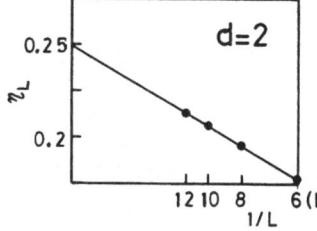

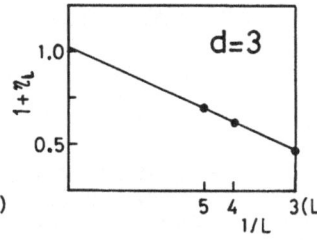

Fig. 1 The plots of η_L versus 1/L for the pure ferromagnetic case

where N=1200(700) for L=3(4). As the pure ferromagnetic case,we assume $f(L,r)$ takes the form like r.h.s. of (2.1), and estimate $1+\eta_L$. We get the configurational average of 50 samples. Consequently,it yields

$$1+\eta_L = \begin{cases} 0.406\pm0.008 & \text{for L=3} \\ 0.492\pm0.009 & \text{for L=4.} \end{cases} \tag{1.3}$$

Though we have only two points in $1+\eta_L$ versus $1/L$ plots, we assume they are well arranged in the extrapolated line as the pure ferromagnetic case. By the extrapolation to $L\to\infty$, we finally obtain $\eta=-0.250\pm0.059$. This value is in good agreement with $\eta=-0.22\pm0.05$ (MC method[1]) and -0.25 ± 0.17 (high temperature series expansion[2]).

2. Verticality of the SG-F Phase Boundary

As for the asymmetric distribution case $(0.5<p\leq1)$, Nishimori[3] proved that the phase boundary between ferromagnetic and spin glass phases is vertical or reentrant. He also argued that the tricritical point is on the so-called cross-over line where $\exp(-2J/T)=(1-p)/p$ is satisfied, and estimates $p_{t_C}=0.767\pm0.004$ and $T_{t_C}=1.68\pm0.025$[4]. In this section, we obtain SG$\leftrightarrow$F phase transition points at some temperatures, T=1.5, 1.0 and 0.5, to clarify whether the phase boundary is vertical or not. We consider $L\times L\times(n+N+n)$ lattices for L= 3 and 4 where $N=1.8\times10^4$ and 10^4, respectively. We calculate the uniform susceptibility $X_L(T_c,p)$, which can be written as follows:

$$X_L(T_c,p)\propto L^{2y_H-d}. \tag{2.1}$$

Therefore, from the values of $X_3(T_c,p)$ and $X_4(T_c,p)$, it yields

$$2y_H-d=\log(X_4(T_c,p)/X_3(T_c,p))/\log(4/3). \tag{2.2}$$

Now we assume that the weak universality of the ferromagnetic transition is valid along the critical line. Then we obtain $2y_H-d=\gamma/\nu=2-\eta$. For the three-dimensional case, the most reliable value of $2y_H-d$ is 1.97. On the other hand, for the pure ferromagnetic case, $\log(X_4/X_3)/\log(4/3)$ at T_c equals 1.96. This value is very close to the most reliable one. Therefore, for the random case, we also judge that the point in T-p plane is in ferromagnetic phase when the value of $2y_H-d$ from L=3 and 4 is above 1.97. We calculate $X_L(T,P)$ at p=0.756 and 0.768 for 20 samples and obtain the data(Table 1). As is seen from Table 1, every p_c at T=1.5, 1.0 and 0.5 is in the region $p_c=0.762\pm0.006$. This value is also in good agreement with $p_{t_C}=0.767\pm0.004$[4]. Thus we conclude that the phase boundary between SG and F phases is vertical.

Table 1. The values of $2y_H-d$

p\T	1.5	1.0	0.5
0.756	1.78±0.07	1.86±0.08	1.74±0.18
0.768	2.17±0.06	2.26±0.10	2.15±0.18

References

1. A.T. Ogielski: Phys. Rev. B32, 7384 (1985)
2. R.P.P. Singh and S. Chakravarty: Phys. Rev. Lett. 57, 245 (1986)
3. H. Nishimori: Prog. Theor. Phys. 66, 1169 (1981)
4. Y. Ozeki and H. Nishimori: J. Phys. Soc. Jpn. 56, 1568 (1987)

Application of the Coherent Anomaly Method to d-Dimensional Ising Spin Glasses

S. Fujiki

Department of Engineering Science, Faculty of Engineering, Tohoku University, Sendai 980, Japan

The spin-glass susceptibility is analyzed by the Coherent Anomaly Method (CAM) [1] applied to a series of closed-form approximations in the Cluster Variation Method (CVM) [2] for the random-bond Ising model on the square, cubic and hyper-cubic lattices. No spin-glass transition is obtained for the square lattice, while for the cubic lattice and higher dimensional lattices spin-glass transitions are obtained; the transition temperature and the exponent of the spin-glass susceptibility obtained for the cubic lattice are in good agreement with those by Monte Carlo simulations [3].

The spin-glass susceptibility — it is defined as a response of Edwards and Anderson's order parameter [4] to the variance of the applied random field and is proportional to the non-linear susceptibility [5] — has been calculated in several levels of approximations in the Cluster Variation Method (CVM) [2,6]: Bethe, cactus square and full square approximations, the last of which is equivalent to Kikuchi's second approximation [7] in the pure case. In the CVM scheme the better estimate of the transition temperature is obtained by the larger cluster taken as a basic cluster. Up to the full square approximation we obtained the spin-glass transition even for the square lattice. The question, however, remains whether the spin-glass transition is obtained by further advanced approximations in the CVM scheme.

Suzuki's Coherent Anomaly Method (CAM) may give an answer to the above question. By the CAM theory we can extract the true critical temperature and the non-classical exponent from at least three different levels of approximations in a certain approximation series. By CVM we have the susceptibility in the classical form,

$$\chi(m) = \frac{C(m)}{(T - T(m))/T(m)}, \tag{1}$$

where $\chi(m)$ and $T(m)$ refer to the susceptibility and the transition temperature by the m-th level approximation. By the CAM scheme, the coefficient, $C(m)$, is related to the corresponding transition temperature, $T(m)$, by

$$C(m) = \frac{\Psi}{(T(m) - T_c)^\psi}, \tag{2}$$

where T_c is the true critical temperature. The exponent, ψ, is related to the non-classical exponent of the susceptibility, γ, by $\psi = \gamma - 1$. We thus determine the three unknown parameters, T_c, ψ and Ψ in eq.(2) by a given set of $T(m)$ and $C(m)$.

To check the validity of the combination of CVM and CAM we first report the pure ferromagnetic case. Table I shows the critical temperatures, $kT_f(m)$, and the coefficients of the uniform susceptibility, $C_f(m)$, by the three different CVM approximations for $d = 2$ and 3. By applying CAM to the three approximations we have $\psi = 0.7738$ and kT_c/J

Table I. The ferromagnetic transition temperatures, $kT_f(m)$, the coefficients of the uniform susceptibility, $C_f(m)$, the spin-glass transition temperatures, $kT_g(m)$, and the coefficients of the spin-glass susceptibility, $C_g(m)$, for the square and cubic lattices by the Bethe ($m = 1$), cactus square ($m = 2$) and full square ($m = 3$) approximations in the Cluster Variation Method (CVM) [2,6].

m	approximation	$kT_f(m)/J$	$C_f(m)$	$kT_g(m)/J$	$C_g(m)$
	$d = 2$ (square lattice)				
1	Bethe	1.4427	0.5000	0.7593	0.3802
2	cactus square	1.3854	0.5854	0.6821	0.7418
3	full square	1.2128	1.4170	0.4824	4.5614
	$d = 3$ (cubic lattice)				
1	Bethe	2.4663	0.2500	1.0390	0.1614
2	cactus square	2.4464	0.2581	1.0165	0.1871
3	full square	2.3049	0.4000	0.8263	1.1903

$= 1.132$ for the square lattice, which are in very good agreement with the exact values, $\psi = 0.75$ and $kT_c/J = 1.1346$.

The coefficients of the spin-glass susceptibility, $C_g(m)$, and the spin-glass transition temperatures, $kT_g(m)$, are also listed in Table I. Analyzing these values by the same way as in the ferromagnetic case, no spin-glass transition is obtained for the square lattice. For the cubic lattice, we have a spin-glass transition at $kT_g/J = 0.663$ with $\psi = 2.39$, which are in good agreement with the estimates by Monte Carlo simulations [3], $kT_g/J = 0.5 \sim 0.65$ with $\psi = 1.6 \sim 2.4$.

I wish to thank Prof. T. Morita, Prof. M. Suzuki and Dr. M. Katori for their valuable comments.

References

1. M. Suzuki, *J. Phys. Soc. Jpn.* **55** (1986) 4205.
   M. Suzuki, M. Katori and X. Hu, *J. Phys. Soc. Jpn.* **56** (1987) 3092.
2. T. Morita, *J. Math. Phys.* **13** (1972) 115.
3. A.T. Ogielski, *Phys. Rev.* **B 32** (1985) 7384.
   R.N. Bhatt and A.P. Young, *Phys. Rev. Lett.* **54** (1985) 924.
4. S.F. Edwards and P.W. Anderson, *J. Phys. F* **5** (1975) 965.
5. S. Fujiki and S. Katsura, *Prog. Theor. Phys.* **65** (1981) 1130.
   M. Suzuki, *Phys. Lett.* **A 127** (1988) 410.
6. S. Katsura and S. Fujiki, *J. Phys. C* **13** (1980) 4711, 4723.
7. R. Kikuchi, *Phys. Rev.* **81** (1951) 988.

The Frequency Dependence of the AC Susceptibility in Spin Glasses

T. Shirakura, S. Inawashiro, and M. Suzuki

Department of Applied Physics, Tohoku University, Sendai, Japan

We investigate the behavior of spin glasses in an external AC field by means of the dynamical 'naive' mean field (NMF) equations

$$\tau(dm_i/dt) = -m_i + \tanh(\beta\Sigma_j J_{ij} + \beta h_{ac}\cos(\nu t)) \tag{1}$$

with infinitely long-ranged interactions whose distribution is Gaussian. The mean is zero and the variance $J/N^{1/2}$, where N is the number of spins.

We solve (1) numerically and examine the following for stationary solutions: the static magnetization $[m_0]_s$, the AC susceptibilities $\mathrm{Re}\chi_1 = [m_{1c}/h_{ac}]_s$, $\mathrm{Im}\chi_1 = [m_{1s}/h_{ac}]_s$, and the generalized EA order parameter in the external AC field $q = [\Sigma_i \langle m_i(t)\rangle_c^2/N]_s$, where m_0, m_{1c} and m_{1s} are defined by $M(t) \equiv \Sigma_i m_i(t)/N \cong m_0 + m_{1c}\cos(\nu t) + m_{1s}\sin(\nu t) + \cdots$, and $\langle\cdots\rangle_c$ and $[\cdots]_s$ denote the averages over one cycle of the AC field and over many samples, respectively. In the numerical analysis the following parameters were used: 35 samples with N=200 and $\Delta t/\tau = 1/4$, Δt being the unit of the time discretization. The stationarity of a solution was checked by $|\mathrm{Re}\Delta\chi_1/\mathrm{Re}\chi_1| + |\mathrm{Im}\Delta\chi_1/\mathrm{Im}\chi_1| < 10^{-5}$, where $\Delta\chi_1$ is the difference in χ_1 by one cycle. The transition temperature T_c is determined as the temperature where q increases significantly from zero. The results are shown in Fig.1. There the solid line is the transition line in the external static field (the AT line in the NMF model) obtained by Bray et al [1]. The broken lines are the results obtained by the perturbational calculation up to the second order in h_{ac}. The numerical results are in agreement with the analytic calculation when h_{ac} is small. When

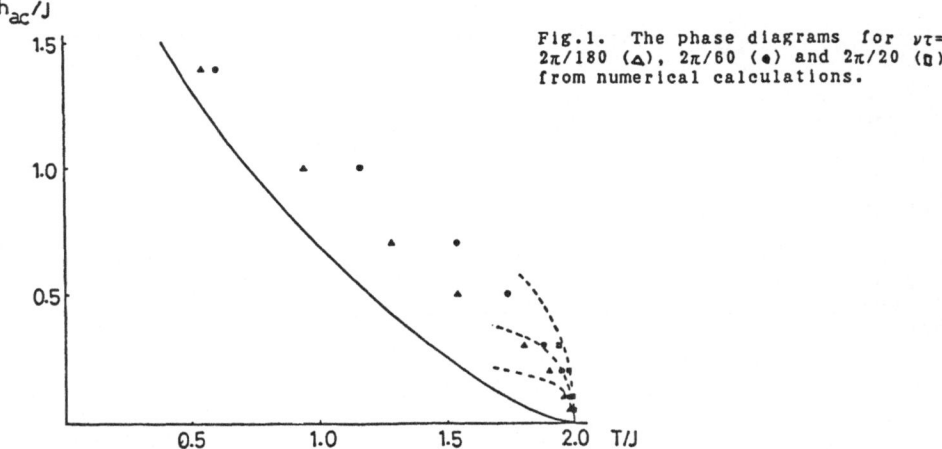

h_{ac}/J

Fig.1. The phase diagrams for $\nu\tau = 2\pi/180$ ($\triangle$), $2\pi/60$ ($\bullet$) and $2\pi/20$ ($\square$) from numerical calculations.

h$_{ac}$ becomes stronger, we see that the transition lines approach the AT line.

The transition lines obtained above characterize the freezing of spins. Similar transition lines are also obtained in the mean field model of a ferromagnet in an external AC field [2]. In a ferromagnet the uniform magnetization can follow the AC field, if the amplitude of the latter is large enough and/or its frequency is low enough. When the period of the AC field is short, the system is trapped in one of the two states symmetric under time reversal, resulting in non-vanishing averaged magnetization. The same argument holds for spin glasses, which explains the ν- and h$_{ac}$-dependent T$_c$ in Fig. 1.

For the spin glass of present interest, however, each state symmetric under time reversal consists of many states. Therefore we may expect another transition temperature T$_h$ where the system becomes trapped in one of these states and so irreversibility sets in even under the AC field. In order to confirm this expectation, we have performed the following numerical calculation: after cooling the system in the external AC field, a small (ideally infinitesimally small) static field is put on, and the static magnetization is measured in the heating process and then in the cooling process (we call them the ZFC and FC magnetizations, respectively). The results for h$_{ac}$/J=0.3, h$_{dc}$/J=0.1 and various $\nu\tau$ are shown in Fig. 2. The transition temperature T$_h$ is specified at which the ZFC magnetization starts to differ from the FC magnetization. We can see that T$_h$ is significantly lower than T$_c$; for example, T$_h$/J≈1.52 and T$_c$/J≈1.88 for $\nu\tau$=2π/60, and it becomes higher with increasing ν as expected.

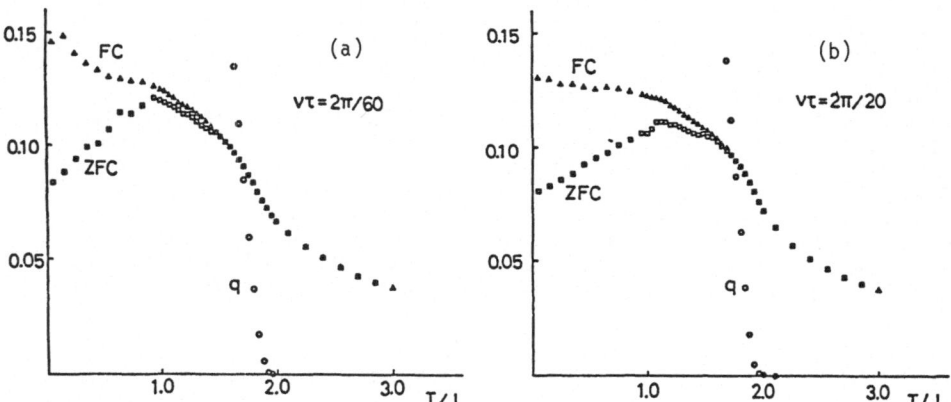

Fig.2. Behavior of m$_{FC}$, m$_{ZFC}$ and q against T/J for (a) $\nu\tau$=2π/60 and (b) $\nu\tau$=2π/20 when h$_{ac}$/J=0.3 and h$_{dc}$/J=0.1.

References

1. A. J. Bray, H. Sompolinsky and C. Yu: J. Phys. C19, 6389 (1986).
2. T. Shirakura, H. Kajitani and S. Inawashiro: J. Phys. C20, 6061 (1987).

Dynamics of Nonlinear Tenuous Structures

A. Jagannathan, R. Orbach, and O. Entin-Wohlman *

Department of Physics, University of California, Los Angeles, CA 90024, USA

I. Abstract

The vibrational dynamics of tenuous structures is developed using extended and localized states. A phonon-fracton model is used for explicit computations. It is
shown that the localized states can contribute to thermal transport via hopping
processes, assisted by absorption of extended state vibrations. This "phonon
assisted-fracton hopping" is the vibrational analog of Mott's variable range hopping contribution to electical conductivity. Comparison with thermal conductivity
experiments on amorphous materials allows an estimate of the magnitude of the anharmonicity coefficient. It is found to be (roughly) two orders of magnitude larger than the acoustic limit. This enhancement is attributed to the "openness" of
the geometry for fracton excitations, and could have profound effects on the vibrational and conformational excitations of tenuous structures.

II. Introduction

The dynamics of tenuous structures has been examined previously by us [1] using
fractal geometry as a basis for detailed analytic calculations. We have stressed
that the conclusions of these treatments are, in fact, broader, and can encompass
the dynamics of structures where a mobility edge exists at an energy ω_c, such that,
for the case of vibrational excitations, for excitation energies less than ω_c, the
states are extended; and for energies greater than ω_c, the states are localized.
Using this structure, we have calculated the the one- and two-vibration-induced
spin-lattice relaxation rate for localized [2] and extended [3] electronic states.
We have shown that the localized character of the states above ω_c has a profound
effect on the time and temperature dependence of the relaxation rate [2,3,4]. Using
the concept of localization above a mobility edge, we have also considered the
thermal conductivity, κ, of a tenuous material. We showed that these considerations
led naturally to a plateau in a plot of κ versus temperature, at temperatures $T_p \sim$
$\hbar\omega_c/k_B$. Because of vibrational localization for energies above ω_c, harmonic considerations alone cannot account for the observed increase in κ at temperatures above
the plateau. We introduced lattice anharmonicity [5] to generate a hopping mechanism for localized vibrational excitations, and found a contribution to κ which was
linear in T.

The purpose of this contribution is to compute quantitatively the anharmonic contribution to the scattering rates for extended and localized excitations of a tenuous structure. The associated inelastic lifetimes will be important for transport,
and for the very concept of vibrational eigenstates when the inelastic scattering
rate becomes comparable to the eigenfrequency.

The next Section will introduce the quantities with which we shall formulate and
compute the effects of anharmonicity. We shall specialize our treatment to fractal
geometry in order to obtain quantitative results, but we again emphasize that our
treatment applies to systems where vibrational localization is present in general.

---

* Permanent address: School of Physics and Astronomy, Tel Aviv University, Ramat
   Aviv, Tel Aviv, Israel

Springer Series in Synergetics, Vol. 43   **Cooperative Dynamics in Complex Physical Systems**
Editor: H. Takayama   © Springer-Verlag Berlin, Heidelberg 1989

Sec. IV presents our results for the phonon and fracton inelastic scattering rates as a function of vibrational frequency and temperature. We derive the phonon-assisted fracton-hopping contribution to κ in Sec. V. Sec. VI applies our results to experiment, and Sec. VII describes other effects associated with anharmonicity. We summarize our results and suggest directions for future research in Sec. VIII.

III. The Model

We use concepts which apply to the dynamics of fractal materials as a specific realization of tenuous structures. An example would be silica aerogels [6], though we have suggested previously that a much broader class of amorphous materials might obey similar dynamics [7].

In particular, we assume a characteristic length, ξ, exists such that, for length scales larger than ξ, the structure appears homogeneous (Euclidean) and the vibrational excitations are extended (phonons). For length scales less than ξ, the system is hypothesized to appear fractal either because of a fractal mass distribution and/or a fractal character to the force constant network, and the vibrational excitations are localized (fractons). We wish to emphasize that, as for a percolating network, the overall mass density need not be fractal, but that pertaining to the infinite network can be. This suggests that for amorphous structures with normal mass densities, it is the force constant network which can have fractal geometry, leading to fracton vibrational excitations for length scales less than ξ.

A fractal network is specified by the Hausdorff dimensionality D for the mass density [7], and the exponent θ for the range dependence of the force constant [1]. This leads to the definition of the fracton dimensionality $\bar{d} = 2D/(2+\theta)$. The vibrational density of states crosses over from

$$N_{ph}(\omega) = Nd\left(\omega^{d-1}/\omega_c{}^d\right) \quad , \quad \omega < \omega_c \quad , \tag{1}$$

where N is the number of cells of size ξ in the total volume V, $\omega_c \propto \xi^{-[1+(\theta/2)]}$, and d is the embedding or Euclidean dimension, to

$$N_{fr}(\omega) = N\bar{d}\left(\omega^{\bar{d}-1}/\omega_c{}^{\bar{d}}\right) \quad , \quad \omega > \omega_c \quad . \tag{2}$$

The fractons are collective excitations, with a dispersion law,

$$\ell_{fr}(\omega) = \xi\,(\omega/\omega_c)^{-\bar{d}/D} \quad , \quad \omega > \omega_c \quad , \tag{3}$$

relative the localization length of the fracton, $\ell_{fr}(\omega)$, to the fracton eigenfrequency. The "superlocalized" [8] form of the fracton wave function is

$$\phi_\omega(\vec{r}) \propto r^{-(d-D)/2} \exp\{-[r/\ell_{fr}(\omega)]^{d_\phi}\} \quad , \quad \omega > \omega_c \quad , \tag{4}$$

where d_ϕ is the superlocalization exponent (a matter of some controversy [8]). The third order vibrational anharmonic Hamiltonian is written as the usual form:

$$\mathcal{H}_{anh} = C_{eff} \int d^d r \left[\vec{\nabla}\cdot\vec{u}(\vec{r})\right]^3 \quad . \tag{5}$$

The theory of finite lattice strain shows that, in general, C_{eff}, contains terms which are of the same magnitude as the harmonic elastic constants. This then gives us a lower bound on C_{eff}. In fact, as noted above, for length scales less than ξ, the value of C_{eff} extracted from experiment will be at least two orders of magnitude larger than the harmonic elastic constant. We express the vibrational excursion $\vec{u}(\vec{r})$ of the atom at site $\vec{r}$ in terms of second quantized operators b_α and $b_\alpha{}^\dagger$, where α specifies the particular vibrational state with frequency ω_α,

$$\vec{u}(\vec{r}) = \sum_\alpha \left(\hbar/2\rho\omega_\alpha\right)^{1/2} \hat{e}_\alpha \left[\phi_\alpha(\vec{r})b_\alpha + \phi_\alpha{}^*(\vec{r})b_\alpha{}^\dagger\right] \quad , \tag{6}$$

where $\hat{e}_\alpha$ is the polarization vector for the state α and ρ is the overall structure mass density, $\rho = NN_{at}/V$ where $N_{at} = (\xi/a_0)^D$, the number of atoms within a cell of

184

size ξ, and a_0 is the fundamental lattice constant upon which the structure is built.

We now express (5) in terms of the matrix element for creation or destruction of vibrational states. One finds in second quantized form,

$$\mathcal{H}_{anh} = C_{eff} \sum_{\alpha,\alpha'',\alpha'} \left[A_{\alpha,\alpha'',\alpha'} \, b_{\alpha'}{}^{\dagger} b_{\alpha''} b_{\alpha} + h.c. \right] , \qquad (7)$$

where the b and $b^{\dagger}$ operators create and annihilate phonons or fractons, depending on whether the index α refers to modes with frequencies less than or greater than the crossover frequency ω_c.

IV. Calculation of the Phonon and Fracton Lifetimes

We shall calculate the consequences of two processes:

(a) phonon + phonon $\leftrightarrow$ fracton $\qquad\qquad\qquad\qquad\qquad\qquad$ (8a)

(b) phonon + fracton $\leftrightarrow$ fracton $\qquad\qquad\qquad\qquad\qquad\qquad$ (8b)

For phonon attenuation, the first requires the sum of the phonon frequencies be greater than ω_c, the latter that a fracton state be occupied. These requirements will generate different temperature dependences for the relaxation processes, depending upon whether $T \gtrless \hbar\omega_c/k_B$. For process (a), the lifetime of the phonon with frequency ω_{ph} is calculated to be [9]

$$1/\tau_{ph}{}^{(a)} = [2\pi C_{eff}{}^2 d\bar{\bar{d}}/8\rho^3 v_s{}^3\omega_c{}^3\xi^6] \, (\hbar^2\omega_{ph}{}^3/k_BT) \, \exp(-\beta\omega_c) \quad , \; T < \hbar\omega_c/k_B \; ; \quad (9a)$$

$$1/\tau_{ph}{}^{(a)} = [2\pi C_{eff}{}^2 d\bar{\bar{d}}/8\rho^3 v_s{}^3\omega_c{}^5\xi^6] \, \omega_{ph}{}^3 k_BT \qquad\qquad , \; T > \hbar\omega_c/k_B \; . \quad (9b)$$

The $\omega_{ph}{}^3$ frequency dependence has been observed by DIETSCHE and KINDER [10] for inelastic phonon scattering in vitreous silica at low temperatures.

The (b) process allows for phonon-assisted fracton-hopping. We use an argument equivalent to Mott for variable range hopping for electronic states to find the most probable fracton hop distance. We require that there be a probability unity to find the final-state fracton at a distance $R(\omega_{fr})$ away from the position of the initial fracton with energy ω_{fr}. This distance is given by the relation

$$N_{fr}(\omega_{fr})\omega_c[R(\omega_{fr})]^D \sim 1 \quad . \qquad (10)$$

Using the dispersion relation for the fracton characteristic length $\ell_{fr}(\omega)$, one find that $R(\omega_{fr})/\ell_{fr}(\omega_{fr}) \sim (\omega_{fr}/\omega_c)^{1/D} > 1$ so that the fracton hops a distance greater than its localization length. Using (10), the lifetime of a phonon of frequency ω_{ph} due to fracton hopping is

$$1/\tau_{ph}{}^{(b)} = [\pi C_{eff}{}^2 \bar{\bar{d}}^2 d_\phi{}^3\Gamma(d/d_\phi)/2\rho^3 v_s{}^2\xi^7\omega_c{}^3] \, \omega_{ph}{}^2 \, (\hbar^2/k_BT) I(\beta) \; , \; T < \hbar\omega_c/k_B \; ; \quad (11a)$$

$$1/\tau_{ph}{}^{(b)} = [\pi C_{eff}{}^2 \bar{\bar{d}}^2 d_\phi{}^3\Gamma(d/d_\phi)/2a\rho^3 v_s{}^2\xi^7\omega_c{}^5] \, \omega_{ph}{}^2 \, k_BT \qquad , \; T > \hbar\omega_c/k_B \; . \quad (11b)$$

Here, $\Gamma(x)$ is the gamma function, $a = 5-2\bar{\bar{d}}-(\bar{d}/D)$, and $I(\beta)$ is $(\beta = 1/k_BT)$,

$$I(\beta) = \int_1^\infty dx \; x^{2\bar{\bar{d}}+(\bar{d}/D)-4} \, \exp(-\beta\omega_c x) \quad . \qquad (12)$$

Using (10), the hopping rate for a fracton of frequency ω_{fr} is

$$1/\tau_{fr}{}^{(b)} = [\pi C_{eff}{}^2\bar{\bar{d}}d_\phi{}^4/2\rho^3 v_s{}^2\xi^7] \, [k_BT/\omega_{fr}{}^3] \, (\omega_{fr}/\omega_c)^{\bar{\bar{d}}+4(\bar{d}/D)} \qquad (13)$$

$$\times \exp\left[-(\omega_{fr}/\omega_c)^{d_\phi/D}\right] \quad , \quad T > \hbar\omega_c/k_B \quad .$$

This expression gives the fracton hopping rate at temperatures above the plateau temperature.

V. Calculation of the Fracton-Hopping Contribution to κ

The kinetic formula for the thermal conductivity κ gives

$$\kappa_{hop}(T) = [k_B/3V] \sum_\alpha R^2(\omega_\alpha)/\tau_{fr}{}^{(b)} \quad . \tag{14}$$

The fracton hopping contribution to $\kappa_{hop}(T)$ is then found by inserting (13) into (14):

$$\kappa_{hop}(T) = [\pi C_{eff}{}^2 \bar{\bar{d}}^2 d_\phi{}^4 I(\beta)/6\rho^3 v_s{}^2 \xi^8 \omega_c{}^3] k_B{}^2 T \quad , \qquad T > \hbar\omega_c/k_B \quad . \tag{15}$$

This is a fully quantitative version of the result derived first by ALEXANDER et al.[5]. It is expressed in terms of microscopic quantities, and predicts a linear temperature dependence for κ above the plateau temperature. It is an additive channel, and so must be added to the contribution of the extended state phonons to κ. The integral in (14) is such that the principal contribution to $\kappa_{hop}(T)$ arises from fractons near ω_c. This means that $R(\omega_{fr}) \sim \xi$, and that the higher energy fractons do not appreciably contribute to κ. It is for this reason that we believe that (15) should apply universally to tenuous structures, with no significant dependence upon the density of vibrational states for energies above ω_c. Experiments [11] appear to be consistent with this prediction. It is instructive to note that the coefficient in (15) looks very much like that appearing in (11) for $1/\tau_{ph}{}^{(b)}$. We make use of this similarity to write (15) at temperature $T > \hbar\omega_c/k_B$ in terms of the phonon mean free path $\Lambda_{ph}{}^{(b)}(T') = v_s\tau_{ph}{}^{(b)}$ measured at temperature $T' < \hbar\omega_c/k_B$. In doing so, we can make use of the estimates of $\Lambda_{ph}(\omega_{ph} = 1.6k_BT/\hbar)$ derived from thermal conductivity experiments using the dominant phonon approximation [11]. We find

$$\kappa_{hop}(T) = [ad_\phi I(\beta)/3\Gamma(d/d_\phi)](k_BT/T')(\omega_c{}^2 v_s/\xi)[1/\Lambda_{ph}{}^{(b)}(T')] \quad , \tag{16}$$

with $T' < \hbar\omega_c/k_B < T$.

VI. Comparison with Experiment

The beauty of (16) is that there are few undetermined parameters. Thus, ROSENBERG et al.[11] find that κ exhibits a linear temperature dependence at temperatures greater than the plateau temperature for epoxy resin with hardener amounts equal to, and twice stoichiometry. Using measured sound velocities, and taking the crossover frequency from Raman scattering data, he can compare experiment with theory using (16). He finds

	$\hbar\omega_c/k_B$(K)	ξ(cm)	$\kappa_{measured}$(Wm^{-1}K^{-2})	$\kappa_{Eq.(16)}$(Wm^{-1}K^{-2})
EDA1X:	27.9	3.15×10^{-7}	1.21×10^{-3}	1.7×10^{-3}
EDA2X	36.0	2.52×10^{-7}	1.30×10^{-3}	2.0×10^{-3}

Given the nature of this class of calculation, the agreement is striking: in the right direction and nearly exact. Working backwards to obtain C_{eff} from (11), one finds a value $\sim 10^{14}$ erg/cm^3, or about two orders of magnitude larger than in a regular solid with the same value of sound velocity. This result may be profound in that it suggests that C_{eff} may be much larger at length scales less than ξ than one would find in conventional non-linear acoustic experiments. This means that $\mathcal{H}_{anh}$ may have much more importance in tenuous structures than heretofore appreciated. The reasons are probably associated with the more open character of the structure at short length scales, allowing much larger scale motions than one would find in the acoustic regime where an equivalent of "steric hindrance" limits atomic motion. Finally, it should be noted that anharmonicity is "standing on its head" in tenuous structures. Whereas it acts to limit thermal transport in ordered structures, it is the origin of vibrational heat conduction in tenuous structures. Without anharmonicity, the thermal conductivity would never rise above its value at the plateau temperature.

VII. Other Effects of Anharmonicity

We have followed the golden rule in deriving our results above with no attention
paid to the validity of its use. We explore the consequences of anharmonicity in
this Section, recognizing that our treatment is only limited to third order, and
that a full diagrammatic analysis will eventually be required.

The first condition we must examine is what we shall refer to as the Simons condition [12]. He recognized that anharmonic processes involving low energy phonons
of frequency ω scattering off thermally excited phonons with energy k_BT was altered
when $\omega\tau_{k_BT} < 1$. This caused the attenuation α_{ph} of microwave frequency phonons to
crossover from the hydrodynamic regime with $\alpha_{ph} \propto T^4$ to the diffusive regime with
$\alpha_{ph} \propto$ constant. A similar effect occurs for $\kappa_{hop}(T)$. Forming the product $\omega_{ph}\tau_{fr}$
from (13) with ω_{fr} taken to be $\hbar\omega_c/k_B$, one finds it approaches unity at temperatures
well above the plateau region for vitreous silica, but close to the plateau region
for the aerogels. Its effect is to reduce the effectiveness of phonon-assisted
fracton-hopping. Calling $\tau_{ph\infty}$ that value of (11) for an infinite fracton lifetime,
following the method of SIMONS [12] (replacing the delta function in the golden rule
by a Dirichlet integral with complex fracton frequency $\omega_{fr} = \omega_{fr} + i/\tau_{fr}$), one
obtains an expression for the phonon lifetime in the presence of finite fracton
lifetime:

$$1/\tau_{ph}^{(b)} = 1/\tau_{ph\infty}^{(b)}[1 - \text{constant} \times (1/\omega_c\tau_{fr})] \quad . \tag{17}$$

This result in combination with (13) shows that the phonon relaxation rate is diminished with increasing temperature, or that $\Lambda_{ph}^{(b)}(T)$ increases with increasing
temperature when $\omega_c\tau_{fr}$ becomes comparable to unity. From (16), this means that
$\kappa_{hop}(T)$ will diminish from the value extrapolated from the linear temperature dependence predicted for $\kappa_{hop}(T)$ by (15). This "curling over" of κ at high temperatures is found experimentally [11]. We attribute it to anharmonicity rather than
zone boundary effects as suggested by CAHILL AND POHL [11].

The curling over of κ has been recently noted by POHL [13] to lead to a value of
κ at melting which is not far from the value obtained coming from the liquid state.
In the liquid, particle collisions are certainly the mechanism for thermal transport. As the fracton frequency increases, the fracton localization length diminishes by virtue of the fracton dispersion law. Phonon assisted fracton hopping can
therefore be thought of as particle motion, the more so the shorter the localization length of the fracton. The similarity in magnitude between κ in the solid near
melting and the value in the liquid might be due to the increasingly particle-like
nature of the fracton as ω_{fr} increases. Anharmonicity will certainly be important
in the formulation of thermal transport near the melting point, so that a full
treatment of its effects would be necessary to establish a meaningful analysis.

It is also intriguing to evaluate $\omega_{fr}\tau_{fr}$ from (13) for process (b) (a fracton
combines with a phonon to produce a fracton) and for process (a) (a fracton splits
into two phonons). One finds that this quantity is well above unity for vitreous
silica, but dangerously close to unity for the aerogels. This suggests that the
more open the structure, the stronger the inelastic scattering, and the greater
danger that anharmonicity may destroy the fracton character of the vibrational
eigenstate. Under such a condition, the golden rule breaks down and one must employ
a fully self-consistent diagrammatic approach.

VIII. Summary

We have shown how anharmonicity can lead to thermal transport for localized vibrational excitations. We have calculated the temperature dependence and magnitude of
the associated thermal conductivity, κ. By using values for phonon mean-free-paths
extracted from thermal conductivity experiments at temperatures below the plateau
regime, one can obtain a quantitative estimate of the hopping contribution to thermal transport at temperatures above the plateau regime. Comparison with experiment
is quantitative. Using the experimental values for κ above the plateau, and work-

ing backwards to obtain the anharmonic coupling constant, one finds that it exceeds the acoustic elastic constant by two orders of magnitude. This leads to a suggestion than anharmonicity in tenuous structures at short length scales is much larger than in crystalline materials.

Analysis of inelastic lifetimes for vibrational excitations (via the temperature dependence of their measured width) in the short length scale regime by optical means [6] would serve as a direct test of our approach. It is an exciting prospect that direct vibrational lifetime measurements can lead to an explicit estimate of thermal transport. We also suggest that phonon propagation experiments similar to DIETSCHE and KINDER [10] in, for example, the aerogels, would also provide relationships between the degree of openness of the structure (larger ξ) and the vibrational anharmonicity at short length scales.

On the theoretical side, third order perturbation has clearly been pushed to its limits. A full scale diagrammatic analysis is in order, with phonon and fracton propagators. The high temperature behavior of vibrational excitations may have similarities to excitations within the liquid state, but anharmonicity must be fully explored to formulate a satisfactory picture.

This work has been supported by the National Science Foundation, grant DMR 84-12898. We are indebted to Dr. Jeremy Grace for bringing [10] to our attention.

REFERENCES

1. S. Alexander and R. Orbach: J. Phys. (Paris) Lett. 43, L-625 (1982)
2. S. Alexander, O. Entin-Wohlman, and R. Orbach: J. Phys. (Paris) Lett. 46, L-549 (1985); ibid L-555 (1985); ibid Phys. Rev. B32, 6447 (1985); ibid Phys. Rev. B33, 3935 (1986)
3. O. Entin-Wohlman, S. Alexander, and R. Orbach: Phys. Rev. B32, 8007 (1985)
4. S. Alexander, O. Entin-Wohlman, and R. Orbach: Phys. Rev. B35, 1166 (1987)
5. S. Alexander, O. Entin-Wohlman, and R. Orbach: Phys. Rev. B34, 2726 (1986)
6. R. Vacher, T. Woignier, J. Pelous, and E. Courtens: Phys. Rev. B37, 6500 (1988); E. Courtens, J. Pelous, J. Phalippou, R. Vacher, and T. Woignier: Phys. Rev. Lett. 58, 128 (1987); E. Courtens, R. Vacher, J. Pelous, and T. Woignier: Europhys. Lett. 6, 245 (1988); Y. Tsujimi, E. Courtens, J. Pelous, and R. Vacher: Phys. Rev. Lett. 60, 2757 (1988)
7. S. Alexander, C. Laermans, R. Orbach, and H.M. Rosenberg: Phys. Rev. B28, 4615 (1983)
8. Y.E. Levy and B. Souillard: Europhys. Lett. 4, 233 (1987); A.B. Harris and A. Aharony, Europhys. Lett. 4, 1355 (1987); K. Yakubo and T. Nakayama: submitted for publication
9. A. Jagannathan, R. Orbach, and O. Entin-Wohlman: submitted for publication
10. W. Dietsche and H. Kinder: Phys. Rev. Lett. 43, 1413 (1979)
11. R.C. Zeller and R.O. Pohl: Phys. Rev. B4, 2029 (1971); D.G. Cahill and R.O. Pohl: Phys. Rev. B35, 4067 (1987); S. Kelham and H.M. Rosenberg: J. Phys. C14, 1737 (1981); R. Orbach and H.M. Rosenberg: In Proceedings of the 17th International Conference on Low Temperature Physics, ed. by U. Eckern, A. Schmid, W. Weber, and H. Wuhl (Elsevier, Amsterdam, 1984) p.375; J.E. de Oliveira, J. Page, and H.M. Rosenberg: submitted for publication.
12. S. Simons: Proc. Phys. Soc. 83, 749 (1964)
13. R.O. Pohl: private communication

Random Field Phenomena at Magnetic Surfaces and Interfaces

A.P. Malozemoff

IBM T.J. Watson Research Center, Yorktown Heights, NY 10598, USA

Abstract

A new class of problems in disordered magnetism concerns interfacial or surface disorder coupled to structurally uniform adjacent layers. One example may be exchange anisotropy in ferromagnetic-antiferromagnetic sandwiches whose hallmark is a field-offset in the hysteresis loop. The phenomenon can arise from interfacial random fields causing the antiferromagnet to break up into Imry-Ma domains. A possibly related phenomenon is that of magnetically dead layers in systems like Fe on GaAs. In this case, spin-glass disorder and a domain structure it induces may create the appearance of a dead layer.

INTRODUCTION

There has been great interest over the years in low-dimensional magnetic systems, and in structurally disordered ones. But the recent experimental progress in studying magnetic surfaces and interfaces is beginning to focus attention on a new class of magnetic phenomena involving structural disorder at a surface or interface which couples magnetically to structurally uniform material on either side. In other words, this is the case where two-dimensional disorder competes with three-dimensional order.

Far from being an abstruse corner of statistical mechanics, this class of problems is potentially relevant to a great variety of experiments, especially since structural disorder at surfaces and interfaces is very much the rule rather than an exception.

This paper considers a subset of these problems, in which the structural disorder leads to random <u>exchange</u> interactions at the surface or interface. One example involves ferromagnetic- antiferromagnetic sandwiches which are known experimentally to exhibit a peculiar effect called exchange anisotropy.[1,2] A brief review of this effect and how random interface exchange helps explain it[3,4] is given below. Another possible example involves ferromagnetic layers on substrates where interdiffusion creates competing ferromagnetic and antiferromagnetic interactions. These new and speculative ideas are compared to recent experiments[5,8] on bcc iron grown on GaAs. This is the problem of ferromagnetic "dead layers".

EXCHANGE ANISOTROPY

While there are a number of different manifestations of exchange anisotropy in ferromagnetic-antiferromagnetic sandwiches, perhaps the most characteristic is a ferromagnetic hysteresis loop whose center is shifted along the applied field axis by an offset or "exchange" field H_E. Experiments with nonmagnetic spacers indicate that the effect arises from interfacial exchange coupling with the antiferromagnet.[1,2]

The time-reversal-symmetry-breaking magnetic field must be injected by some kind of preliminary "writing" process. For example, the sandwich can be cooled through the antiferromagnetic Neel temperature in the presence of an applied field which single-domains the ferromagnetic layer. Alternatively, the antiferromagnetic layer can be grown directly on the single-domained ferromagnetic layer, at a temperature below both the Neel and Curie temperatures.[9]

Given the likelihood of roughness on an atomistic scale at the interface, a random polarity of coupling between the ferromagnet and antiferromagnet is to be expected since every interfacial atomic step up or down typically reverses the sign of the antiferromagnetic sublattice which couples to the ferromagnet. This can be modeled as a random nearest neighbor exchange $\pm J$ along a hypothetical perfect interface.[3,4] Thus a single-domained ferromagnet with spin S_F exerts, in effect, a random field $\propto \pm J S_F$ on the top atomic layer in the antiferromagnet with spin S_A.

As in the classic Imry-Ma random field theory,[10] the effect of this random field will be to induce domain formation in the antiferromagnet.[3] In a domain encompassing N interface atoms, the average exchange per site will thus be reduced statistically by a factor $1/\sqrt{N}$. The antiferromagnet can reduce its energy by following this net exchange field from region to region. But the domain walls which must inevitably separate the domains have a positive energy which resists domain formation. Since these domain walls will stand perpendicular to the interface, their energy will grow relative to the interfacial stabilization as the anitferromagnetic layer thickness is increased. Thus it is reasonable to expect critical thickness above which the uniform antiferromagnetic state will become stable.[3,4] Nevertheless, kinetic barriers may prevent attaining the uniform state if the domain state is established first, either during a low temperature growth of the antiferromagnetic layer on the ferromagnet or because of some reduction in the Neel temperature away from the interface.

The size of the domains has been deduced from a simple extension[3,11] of the Imry-Ma domain theory to treat a Heisenberg antiferromagnet with exchange stiffness A and a uniaxial anisotropy K. Below a critical thickness t_{crit} of order $\sqrt{A/K}$, this domain size is about $\pi\sqrt{A/K}$, and it decreases yet further below $t_{crit}/2$.

One more concept is required to account for exchange anisotropy in this context. Once the antiferromagnetic domains are formed, the domain walls must be frozen in place by coercivity arising from defects in the antiferromagnet.[3,4,12] Then when the applied field polarity is reversed during measurement of a hysteresis loop, the interfacial couplings, initially optimized by the domain formation process, now become unfavorable. That is, the initially negative interfacial energy $-JS_FS_A/a^2\sqrt{N}$ changes sign (here a is the lattice parameter of an assumed simple cubic grid.) Thus one can speak of an effective interfacial energy difference $\Delta\sigma = 2JS_AS_F/\pi a\sqrt{A/K}$. Now the offset field is just $\Delta\sigma/2M_Ft_F$, from simple force balance considerations (here M_F and t_F are the ferromagnetic magnetization and layer thickness respectively). Combining these expressions one derives an offset field which agrees with the experiment for reasonable (though imperfectly known) parameters.[3,4]

Recent experiments have revealed a distribution of activation energies which characterize thermal hysteresis of the exchange anisotropy.[13] In the above context, these activation energies might be interpreted as the potential barriers which pin the antiferromagnetic domain walls.

Let us next apply these concepts to a different and perhaps more general problem.

FERROMAGNETIC DEAD LAYERS

A number of recent experiments have revealed rather substantial layers of reduced net magnetization near ferromagnetic surfaces or interfaces. An example is single crystal $\alpha - Fe$ grown by molecular beam epitaxy on GaAs [100] and [110] faces.[5,6] While As has been found to diffuse into the iron to a depth of up to 10 Å, the "dead" layer extends 50 to 100 Å, well into the pure iron.[7,8]

A possible key to understanding this phenomenon is to recognize that the dead layer could consist of domains randomly oriented so that their net magnetization is zero. These domains could have an origin analogous to that of the antiferromagnetic domains of the exchange anisotropy problem, as can be seen in the following way.

It is known that various iron-arsenic compounds are antiferromagnetic with Neel temperatures well above room temperature. In the FeAs interdiffusion layer, alloys disorder could thus lead to competing ferromagnetic and antiferromagnetic interactions, hence to spin glass behavior.

If the pure iron layer now grows at a temperature below the spin glass ordering temperature, it will experience random fields from the already formed spin-glass. As in the Imry-Ma theory,[10] it should then break up into domains. Since the domain moment orientations are random, they will in general entail dipolar energies of order $2\pi M^2$, which were absent in the antiferromagnetic case considered earlier. So in this case one can, as a first approximation, expect a domain scale $\pi\sqrt{A/2\pi M^2}$ and a corresponding critical thickness. For bcc iron, this works out to 50 Å, just about the observed dead layer thickness.[8]

This suggestive result calls for a more convincing experimental test. One consequence of this model is that at larger layer thicknesses, the domain state is metastable, and so it should be possible to recover the full magnetization by an appropriate annealing or field treatment. However, because of possible topological defects,[4] the required fields may be quite high. Magnetic annealing temperatures may be limited by increasing interdiffusion. Iron deposition in a large magnetic field is another possible way to limit the domain formation.

Perhaps the greatest problem for this model is that the optimal growth temperatures for Fe on GaAs are somewhat above 100°C. Since spin glass ordering usually occurs at temperatures much below those of the component ferromagnetic and antiferromagnetic ordering, the experimental growth temperature may be too high to permit spin glass ordering during growth.

An alternative model based on random anisotropy has been proposed earlier.[7,8] While also predicting a disordered magnetic state, the model neglects magnetostatic effects which will reduce the size of the canting. In general exchange fields are larger than anisotropy fields, and the time-reversal-symmetry-breaking character of the random fields is much more effective in nucleating local domain structure. Nevertheless, a more definitive test to distinguish these models is still needed.

In summary, random exchange effects at interfaces lead to a variety of novel phenomena which will surely receive increasing attention as experimental work on magnetic surfaces and interfaces accelerates in coming years.

The author thanks V. Speriosu, J. Cullen, K. Hathaway and G. Prinz for valuable discussions.

REFERENCES

1. A. Yelon, in Physics of Thin Films, ed. M. Francombe and R. Hoffman, (Academic Press, New York 1971), Vol. 6, P. 205.
2. N.H. March, P. Lambin and F. Herman, J. Magn. Mater. 44, 1 (1984).
3. A.P. Malozemoff, Phys. Rev. B 37, 7673 (1988).
4. A.P. Malozemoff, J. Appl. Phys., 63, 3874 (1988)
5. G.A. Prinz and J.J. Krebs, Appl. Phys. Lett., 39, 397 (1981).
6. G. A. Prinz, Phys. Rev. Lett., 54, 1051 (1986).
7. J.R. Cullen, K.B. Hathaway and J.M.D. Coey, J. Appl. Phys., 63 (1988).
8. J.R. Cullen and K. B. Hathaway, J. de Physique, Proceedings of the International Conference on Magnetism, Paris, July 25-29, 1988, to be pubished.
9. C. Tsang, N. Heiman and K. Lee, J. Appl. Phys., 52, 2471 (1981); T. Tsang and K. Lee, J. Appl. Phys. 53, 2605 (1982).
10. Y. Imry and S.K. Ma, Phys. Rev. Lett., 35, 1399 (1975).
11. Y.Y. Goldschmidt and A. Aharony, Phys. Rev. B, 32, 264 (1985).
12. L. Neel, Annales de Physique 2, 61 (1975).
13. V. Speriosu and S.S.P. Parkin, J. de Physique, Proceedings of the Intrenational Conference on Magnetism, Paris, July 25-29, 1988, to be published.

Spin Glasses and High-T_c Superconductivity

I. Morgenstern, K.A. Müller, and J.G. Bednorz

IBM Research Division, Zurich Research Laboratory, CH-8803 Rüschlikon, Switzerland

This publication deals with the current controversy concerning flux creep or glass behavior in high-T_c superconducting single crystals. It is shown that the flux creep picture is merely a phenomenological approach to the glass behavior for relatively short times and low temperatures. Glassy effects are predicted for temperatures between 70% and 95% T_c and magnetic field in the range of 300 G to 2000 G. The glass concept can be understood as a generalization of the traditional flux creep picture.

At the beginning we want to clarify that in the case of the macroscopic glass picture, we consider glass behavior inside a single grain or single crystal, and not in a ceramic. Figure 1 shows the situation. On level 1 we have a disordered array of grains, a ceramic. Inside a grain we see the macroscopic glass consisting of weakly coupled two-dimensional (2-D) planes. Each plane (level 3) consists of an array of domains, probably due to the heavy twinning of the system. The resulting weak links are essential for the glass behavior. Level 4 inside the domain describes the microscopic mechanism. It should be noted that the macroscopic glass behavior is not the origin of high-T_c superconductivity and therefore we do not consider level 4.

The first experiment stating glassy behavior in high-T_c materials was carried out by MÜLLER et al. [1]. In the spirit of spin glass experiments [2, 3], they considered zero-field cooling (z.f.c.) versus field-cooling behavior for a ceramic La-Ba-Cu-O system. Measuring the susceptibility (Fig. 2), they found reversible behavior after field cooling (f.c.). In sharp contrast, metastability showed up in the irreversible behavior after z.f.c. The two curves met at a temperature further on denoted as $T_c^*(H)$, above which only reversible behavior was detected.

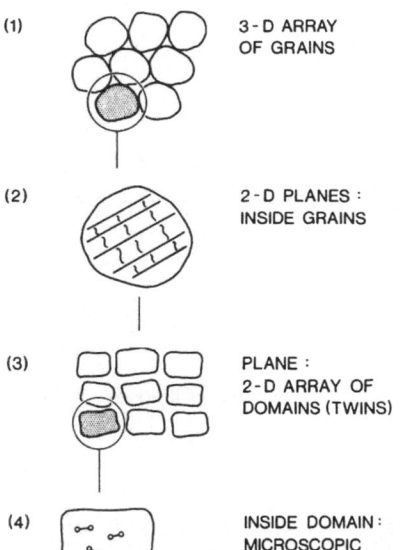

(1) 3-D ARRAY
OF GRAINS

(2) 2-D PLANES :
INSIDE GRAINS

(3) PLANE :
2-D ARRAY OF
DOMAINS (TWINS)

(4) INSIDE DOMAIN :
MICROSCOPIC
MECHANISM

Fig. 1. Schematic display of the physical situation in a ceramic (single crystal-level 2). From I. Morgenstern et al. *Physica C* **153-155**, 59-62 (1988). © Elsevier Science Publishers, B. V.

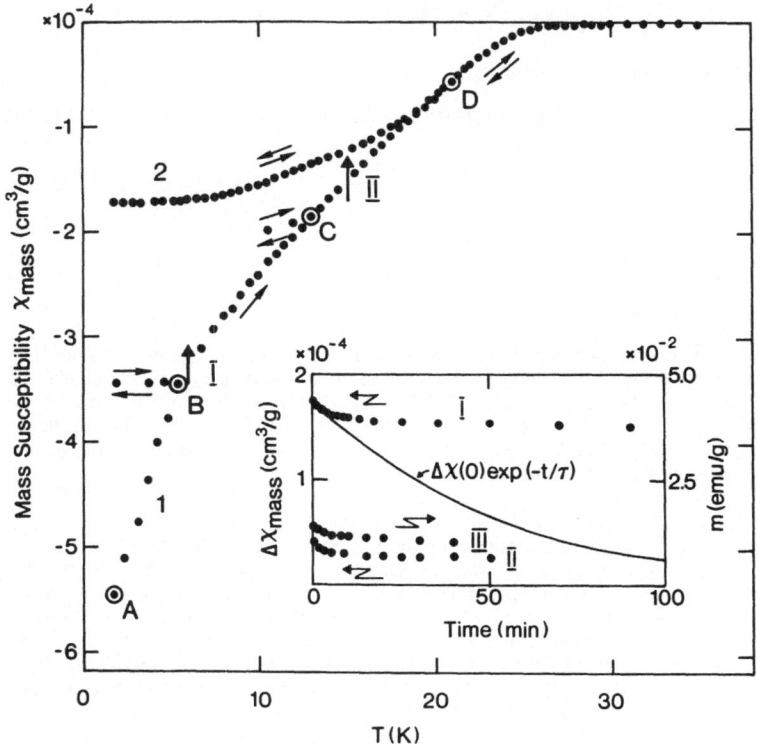

Fig. 2. Susceptibility versus temperature. Experimental situation for La-Ba-Cu-O. Field-cooling and zero-field cooling measurements. Insert: non-exponential time decay. From [1]. © The American Physical Society 1987

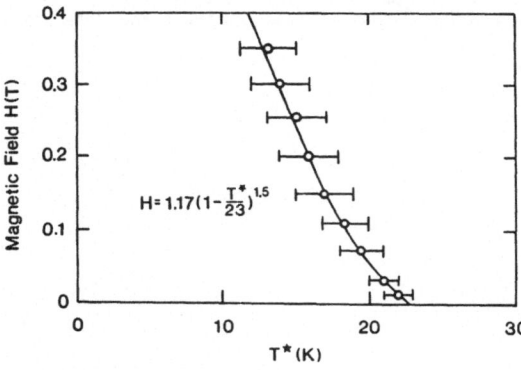

Fig. 3. "Quasi" de Almeida-Thouless line experiment. From [1]. © The American Physical Society 1987

Thus, $T_c^*(H)$ (H − magnetic field) is the temperature below which metastable behavior occurs just as in the corresponding spin-glass experiment.

Measuring in different magnetic fields the analogy to spin glasses was clarified even more. A "quasi" de AlMEIDA-THOULESS line [4] was found, i.e. we have

$$H^{2/3} \sim T_c^*(0) - T_c^*(H) \tag{1}$$

just as in spin glasses (Fig. 3). The addition "quasi" had to be made, as further theoretical work [5] showed that the underlying mechanism for the $H^{2/3}$ behavior is different from that in spin glasses.

Furthermore, experiments [1] also showed nonexponential decay at magnetization or susceptibility (insert of Fig. 1) also reminiscent of spin glasses. Therefore, a natural theoretical approach to the problem had to consider various (spin) glass models. Carrying out numerical simulations from an XY-spin-glass model based on earlier work of EBNER et al. [6], we were able to repeat the experimental findings. Figure 4 shows the f.c. and z.f.c. susceptibilities for various magnetic fields H. The resulting error bars denote $T_c^*(H)$. Figure 5 shows nothing surprising: plotting $H^{2/3}$ as a function of $T_c^*(0) - T_c^*(H)$ results in a straight line. Furthermore, the model was surprisingly successful in describing the features of high-T_c glass experiments [7]. We refer the reader to Ref. [5] for details.

The theoretical model of the XY-spin glass of EBNER et al. [6] can be understood as a disordered array of Josephson junctions. The following Hamiltonian describes the system:

$$-\beta \mathscr{H} = J \sum_{<ij>} \cos(\phi_i - \phi_j - A_{ij}) . \qquad (2)$$

The ϕ_i's (the XY-spins) describe the regions or domains of coherent phases in the system, originally the (physical) grains of the ceramic or the granular superconductor, as considered in [6]. But soon it became clear that our La-Ba-Cu-O ceramic behaved differently — the domains of the glass model had to be located *inside* the grains. DEUTSCHER et al. [8] solved the resulting puzzle. The extremely short coherence length in the system and the existence of twin boundaries lead to Josephson junctions or more generally weak links inside the grains. The resulting disorder is described by the phase factor A_{ij} in the model which is given by the line integral

$$A_{ij} \sim \int_i^j \vec{A}\, d\ell \qquad (3)$$

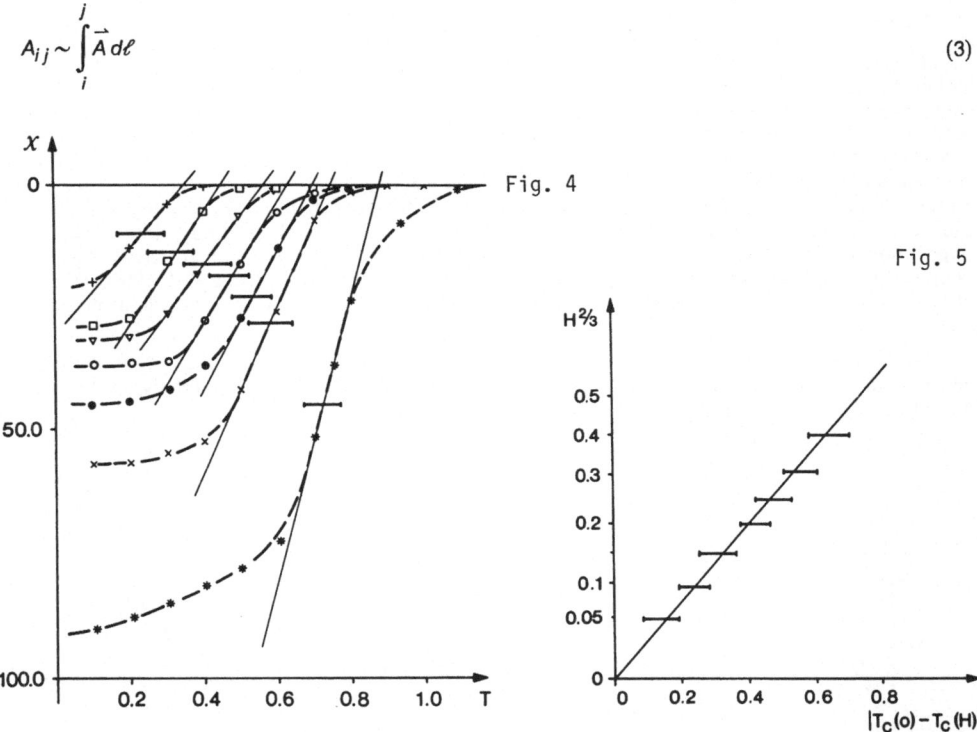

Fig. 4. Susceptibility versus temperature; numerical simulation; various magnetic fields from $H = 0.05$ (lowest curve) to $H = 0.40$ in steps of $\Delta H = 0.005$ in units of $2\pi/\phi_0$. ϕ_0 is the elementary flux quantum. From [5]. © Springer-Verlag 1987

Fig. 5. "Quasi" de Almeida-Thouless line numerical simulation. From [5]. © Springer-Verlag 1987

from site i to site j over the vector potential $\vec{A}$. The sites i and j reflect the positional disorder in the system.

The main feature of Hamiltonian (2) is frustration. Figure 6 just shows the effect well known to spin glass physicists [2]. Spin 4 just does not know whether to follow spin 1 according to phase factor A_{41} or spin 3 due to A_{34}. It is frustrated. The origin of frustration is disorder leading to "randomized" A_{ij}. The interaction in the system via the weak links leads to cooperative effects — the glass behavior. Glass behavior is best described by considering the energy landscape of the system (Fig. 7). Again, it is well known that we have a hierarchy of barriers, as shown in the 2-D cut through the multidimensional phase space (denoted by coordinate P). Most of the behavior results from the fact that the system has to hop over all these little but also those larger barriers, i.e. the hierarchy of barriers in the system.

But soon experimentalists such as MOTA [9] and MALOZEMOFF [10] realized that their results could also be explained in the framework of the traditional flux creep picture [11], just considering a single or only a few barriers originating from the pinning forces acting on the flux lines. It is important to note that these experiments were carried out at low temperatures (in Mota's case about Helium temperature) and only "relatively" short measurements were taken, especially leading to a logarithmic decay of the magnetization with time, as expected for the flux creep picture.

But now back to Fig. 7, our hierarchy of barriers. Just considering the slope of a hill, we know from spin glass research that this slope consists of a "rough surface" [12], shown in Fig. 8. Then it becomes immediately clear that the flux creep theory, based on an identical physical picture on phenomenological grounds, only describes the decay of the system one single slope down. Considering the average size of a barrier U, MALOZEMOFF's experiments [10] show that it is a) relatively too low for the traditional flux creep picture but b) also relatively too high for the glass picture, again in the traditional sense. MALOZEMOFF's reasoning leads to the introduction of the Giant Flux Creep picture, just allowing more flux to creep over the relatively small barriers [10]. But is this picture sufficient to describe the behavior of the system also at higher temperatures, when the hopping probabilities are higher? Or from the glass point of view: is the system even able to experience our nice hierarchy of barriers? The answer is certainly yes! Giant flux creep or the influence of only a few barriers is seen at temperatures below about 70% T_c and relatively short times which can be up to several days or weeks at Helium temperatures. Here the phenomenological approach to the problem, the flux creep picture, is certainly valid, but we always have to keep in mind that we are dealing with a glass, but with such large relaxation times, that they do not show up in ordinary low temperature experiments. But above about 70% T_c glass behavior can be detected on relatively large time scales. The appropriate magnetic fields are roughly in the range of 300 G to 2000 G. A too close approach to T_c destroys the glassy behavior again, as we experience critical effects (Fig. 9).

The general message of this paper is the following: in the regime of about 70%-95% T_c we will find glass behavior and a new rich physics in the field of superconductivity. What do we have to expect and what is already known?

Fig. 6

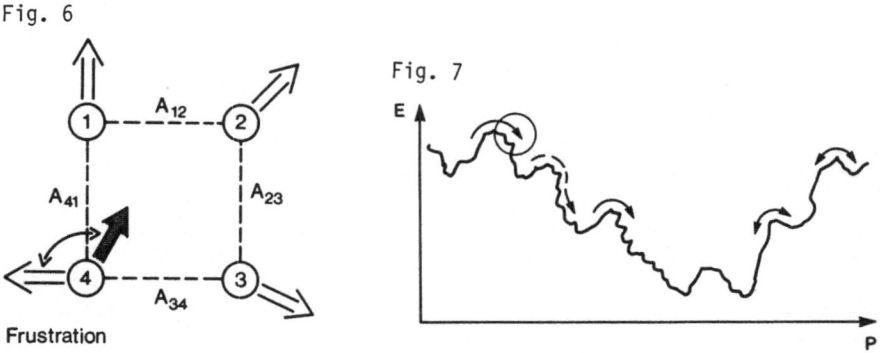

Frustration

Fig. 6. Frustration in theoretical model (see text). From [5] © Springer-Verlag 1987

Fig. 7. Energy landscape of glassy system. Hierarchy of barriers

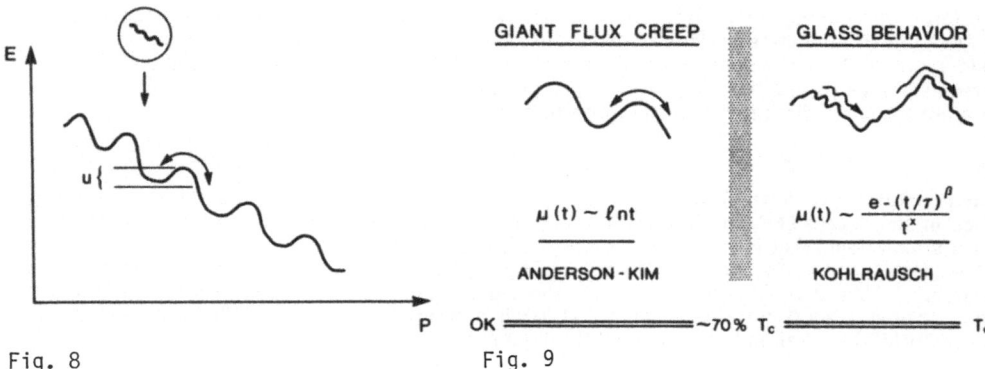

Fig. 8

Fig. 9

Fig. 8. Enlarged slope of energy-hill. Corresponds to giant flux creep picture

Fig. 9. Schematic view of experimental situation

1. Kohlrausch decay.

Decay of the magnetization $M(t)$ should decay as

$$M(t) \sim \frac{e^{-(t/\tau)^{\beta}}}{t^{x}} \tag{4}$$

with $0 < \beta < 1$ and $0 < x < 1$, τ is the relaxation time. It should be noted that the short time limit of (4) results in $\ln t$. Therefore, only very time-consuming experiments can decide on (4), as knowledge of the entire decay is necessary to unambiguously show Kohlrausch behavior. But still, Kohlrausch decay does not prove glass behavior, as it also exists in an Ising-ferromagnet below T_c. It could, however, disprove it. Therefore, we have to look for better, more easily accessible experiments.

2. Aging.

ROSSEL et al. [12] repeated the aging experiment, well known in spin glasses, for a single crystal. After field-cooling the crystal in $H_0 = 500$ G and aging t_w up to 24 hours, the decay of the magnetization measured after switching the field by $\Delta H = 1000$ G showed a characteristic inflection point, just after the same time t_w. Such an effect can be only described by a hierarchy of barriers or the resulting broad distribution of relaxation times [12].

3. Time-Dependent Specific Heat.

This effect was found only in simulations for 2-D spin glasses [13]. It was not as strong as in recent experiments by A. Voronel [14]. This shows that the hierarchy in the high-T_c glasses is even more pronounced than in spin glasses. Furthermore, the strong effect can also be described in terms of a super glass: a system with annealed and quenched disorder at the same time. In terms of our XY-model, it means that the A_{ij} phase factors relax additionally on a time scale, which is comparable to that of the whole system. Physically, we deal with structural transitions leading to a relaxation of the twins, microtwins, etc. in the system. That means our weak links change with time. But they do not disappear. Disorder relaxes to a certain degree of weaker disorder. The resulting large energy fluctuation can explain the strong relaxation effects in the specific heat. We want to emphasize that the super glass concept is currently worked out and will be published separately. Especially for experimentalists, it will open a new world of "superglassy" effects to be explored.

4. "Quasi" de Almeida-Thouless Line.

As we find this line just in the "glass regime", its origin is clearly related to the hierarchy of barriers. Other approaches which take only the flux creep single barrier picture into account have to fail. Here at this point, it becomes clear why the certainly "oversimplified" XY-spin glass model was so successful. It contains the essential physics, the hierarchy of barriers, and therefore reproduces the (universal) $H^{2/3}$ behavior. Both the flux creep approach and the *XY*-model describe different sides of the same coin. The *XY*-model can be considered a rather basic approach to the Landau-Ginzburg theory, and which just keeps the essential physics for a more qualitative description of the glass phase. Further intensive work is certainly necessary to obtain a quantitative theory for the glass behavior.

5. Decoration Experiment.

A major point of criticism against a glass picture was the occurrence of an Abrikosov vortex lattice in single-crystal decoration experiments at Bell Labs [15]. But just in the spirit of small barriers experienced at lower temperatures, we should also expect this phenomenon according to the glass picture. At higher temperatures, when the hierarchy dominates, the effect should disappear. This was actually the case in the experiment which showed an Abrikosov lattice with a relatively short correlation length, at low T and small H. This situation was reproduced by T. SCHNEIDER et al. in their approach to the *XY*-model [16].

Summarizing at this point, we want to emphasize that a whole new world of glassy effects is waiting to be discovered in high-T_c experiments. It should be noted that in view of applications, when we deal with nitrogen temperature, glass behavior becomes dominant and has to be understood. Especially in view of reaching higher critical currents in (ceramic) superconductors it obviously plays an important role. The current analysis of magnetization measurements may lead to critical currents which are much too low at higher temperatures.

From the theoretical point of view our current situation is the following. We do not have a real contradiction between the (giant) flux creep and the glass picture. Flux creep is just a phenomenological approach to the glass behavior for low temperatures and relatively short times. Glass behavior at higher temperatures and then relatively longer times could also be understood in terms of a flux glass.

Concluding, we want to emphasize that glass behavior plays an important role, especially considering future applications in the field of high-T_c superconductivity.

References

1. K. A. Müller, M. Takashige, J. G. Bednorz. *Phys. Rev. Lett.* **58**, 1143 (1987).
2. For a recent review, see: J. L. van Hemmen and I. Morgenstern (eds.) *Lecture Notes in Physics*, **275**, Heidelberg Colloquium on Glassy Dynamics, Springer-Verlag, 1987.
3. For a recent review on spin glasses in particular, see: K. Binder and A. P. Young. *Rev. Mod. Phys.*, **58**, 801 (1986)
4. J. R. L. de Almeida and D. J. Thouless. *J. Phys.*, A **11**, 983 (1978)
5. I. Morgenstern, K. A. Müller and J. G. Bednorz. *Z. Phys.*, B **69**, 33 (1987)
6. C. Ebner and D. Stroud. *Phys. Rev.*, B **31**, 165 (1985)
7. H. Keller, B. Pümpin, W. Kündig, W. Odermatt, B. D. Patterson, J. W. Schneider, H. Simmler, S. Connell, K. A. Müller, J. G. Bednorz, K. W. Blazey, I. Morgenstern, C. Rossel and I. M. Savic. *Physica C*, **153**, 71 (1988)
8. G. Deutscher and K. A. Müller *Phys. Rev. Lett.*, **59**, 1745 (1987)
9. A. C. Mota, A. Pollini, P. Visani, K. A, Müller and J. G. Bednorz, *Physica C*, **153**, 67 (1988)
10. L. Krusin-Elbaum, A. P. Malozemoff, Y. Yeshurun, D. C. Cronemeyer and F. Holtzberg. *Physica C*, **153**, 1469 (1988) and references therein.
11. See for example M. Tinkham "Introduction to Superconductivity", McGraw-Hill Inc., New York (1975)
12. C. Rossel, H. Maeno, I. Morgenstern, K. A. Müller and F. Holtzberg, preprint
13. I. Morgenstern in: *Lecture Notes in Physics*, **192**, Heidelberg Colloquium on Spin Glasses, Springer-Verlag, 1983
14. A. Voronel, private communication.
15. P. L. Gammel, D. J. Bishop, G. J. Dolan, J. R. Kwo, C. A. Murray, L. F. Schneemeyer and J. V. Waszczak. *Phys. Rev. Lett.*, **59**, 2592 (1987)
16. T. Schneider and R. Hetzel, to be published.

Magnetic Frustration and Pairing in Doped Lanthanum Cuprate

A. Aharony[1-3], *R.J. Birgeneau*[2], *A. Coniglio*[1], *M.A. Kastner*[2],
and H.E. Stanley[1]

[1]Center for Polymer Studies and Department of Physics, Boston University,
  Boston, MA 02215, USA
[2]Physics Department, M.I.T., Cambridge, MA 02139, USA
[3]School of Physics and Astronomy, Tel Aviv University, Tel Aviv 69978, Israel

We discuss the temperature-concentration $(T-x)$ phase diagram of $La_{2-x}(SrBa)_x CuO_4$. The magnetic interactions of the hole spins with the Cu spins yield frustration, explaining the fast decrease in the Néel temperature and yielding a new spin glass phase. The same interactions yield a strong attractive hole-hole potential, which can lead to pairing and superconductivity.

1. Introduction

Both $La_{2-x}(Sr_1Ba)_x CuO_{4-\delta}$ and $YBa_2Cu_3O_{6+\delta}$ exhibit antiferromagnetism—with high Néel temperatures—at low doping and superconductivity (SC) at higher doping. In both cases, there exist strong AF exchange interactions ($J \sim 1100$ K) between the Cu spins in the CuO_2 planes. The possible relevance of magnetism to the SC is thus a topic of much current research.

Since we believe the physics of the two classes of high T_c superconductors is the same, we discuss the simpler case of $La_{2-x}Sr_x CuO_4$, which exhibits the $T - x$ phase diagram shown (schematically) in Fig. 1.

Given the experimental fact that for $x \leq 0.05$ the holes are localized[1] on the O^- ions,[2,3] we show in Sec. 2 that each hole generates a strong local ferromagnetic (F) Cu-Cu interaction, which competes with the otherwise AF exchange. The consequences of this frustration on the $T - x$ phase diagram are then discussed in Sec. 3.

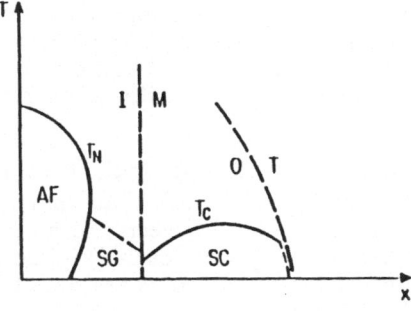

Fig. 1: Temperature-concentration phase diagram (AF = antiferromagnetic, SG = spin glass, I = insulator, M = metal, O = orthorhombic, T = tetragonal, SC = superconducting). Model fractal.

2. Frustration

For small x, the random potential localizes the extra holes within a localization length ℓ_o of order $(2\text{-}3)a$.[1] ℓ_o probably grows gradually as x approaches the I-M transition. There is also evidence that the holes are on the O^- ions.[2,3]

Consider first an instantaneous configuration with the hole on one O^- site. The spin of the hole $\vec{\sigma}$ will have strong exchange interactions with the two neighboring Cu spins $\vec{S}_1$ and $\vec{S}_2$. Writing $H_\sigma = -J_\sigma \vec{\sigma} \cdot (\vec{S}_1 + \vec{S}_2)$, it is intuitively clear that, regardless of the sign of J_σ, the ground state of H_σ prefers $\vec{S}_1 \parallel \vec{S}_2$. Quantum mechanically, the exact ground state of H_σ (neglecting the couplings to other Cu spins) indeed has $S_{12} = 1$ (where $\vec{S}_{12} = \vec{S}_1 + \vec{S}_2$), i.e., $\langle \vec{S}_1 \cdot \vec{S}_2 \rangle = 1/4$.[4] It is thus reasonable to replace H_σ by an F interaction, $\tilde{H}_\sigma = -K(\vec{S}_1 \cdot \vec{S}_2)$, where $K = O(|J_\sigma| \gg |J|)(|J_\sigma| \gg |J|$ because the Cu-Cu distance is twice that of Cu-O. This replacement is exact for classical spins at low temperatures.[4]

Since a strong F bond in the CuO_2 plane destroys the local AF order, it also influences the coupling to the neighboring planes. The Cu spins thus feel competing AF and F interactions. Each F interaction arises from one hole sitting on a Cu-O-Cu bond. In the extremely localized case, the concentration of these F bonds would be x. However, for a finite localization length ℓ_o, the holes are shared by $(\ell_o/a)^2$ bonds, hence the F-bond concentration is of order $x(\ell_o/a)^2$. This is of order $10x$ for small x, and increases as the I-M transition is approached.

3. Phase Diagram

Competing AF and F interactions are known to yield a sharp decrease in T_N, a spin glass (SG) phase[5] and a re-entrance from the AF to the SG phase upon cooling, because of frozen random local moments.[6] This yields the magnetic parts of Fig. 1. In the isostructural $K_2Cu_xMn_{1-x}F_4$, the Cu ferromagnetism is lost at $x \simeq 0.8$,[7] corresponding to a concentration 0.36 of the very weak Cu-Mn and Mn-Mn AF bonds. Renormalizing this by $(\ell_o/a)^2$, and remembering that we have $K \gg |J|$, explains why in $La_{2-x}Sr_xCuO_4$ the SG phase appears at $x \simeq 0.02$.[8,9]

Both the re-entrance[10] and the existence of frozen spins at low T for $0.02 \le x \le 0.05$[8,9] have now been confirmed experimentally.

4. Pairing Potential

A strong F bond between two Cu spins turns them parallel, against the AF coupling to the other Cu spins. The details of the resulting spin configuration depend on the symmetry of the spins. At low temperatures, the spins order along the orthorhombic c axis, indicating a weak Ising anisotropy.[11] Assuming this anisotropy dominates the ground state, the K-bond will simply flip one of its spins (Fig. 2a), with an energy gain of $(K - 7|J|)S^2$ (compared to the AF state without the hole).

When two K bonds are placed next to each other (Fig. 2b), flipping the central spin yields a gain of $(2K - 6|J|)S^2$, which is larger by $8|J|S^2 = 2|J|$ than that of two isolated holes. This implies an attractive potential energy between the holes. Similarly, a gain of $4|J|S^2 = |J|$ results for next nearest neighbor bonds (Fig. 2c). Comparison of Figs. 2b and 2c shows, however, that the two hole spins are parallel (triplet) in the former, and antiparallel (singlet) in the latter.

In the Ising case, similar arguments can be applied to each of the 22 neighboring bonds shown in Fig. 3. The singlet state is unfavorable for the six bonds denoted

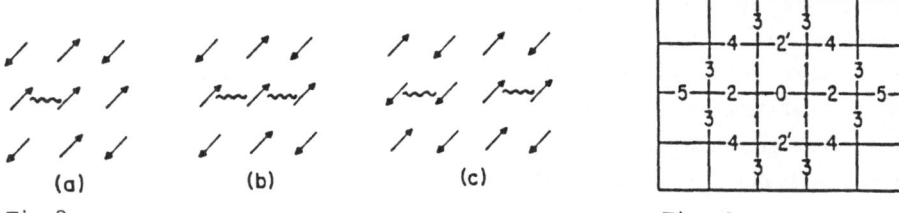

Fig.2   Fig. 3

Fig. 2: Ising ground state with (a) one K bond (wiggly line), (b) two nn K bonds, (c) two nnn K bonds.

Fig. 3: Neighboring K bonds: All except 1,2 attracted to 0.

1,2, and favorable (with energy gain $|J|$) for the remaining 16 bonds. A similar, albeit weaker, attraction will occur between neighboring planes.

We next consider the Coulomb repulsion. Using a Thomas-Fermi estimate, we find a bare screening length of 1-2Å, yielding a repulsion of ~ 0.04eV at 6Å, assuming a dielectric constant ~ 10. This is much smaller than the effective attraction there, $|J| \sim 0.12$eV.

So far, we have calculated the attractive energy of K bonds in a completely ordered AF background. In fact, for $x > 0.05$ the AF correlation length decays as $\xi \sim 3.78x^{-1/2}$.[12] Thus, the above attractive interaction is reduced by a factor $\exp(-r/\xi)$, causing a decrease of the attractive energy.

An attractive potential for singlet pairing can also be derived when the spins have XY or Heisenberg symmetry. Instead of the finite range potential derived above, one obtains a dipole-dipole potential, which decays as $1/r^2$, with an oscillation $\exp(i\vec{Q}\cdot\vec{r})$, where $\vec{Q} = (\pi/a, \pi/a)$.[4] Since the factor $\exp(-r/\xi)$ eliminates the distant K-bonds, the resulting potential is qualitatively similar to the one discussed above, i.e., repulsion at bonds 1, 2 and attraction at bonds 2', 3, 5.

5. Superconductivity

Having established a strong attractive potential between the holes, one can then find either real space bound pairs, which undergo Bose condensation, or correlated BCS pairing. T_c should grow from zero above the I-M transition, with the number of mobile holes, and then decrease due to the decrease in ξ. This qualitatively agrees with the shape of $T_c(x)$ in Fig. 1. A more detailed discussion of the consequences of our model to superconductivity will be given in Ref. [12].

Acknowledgements

This work was supported by the NSF and ONR (at Boston University), the NSF (at MIT) and the US-Israel BSF and the Israeli Academy (at Tel Aviv).

1. M. A. Kastner et al., Phys. Rev. B **37**, 1329 (1987).

2. V. J. Emery, Phys. Rev. Lett. **58**, 2794 (1987).

3. J. M. Tranquada et al., Phys. Rev. B **36**, 5263 (1987); N. Nücker et al. (unpublished).

4. A. Aharony et al., Phys. Rev. Lett. **60**, 1330 (1988).

5. K. Binder and A. P. Young, Rev. Mod. Phys. **58**, 801 (1986).

6. G. Aeppli et al., Phys. Rev. B **25**, 4882 (1982).

7. Y. Kimishima et al., J. Phys. Soc. Japan **55**, 3574 (1986).

8. J. Budnick et al., Europhys. Lett. (in press); D. W. Harshman et al. (unpublished).

9. Y. Kitaoka et al. (unpublished).

10. Y. Endoh et al. (unpublished).

11. D. Vaknin et al., Phys. Rev. Lett. **58**, 2802 (1987).

12. R. J. Birgeneau et al. (unpublished).

Glassy Dynamics in Proteins

R.D. Young

Department of Physics, University of Illinois at Urbana-Champaign,
1110 West Green Street, Urbana, IL 61801, USA, and
Illinois State University, Normal, IL 61761, USA

1. INTRODUCTION

The conformational energy surface of a protein is complex -- the ground state of a protein is highly degenerate and consists of a large number of energy minima separated by energy barriers (1,2,3). The energy minima correspond to conformational substates (CS) of a protein. The CS have the same overall structure, but differ in detail; they perform the same biological function, but possibly with different rates (1,4). Moreover the energy surface (and CS) may be hierarchical and contain minima within minima (2,3,4). Many, perhaps all, complex systems are characterized by a highly degenerate and hierarchically arranged ground state (5). Among such systems are glasses (6,7), spin glasses (8), evolution (9), neural networks (10) and, now, proteins.

We study the infrared CO stretch bands in carbonmonoxymyoglobin (MbCO) as functions of solvent, temperature, and pressure (3,11,12,13). Quasistatic and kinetic studies prove that MbCO experiences a glass transition which is slaved to the solvent (14). Temperature and time dependences of the relaxation of the CO stretch bands in MbCO in 75% glycerol/water are similar to those of the α relaxation in glasses (14).

2. PROTEINS AND COMPLEXITY

2.1 The Multi-Valley Energy Surface and Protein Structure

Proteins are complex systems as a result of their structure (15). The primary structure of a protein consists of linear chains of amino acids linked by strong covalent bonds. Weak hydrogen and sulfur-sulfur bonds link different parts of the primary polypeptide chain and thus stabilize the biologically active, folded, tertiary structure. Complexity occurs because no periodicity condition forces the macro-molecule into a unique structure. In a protein hydrogen bonds assume a variety of positions. Side chains of amino acids can be in slightly different positions in different proteins. Therefore the conformational energy surface of a protein exhibits many energy minima which are separated by energy barriers (Fig. 1). The energy minima are the CS of a protein.

2.2 Ligand Binding to Heme Proteins and Complexity

H. Frauenfelder and collaborators using laser-induced flash photolysis at low temperature developed the first and some of the most dramatic evidence for complexity in proteins (1). They observed non-exponential time dependence for the rebinding of a small ligand such as carbon monoxide (CO) to the iron ion of a heme protein such as myoglobin (Mb) after photodissociation below about 160K. The nonexponential time course for rebinding is explained by postulating that the barrier for bond formation between the ligand and heme iron does not have a single value, but varies from protein to protein. The resulting distribution of barriers is inhomogeneously broadened since it results from proteins being stuck in different energy minima or CS[1] with different barriers for rebinding (1,4). We call these conformational substates resulting in the nonexponential kinetics CS[1] in anticipation of the hierarchy of CS.

2.3 Hierarchy of Conformational Substates

A number of experiments (2-4) have also implied that there is a hierarchy of CS or, equivalently, that the conformational energy surface is hierarchical with minima within minima (Fig. 1). Tier 0 (CS[0])

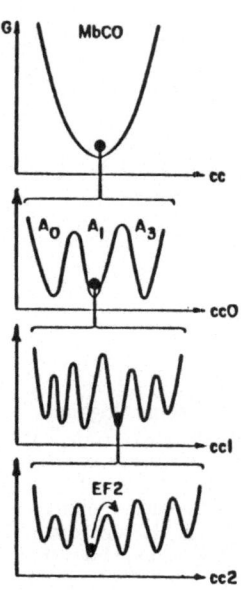

Fig. 1 Conformational Gibbs energy of a protein (MbCO) as function of a conformational coordinate. The various minima correspond to conformational substates (CS). Hierarchical arrangement of CS is indicated through three tiers. EF represent equilibrium fluctuations which are connected to FIMs by fluctuation-dissipation theorems (2,4).

comprises a small number of substates which we characterize below by the infrared CO stretch bands of MbCO. Nonexponential rebinding in each substate of tier 0 implies that each CS^0 is furcated into a large number of substates of the tier 1 (CS^1). Each CS^1 may again be further split. Evidence from low temperature flash photolysis, X-ray diffraction, Mössbauer effect, and Fourier-Transform infrared spectroscopy (FTIR, summarized below) suggest that the furcation continues for at least three tiers.

3. GLASSY DYNAMICS IN PROTEINS

3.1 Experimental Techniques

MbCO exists in three major conformational substates (A_0, A_1, A_3) of tier 0 which are characterized by different angles θ and center frequencies ν of the infrared CO stretch bands (Fig. 2 inset). The IR bands are excellent markers with which to probe effects of external parameters like solvent, temperature (T), and pressure (P). We measure IR spectra using FTIR spectroscopy. The areas (A_i), center fre-

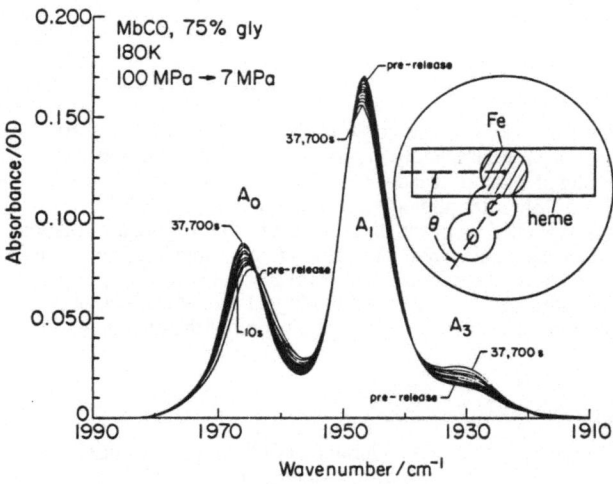

Fig. 2 Time evolution of the absolute absorption spectra of the CO stretch bands for the A substates in MbCO during a P-release experiment. Inset: Cross section through the heme showing the angle θ between the CO dipole and the heme plane.

quencies (ν_i), and widths (Γ_i) of the A bands are fit to Voigtian line shapes (3,13). We also measure the rate of heat absorption using differential scanning calorimetry (DSC).

3.2 Slaved Glass Transition in Proteins (3,14)

In quasistatic experiments we measure the IR spectra from 1910 to 1990 cm^{-1} in several solvents from 140 to 300K for pressures up to 100 MPa (1 kbar). Several important aspects of the data are summarized in Fig. 3 which gives the temperature dependence of the ratio $r_0 = A_0/A_1$ at 7 MPa. The ratio r_0 exhibits six different regions denoted by i to vi: (i) Glass Region. The protein is in a nonequilibrium state in which large-scale motions are frozen on any reasonable experimental time scale. Each protein is frozen into a single CS and the system is nonergodic (7). (ii) Slaved Glass Transition. The bend in Fig. 3 at T_{sg} shows the transition from an equilibrium region well above T_{sg} to a frozen, nonergodic state well below. A similar bend occurs in the width $\Gamma_0(T)$. The transition can also be observed in the DSC data. The midpoints of the transitions define glass temperatures with the following properties: (a) The glass temperature T_{sg} of the protein depends crucially on the glass temperature T_g of the solvent so we refer to the protein transition as a "slaved glass transition". For example DSC data in 75% glycerol/water give $T_g = 175K$ and $T_{sg} = 178K$. A similar pattern is found in other solvents; (b) T_{sg} is higher for the ratio r_0 (195K in 75% glycerol/water) than for the width Γ_0 (175K in 75% glycerol/water). The T_{sg} from the DSC coincides better with that from the width Γ_0. (iii)-(v) Equilibrium Regions. Transitions among the A substates occur faster than the time of observation so that the protein is in thermal equilibrium. (vi) Unfolding. The protein begins to unfold. Although interesting phenomena occur in Regions (iii) to (vi) they are not the focus of this work.

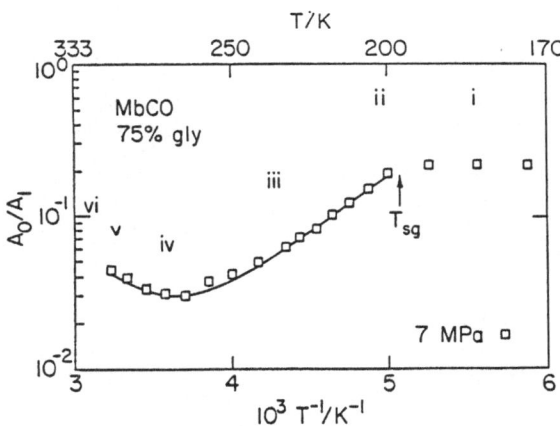

Fig. 3 Plot of $r_0 = (A_0/A_1)$ versus $10^3/T$ at 7 MPa. Roman numerals signify the six regions discussed in the text. The transition temperature T_{sg} is indicated. Regions (iii) to (v) are thermal equilibrium regions. T and P dependences of the ratios r_i determine the relative energies, entropies, and volumes of the A substates in the equilibrium regions (13,14).

3.3 Glassy Relaxation in Proteins (14)

In a typical kinetic experiment, 100 MPa is applied at 240K, the sample is cooled to the desired temperature, a reference IR spectrum is measured (labelled pre-release in Fig. 2), and the pressure is then released to 7 MPa in a few seconds. IR spectra are then taken at approximately exponentially increasing times from about 10 to more than 10 ks. The behavior of the A bands after P release reveals several relaxation processes which can be interpreted in terms of a hierarchy of substates. The data show three characteristic temperature regions: (a) 140-165K. The A bands shift in peak frequency by a constant amount faster than our shortest measuring time (10 s), but no changes in relative intensities occur up to more than 10 ks. We interpret this behavior to mean that each protein remains in the same CS^0 and CS^1; (b) 170-190K. The area of band A_0 remains constant but the peak frequency ν_0 shifts from its high-P toward its low-P value. See Fig. 2. This shift is interpreted as FIM 1, the internal redistribution of the substates CS^1 within A_0. FIM is an acronym for "functionally important motion". FIM 1 is nonexponential in time over several decades and occurs within a narrow T range between 165 to 195K for our experimental time range. The shift in ν_0 can be described by a relaxation function $\phi_1(t)$ where t is the time after P release. The relaxation function for Γ_0 has essentially the same time course as that for ν_0. A_1 and A_3 also interconvert in this temperature region with A_1 decreasing and A_3 increasing with nonexponential time dependence. The barrier between A_1 and A_3 is smaller than that between A_0 and the other A substates; (c) <u>195-210K</u>. FIM 1 and the interconversion between $A_1$and A_3 are faster than 10 s so that the peak positions, widths, and ratio A_1/A_3 after P release have the values obtained in the quasistatic experiment at 7 MPa. The slow exchange of A_0 with $(A_1 + A_3)$, called FIM 0, occurs in a narrow T interval of about 25K. We do not give any more details regarding FIM 0 (14).

A glass-forming liquid forms a disordered phase on cooling to T_g. Near T_g relaxation is dominated by the α process which is nonexponential in time (6,16). The average rate coefficient k_α can be approximated by an Arrhenius temperature dependence near T_g, $k_\alpha = A_\alpha \exp(-E_\alpha/k_B T)$. The parameters for glycerol (17,18) near 200K are $E_\alpha \approx 1.6$ eV and $A_\alpha = 10^{41}$ to 10^{42} s^{-1}. The large A_α for the α relaxation can be interpreted as a signature of a cooperative process and means that the Arrhenius temperature dependence is not physically meaningful. In fact it is well known that the Arrhenius relation fits k_α only over a narrow temperature range while other semi-empirical expressions like the Vogel-Fulcher relation or a power law describe k_α over a wide temperature range (6,16). We have chosen a two-parameter approximation

$$k_\alpha = k_0 \exp\{-(T_0/T)^2\} \tag{1}$$

derived by Baessler (19) and Zwanzig (20). Eq. (1) with $k_0 = 3 \times 10^{19}$ s^{-1}, $T_0 = 1320$K fits the glycerol data well (17,18,19). W. Doster (14) measured k_α in 75% glycerol/water using specific heat spectroscopy (17) giving $k_0 = 10^{18}$ s^{-1}, $T_0 = 1130$K. We evaluate the MbCO relaxation FIM 1 in the range 170-190K by treating the temperature dependence of $\phi_1(t)$ phenomenologically using $k_e = 1/t_e$ where t_e is the time where $\phi_1(t_e)$ has fallen to 1/e. An Arrhenius fit to $k_e = A\exp\{-E/k T\}$ gives $E \approx 1.5$ eV, $A \approx 10^{40}$ s^{-1}. The large preexponential A implies again that the relaxation is cooperative and that the Arrhenius relation is inappropriate. We treat the nonexponential time and non-Arrhenius temperature behavior together by a power law

$$\phi_1(t) = \{1 + k(T)t\}^{-n} \tag{2}$$

with $k(T)$ given by Eq. (1). A fit to the relaxation function ϕ_1 at four temperatures from 168 to 183K yields $k_0 = 3 \times 10^{17}$ s^{-1}, $T_0 = 1170$K, $n = 0.26$. These values for k_0 and T_0 are very close to those for 75% glycerol/water solvent given above and support the idea that large-scale motions of the protein are slaved to those of the solvent: Relaxation FIM 1 in tier 1 of the conformational substates, characterized by both ν_0 and Γ_0, is slaved to the α relaxation of the solvent. In addition the temperature-independent exponent $n = 0.26$ suggests scaling with a scaling function $\tau(T) = k(T)t$ yielding a single, master curve $\phi_1(\tau)$ (16,17).

4. CONCLUSIONS, CONNECTIONS AND OUTLOOK

The results sketched above allow several conclusions: (a) Proteins and glasses have similar properties including a transition to a nonequilibrium state as temperature is lowered past a transition temperature. (b) Proteins like glasses also exhibit cooperative dynamics. In particular proteins show a relaxation process for large-scale motions (α relaxation) which is nonexponential in time and does not follow an Arrhenius temperature dependence. (c) The protein relaxation FIM 1 is slaved to solvent motions. (d) The two relaxations, FIM 0 and FIM 1, observed near 200K in MbCO in 75% glycerol/water are very different. FIM 0 is slower than FIM 1 leading to the higher transition temperature T_{sg} for FIM 0. FIM 0 probably involves both solvent motions and internal protein barriers.

Schroedinger referred to proteins as "aperiodic crystals" (21). However proteins may be closer to glasses and other complex systems. Additional evidence for this point of view has come from other studies: A molecular dynamics simulation supports the concept of a degenerate ground state (22). Specific heat data (23,24) provide evidence for two-level states (TLS) as in glasses. If deep connections do exist among different complex physical systems, theories developed for spin glasses, for example, may be applicable to proteins (25), and progress in protein dynamics may have impact on other fields.

I thank Professor H. Frauenfelder for many discussions and fruitful collaborations which have been essential to this work. I also thank the biomolecular physics group at the University of Illinois for a stimulating atmosphere in which to work and discuss, especially I. E. T. Iben who performed the FTIR experiments and Alfons Schulte who performed the DSC measurements. This work was supported in part by the U.S. National Science Foundation Grant DMB88-16476, National Institutes of Health Grant GM 18051, and the Office of Naval Research Grant N00014-86-K-00270.

5. LITERATURE CITED

1. R. A. Austin *et al*: Bioch. **14**, 5355 (1975).
2. A. Ansari *et al*: Proc. Natl. Acad. Sci. USA **82**, 5000 (1985).
3. A. Ansari *et al*: Biophys. Chem. **26**, 337 (1987).
4. H. Frauenfelder, F. Parak and R. D. Young, Ann. Rev. Biophys. Biophys. Chem. **17**, 451 (1988).
5. Chance and Matter, Les Houches Proceedings, Session No. 46 (North Holland, NY, 1987).
6. J. Jaeckle, Rep. Prog. Phys. **49**, 171 (1986).
7. R. G. Palmer, Adv. Phys. **31**, 669 (1982).
8. K. Binder and A. P. Young, Rev. Mod. Phys. **58**, 801 (1986).
9. P. W. Anderson, Proc. Natl. Acad. Sci. USA **80**, 3386 (1983).
10. J. J. Hopfield, Proc. Natl. Acad. Sci. USA **79**, 2554 (1982).
11. M. W. Makinen *et al*: Proc. Natl. Acad. Sci. USA **76**, 6042 (1979).
12. F. G. Fiamingo and J. O. Alben, Bioch. **24**, 7964 (1985).
13. I. E. T. Iben, Ph.D. Dissertation, University of Illinois (1988).
14. I. E. T. Iben *et al*: Phys. Rev. Lett. submitted.
15. L. Stryer, *Biochemistry* (W. H. Freeman Co., San Francisco, 1988).
16. S. Brawer, *Relaxation in Viscous Liquids and Glasses* (American Ceramic Society Inc., Columbus, OH, 1985).
17. N. O. Birge and S. R. Nagle, Phys. Rev. Lett. **54**, 2674 (1985) and Y. H Jeong *et al*: Phys. Rev. A**34**, 602 (1986).
18. P. Kuhns and M. S. Conradi, J. Chem. Phys. **77**, 1771 (1982).
19. H. Baessler, Phys. Rev. Lett. **58**, 767 (1987).
20. R. Zwanzig, Proc. Natl. Acad. Sci. USA **85**, 2029 (1988).
21. E. Schroedinger, *What is Life?* (Cambridge Univ. Press, NY, 1944).
22. R. Elber and M. Karplus, Science **235**, 318 (1987).
23. V. I. Goldanskii *et al*: Dok. Akad. Nauk SSSR **272**, 978 (1983).
24. G. P. Singh *et al*: Z. Phys. B**55**, 23 (1984).
25. D. Stein, Proc. Natl. Acad. Sci. USA **82**, 3670 (1985).

Vulcanization: How Randomly Cross-Linked Macromolecules Form Equilibrium Amorphous Solids

P. Goldbart and N. Goldenfeld

Department of Physics and Materials Research Laboratory,
University of Illinois at Urbana-Champaign, 1110 West Green Street,
Urbana, IL 61801, USA

1. Introduction

As Goodyear discovered [1], when he first vulcanized rubber in 1839, a viscous liquid of macro-molecules becomes solid when a sufficient number of permanent cross-links are introduced. In common with three-dimensional simple atomic solids [2], this equilibrium rigidity is a consequence of the spontaneous breakdown of translational symmetry: monomers no longer explore the container and, instead, become localized and fluctuate around preferred positions. In contrast, however, the formation of a crystalline macromolecular solid is frustrated by the random locations of   the permanent cross-links, together with the impenetrability of the chains. As a result, rigidity is acquired through the formation of one of many equilibrium amorphous solid states.

Many simple pictures have been advanced to caricature the emergence of rigidity in randomly cross-linked macromolecular systems [3]. However, to the best of our knowledge, there has not been any attempt to provide a conceptually sound statistical mechanical theory of the transition to the solid state, reflecting as it should the variety of distinct topologies in which the system may find itself confined and the non-crystalline nature of the emergent ordered states [4]. Recently, we have begun to develop a theoretical description of how solidness, or rigidity, is acquired as macromolecular systems are randomly cross-linked [5]. Here, we shall elucidate some of the physical issues which contribute to the complexity of this issue. We begin with a brief reminder of what we mean by rubber. Then, we shall discuss how our approach addresses the complexity and, thus, may begin to provide a systematic answer to the question: why is rubber solid? Along the way, we shall try to emphasize the similarities, both formal and substantial, between spin glass ordering and the solidification of rubber by vulcanization. We believe this program to be a prerequisite to the formulation of a genuine theory of the remarkable properties of the rubbery state [6].

We have in mind a monodisperse system of long, flexible, linear macromolecules, or chains. These macromolecules interact with each other and with themselves through a weak long-ranged attraction and a strong short-ranged repulsion, or hard core, which prevents any chain segment, or monomer, from passing through any other. They may also be in solution, in which case the only significant contribution of the solvent to the equilibrium properties is a renormalization of the chain-chain interaction. (Of course, the hard core will remain.)

The process of vulcanization causes cross-links to be formed between randomly selected pairs of monomers which, it should be noted, do not break the translational and rotational invariance of the system. Our aim is to develop a theory of the equilibrium properties of vulcanized macromolecular systems. As always, the notion of equilibrium requires the specification of a time scale: we shall be considering the range of times during which the fracture of any cross-link is extremely unlikely. Hence, we shall model these cross-links by a set of permanent, or quenched, random constraints which glue these monomers together. We shall not be considering weak gels, in which cross-links are constantly forming and breaking.

Springer Series in Synergetics, Vol. 43   **Cooperative Dynamics in Complex Physical Systems**
Editor: H. Takayama   © Springer-Verlag Berlin, Heidelberg 1989

In practice, the probability distribution for the number and location of the cross-links depends on the method used to create them. Typical cross-linking methods induce correlations between the cross-link positions. However, for explicit calculations we have chosen a convenient distribution, Poisson in the number of cross-links and independent of their locations, solely because of the simplification it causes. We expect that our conclusions will apply not only to the system described above but also to more general macromolecular systems with, for example, polydispersity, branching and correlations between cross-links.

2. Topology and equilibrium states

Consider two linear macromolecules, each cross-linked to itself once. This small system has many distinct ways to be in equilibrium: some when the chains are not linked, and many others when they are [7]. Each way, or equilibrium state, is defined by a probability distribution for the occupation of a collection of mutually accessible microscopic configurations. Of course, the proliferation of distinct equilibrium states is not caused by the spontaneous breakdown of some symmetry, as would occur, e.g., for an infinite ferromagnet below its critical temperature. Instead, it is caused by the impenetrability of the chains and the permanence of the cross-links. We label the equilibrium states Σ and refer to them as being topologically distinct because to move between them *requires* the intersection of two chains.

It will be useful to introduce $m_{\mathbf{k}}^{\Sigma}(j;s)$, the Fourier transform of the equilibrium density associated with the sth monomer on the jth chain, defined by

$$m_{\mathbf{k}}^{\Sigma}(j;s) \equiv \left\langle e^{-i\mathbf{k}\cdot\mathbf{c}_j(s)} \right\rangle^{\Sigma}.$$

The average, denoted by $\langle \cdots \rangle^{\Sigma}$, is taken over the set of configurations $\{\mathbf{c}_i(s)\}$ consistent with the topology Σ and weighted with the probability distribution which minimizes the free energy per monomer. As the system is finite, this minimization gives a unique, translationally invariant probability distribution, the Gibbs distribution, for each topology Σ and, hence, $\{m_{\mathbf{k}\neq 0}^{\Sigma}(j;s)\} = 0$. The states $\{\Sigma\}$ cannot be distinguished by comparing the monomer densities $\{m_{\mathbf{k}}^{\Sigma}(j;s)\}$.

Unfortunately, it is not possible to restrict the sum over configurations of the system to those with a particular topology. To do so would require the use of an invariant which uniquely characterizes each topology. However, no such invariant is known to exist, even for this finite system [8]. Even if the necessary invariant existed for the classification of an infinite network, its implementation would be technically infeasible.

This finite system possesses distinct equilibrium states which are not related by translations or rotations. The ensemble of configurations which respect the cross-links should, therefore, be decomposed into smaller sets of configurations, each with a particular topology and each representing one of the ways for the system to be in equilibrium. Which state is realized in practice depends on the configuration at the instant at which the cross-links are inserted. Of course, none of the states will be solid. As the system is finite, it will translate and rotate over finite times, thus homogenizing and isotropizing all observables; it will not resist infinitesimal shearing.

3. Infinite networks

Now consider an infinite system of macromolecules, with cross-links imposed between randomly selected pairs of monomers. The space of configurations which respect the cross-links will consist of a collection of disjoint regions, one for each topology. As the topology is unchanged by global rotations and translations each of these sets is invariant under these symmetry operations.

Suppose that the system is prepared in one of the topologies and permitted to evolve dynamically. If, in the course of time, it is ergodic over all the configurations with that topology (or a symmetric subset of them) then the system will be fluid. If, on the other hand, the system remains in a subset which is not invariant under symmetry operations then the symmetry is said to be spontaneously broken and the system will exhibit rigidity. In principle, this issue should be addressed by performing a statistical mechanical average over the configurations associated with one topology and searching for spontaneous symmetry breakdown using the conventional method: apply a small symmetry breaking field which eliminates the contribution from all but one ergodic component and, ultimately, allow the field to vanish. The approach is impossible at both stages. Firstly, the invariant required to select a particular topology does not exist. Secondly, as the solid state may be amorphous, we do not *a priori* know the appropriate small field to apply. As with the spin glass problem, a direct attack is not presently possible.

As an alternative approach, we decompose the configuration space into as yet unknown ergodic components which we label σ. We shall simply require that these components correspond to equilibrium states, in that they each support a probability distribution which minimizes the free energy density and, furthermore, that these states are *extremal* equilibrium states [9] possessing, as such, the cluster property. Physically, these ergodic components can certainly not contain contributions from more than one topology. Moreover, as the system is infinite, there may be degenerate extremal probability distributions for a given topology, each containing only a fraction of the configurations with that topology. If this happens then ergodicity is said to be spontaneously broken. If, in addition, the ergodic components are not fully symmetric then symmetry is also said to be spontaneously broken. Only in this case it is possible for the equilibrium monomer densities

$$m_{\mathbf{k}}^{\sigma}(j; s) \equiv \left\langle e^{-i\mathbf{k}\cdot\mathbf{c}_j(s)} \right\rangle^{\sigma}$$

not to vanish for non-zero $\mathbf{k}$. Whether or not the the state σ corresponds to a liquid, a crystalline solid or an amorphous solid is determined by the nature of the equilibrium distribution through the set $\{m_{\mathbf{k}}^{\sigma}(j; s)\}$, which is incalculable using direct methods.

The analogy with the Heisenberg spin glass is realised in the following way. The degrees of freedom in the macromolecular system are the position vectors $\{\mathbf{c}_j(s)\}$ of the monomers, labeled by chain j and arclength location s. They correspond to the spin vectors $\{\mathbf{S}(\mathbf{x})\}$ at the set of M lattice sites $\{\mathbf{x}\}$ in the vector spin glass. The monomer densities $\{m_{\mathbf{k}}^{\sigma}(j; s)\}$ correspond to the local magnetizations $\{\langle \mathbf{S}(\mathbf{x})\rangle^{\sigma}\}$ in the equilibrium state σ, which vanish in a paramagnetic state and acquire non-zero values in a spin glass or ferromagnetic state.

Although the monomer densities $\{m_{\mathbf{k}}^{\sigma}(j; s)\}$ distinguish between liquids, crystals and amorphous solids, the distinction between the liquid and the amorphous solid is lost in the thermodynamic limit of the normalized density,

$$m_{\mathbf{k}}^{\sigma} \equiv \frac{1}{NL} \sum_{j=1}^{N} \int_{0}^{L} ds \left\langle e^{-i\mathbf{k}\cdot\mathbf{c}_j(s)} \right\rangle^{\sigma},$$

of a system of N chains, each of arclength L. It is zero for both liquid and amorphous solid states because, although $m_{\mathbf{k}}^{\sigma}(j; s) \neq 0$ for the amorphous solid, random phase cancellations cause its sum to vanish. In the spin glass, it is the magnetization density,

$$\mathbf{m}^{\sigma} \equiv \frac{1}{M} \sum_{\mathbf{x}} \langle \mathbf{S}(\mathbf{x})\rangle^{\sigma},$$

which vanishes in both the spin glass and paramagnetic states, even though $\langle \mathbf{S}(\mathbf{x}) \rangle^{\sigma} \neq 0$ in the spin glass state.

How can the random phase cancellation be avoided? For the spin glass problem [10] the self-overlap

$$q_{\mu\nu}^{\sigma\sigma} \equiv \frac{1}{M} \sum_{\mathbf{x}} \langle S_{\mu}(\mathbf{x}) \rangle^{\sigma} \langle S_{\nu}(\mathbf{x}) \rangle^{\sigma},$$

together with the magnetization density, serve to distinguish paramagnetic, ferromagnetic and spin glass states. For macromolecular systems, the self-overlap

$$q_{\mathbf{k}\mathbf{k}'}^{\sigma\sigma} \equiv \frac{1}{NL} \sum_{j=1}^{N} \int_{0}^{L} ds \, \langle e^{-i\mathbf{k}\cdot\mathbf{c}_j(s)} \rangle^{\sigma} \langle e^{-i\mathbf{k}'\cdot\mathbf{c}_j(s)} \rangle^{\sigma},$$

together with the normalized density, will distinguish fluid, crystalline and amorphous solid states.

To become acquainted with the properties of the self-overlap, consider the simple example of an atomic solid in which each of the atoms $\{j = 1 \dots N\}$ is, by some means, identically harmonically localized around a mean position, $\langle \mathbf{c}_j \rangle^{\sigma}$. Then

$$\langle e^{-i\mathbf{k}\cdot\mathbf{c}_j} \rangle^{\sigma} = e^{-\frac{1}{2}k^2\xi^2} e^{-i\mathbf{k}\cdot\langle \mathbf{c}_j \rangle^{\sigma}},$$

where ξ is the common localization length. In this example, the normalized density is given by

$$m_{\mathbf{k}}^{\sigma} = e^{-\frac{1}{2}k^2\xi^2} \frac{1}{N} \sum_{j=1}^{N} e^{-i\mathbf{k}\cdot\langle \mathbf{c}_j \rangle^{\sigma}}.$$

For a crystalline solid, there exists a set of reciprocal lattice vectors, $\{\mathbf{G}\}$, for which all terms in the summation have the same phase, and no cancellations occur. For an amorphous solid, no such vector exists (other than the trivial $\mathbf{k} = 0$), the summation is over a set of random phase factors and, in the thermodynamic limit, the normalized density vanishes. On the other hand, the self-overlap is given by

$$q_{\mathbf{k}\mathbf{k}'}^{\sigma\sigma} = e^{-\frac{1}{2}\xi^2(k^2+k'^2)} \frac{1}{N} \sum_{j=1}^{N} e^{-i(\mathbf{k}+\mathbf{k}')\cdot\langle \mathbf{c}_j \rangle^{\sigma}}$$

which is non-zero for $\mathbf{k}+\mathbf{k}' = 0$, when $\{\langle \mathbf{c}_j \rangle^{\sigma}\}$ corresponds to either a crystalline or an amorphous state. The self-overlap only vanishes for the delocalized, or fluid, system for which $\xi \to \infty$.

4. Overlap distributions

The self-overlap, discussed in the previous section, is no easier to calculate directly than the monomer density $\langle e^{-i\mathbf{k}\cdot\mathbf{c}_j(s)} \rangle^{\sigma}$. As with the spin glass problem, progress comes with the generalization of the self-overlap to the overlap between pairs of extremal equilibrium states,

$$q_{\mathbf{k}\mathbf{k}'}^{\sigma\sigma'} \equiv \frac{1}{NL} \sum_{j=1}^{N} \int_{0}^{L} ds \left\langle e^{-i\mathbf{k}\cdot\mathbf{c}_j(s)} \right\rangle^{\sigma} \left\langle e^{-i\mathbf{k}'\cdot\mathbf{c}_j(s)} \right\rangle^{\sigma'},$$

which, loosely speaking, exhibits the rigidity and compares the configurations of the states σ and σ'.

For the spin glass, one can define a symmetrized overlap [11] which is invariant under the global rotation of either σ or σ'. Then, at low temperatures, similar states have overlaps which differ from unity by small amounts, as do states related by global rotations. This conveniently causes symmetry related sets of states to have a single common overlap; any variety in the value of a set of symmetrized overlaps reflects the presence of states which are not related by symmetry. Similarly, a symmetrized overlap for the amorphous solid can be constructed which eliminates the spurious dependence of the overlaps on relative translations and rotations of the equilibrium states. This construction is discussed in detail in [12].

The central idea is to construct a quantity which is calculable, at least approximately, and which exhibits the solidness of, and relationship between, the equilibrium states of the system. One such quantity is the disorder-average of the probability distribution of the overlaps between equilibrium states,

$$[P_{\{\mathbf{k}\mathbf{k}'\}}(q)] \equiv \Big[\sum_{\sigma\sigma'} w^{\sigma} w^{\sigma'} \delta(q - q_{\{\mathbf{k}\mathbf{k}'\}}^{\sigma\sigma'}) \Big],$$

where $[\cdots]$ denotes an average over the position and number of the random cross-links, $q_{\{\mathbf{k}\mathbf{k}'\}}^{\sigma\sigma'}$ is the symmetrized overlap and w^{σ} is the normalized Boltzmann weight associated with the equilibrium state σ. From the properties of this distribution one can determine the nature of the underlying equilibrium states. In particular,

(a) if $[P_{\{\mathbf{k}\mathbf{k}'\}}(q)] = \delta(q)$ then there may be many equilibrium states, at least one for each topology, all of which correspond to translationally invariant liquid states;

(b) if $[P_{\{\mathbf{k}\mathbf{k}'\}}(q)] = \delta(q - \bar{q})$ then the system possesses at least one family of solid, symmetry related equilibrium states and, potentially, several such families which are microscopically distinct but macroscopically identical;

(c) if $[P_{\{\mathbf{k}\mathbf{k}'\}}(q)]$ is a broad distribution then, along with symmetry-related solid states, there are also solid states which are not only unrelated by symmetry but also differ from each other macroscopically. This means that they cannot be related by a finite amount of local reweaving of the topology. States which can be related by a finite amount of local reweaving are not macroscopically distinct and have common mutual- and self-overlaps.

An exact treatment of our macromolecular system with a small number of random cross-links should generate case (a), corresponding to the existence of one or more fluid states. As the mean number of cross-links is increased, a transition should occur to case (c), when the system becomes one of many equilibrium amorphous solids. This transition is not a conventional liquid-to-crystal transition and may be continuous; this possibility is realised in the present theory.

5. Field theoretic representation

In the previous section we have discussed the overlap distribution because, unlike the overlaps themselves, it is at least approximately calculable. Two principal methods exist: Monte Carlo simulation, on which we intend to report in the future, and replica field theory, which we now discuss.

It is not *a priori* obvious that the overlap distribution will emerge from a replica calculation. The derivation is quite lengthy and is discussed in detail in [5,12,13]. Here we shall simply sketch the main ideas. Using the cluster property, the overlap distribution, $P_{\{kk'\}}(q)$, is related, in an approximation which is exact for mean field theory, to the expectation value of the distribution of microscopic overlaps between the configurations of two uncoupled replicas of the system with identical cross-links. The expectation value is computed by averaging over *all* configurations of *both* systems weighted with one physical Hamiltonian for each system,

$$P_{\{kk'\}}(q) \simeq \langle \delta(q - Q^{12}_{\{kk'\}}) \rangle_2$$

$$Q^{12}_{kk'} \equiv \frac{1}{NL} \sum_{j=1}^{N} \int_0^L ds\, e^{-i\mathbf{k}\cdot\mathbf{c}_j(s) - i\mathbf{k}'\cdot\mathbf{c}_j(s)}.$$

Here $Q^{12}_{\{kk'\}}$ is the symmetrized microscopic overlap obtained by the symmetrization of $Q^{12}_{kk'}$ and $\langle \cdots \rangle_2$ denotes the average over the two uncoupled systems. This expectation value is still not calculable, as it depends on the choice of random cross-link positions; neither is its disorder average, because of the usual problem of denominators. Introducing a further $n-2$ identical uncoupled replicas, taking the limit $n \to 0$ to eliminate the denominators, and averaging over the disorder, yields

$$[P_{\{kk'\}}(q)] \simeq \lim_{n\to 0} [\langle \delta(q - Q^{12}_{\{kk'\}}) \rangle_n].$$

The random constraints enter the replica theory through the factor

$$\prod_{e=1}^{M} \prod_{\alpha=1}^{n} \delta(\mathbf{c}^{\alpha}_{i_e}(s_e) - \mathbf{c}^{\alpha}_{i_e}(s'_e))$$

in the expectation value $\langle \cdots \rangle_n$, where the replicas are labeled $\alpha = 1 \ldots n$. Assuming the Poisson form,

$$\mathcal{P}_M = \frac{1}{M!} \left(\frac{\mu^2}{2NL^2} \right)^M \exp\left(-\frac{1}{2}\mu^2 N \right)$$

for the probability of finding M cross-links connecting arclength positions $\{s_e, s'_e\}$ on chains $\{j_e, j'_e\}$, $e = 1 \ldots M$, and averaging over the disorder, generates the term

$$\frac{-\mu^2}{2NL^2} \sum_{h,j=1}^{N} \int_0^L ds \int_0^L dt \prod_{\alpha=1}^{n} \delta(\mathbf{c}^{\alpha}_h(s) - \mathbf{c}^{\alpha}_j(t))$$

which augments the replicated Hamiltonian. Note that this term, which couples the replicas, is invariant under independent rotations and translations of the replicas, and that the mean number of cross-links per chain is $\mu^2/2$.

The replica-replica interaction may be decoupled in the following way. Introducing the Fourier representation of the Dirac delta function, and assuming the system to be confined in a cubic container of volume V, we obtain

$$\frac{-\mu^2}{2NL^2} \sum_{h,j=1}^{N} \int_0^L ds \int_0^L dt \frac{1}{V^n} \sum_{\hat{k}} \exp(i\hat{k} \cdot (\hat{c}_h(s) - \hat{c}_j(t))),$$

where $\hat{k}$ denotes the set of n d-vectors $\{\mathbf{k}^1 \ldots \mathbf{k}^n\}$, and the inner product $\hat{k} \cdot \hat{c}$ is $\sum_{\alpha=1}^{n} \mathbf{k}^{\alpha} \cdot \mathbf{c}^{\alpha}$. Consider the summation over $\hat{k}$. The replicas are only coupled by terms for which at least two

d-vectors in $\hat{k}$ are non-zero. The $\hat{k} = 0$ term is a constant, which we ignore, and the terms with one non-zero d-vector renormalize the physical Hamiltonian in each replica. Summations with these terms omitted carry an overbar. The replica coupling term is quadratic,

$$\frac{-\mu^2}{2NL^2V^n}\overline{\sum_{\hat{k}}}\left|\sum_{h=1}^{N}\int_0^L ds\ e^{i\hat{k}\cdot\hat{c}_h(s)}\right|^2,$$

and thus may be decoupled in the standard way at the expense of introducing the complex field $\Omega_{\hat{k}}$, whose expectation value is

$$\langle\Omega_{\hat{k}}\rangle_n = \frac{1}{NL}\sum_{j=1}^{N}\int_0^L ds\ \langle e^{-i\hat{k}\cdot\hat{c}_j(s)}\rangle_n.$$

At this stage we formally integrate the macromolecular degrees of freedom to obtain a representation purely in terms of $\Omega_{\hat{k}}$. The effect of the chain interactions are felt through the correlation functions of a ficticious fluid which determine the coefficients in the perturbative expansion of the effective Hamiltonian of the $\Omega_{\hat{k}}$ theory. This ficticious fluid is the uncross-linked macromolecular system with interactions which are renormalized by the contribution from the one-replica sector of the cross-link term. The presence of repulsive interactions in the physical system is crucial, however, to avoid a net attraction due to the cross-link contribution in the replica theory. Simplification arises if the two interactions are supposed to cancel exactly. This is analogous to chosing the spin glass interaction distribution to have zero mean.

In the limit $N \to \infty$ the saddle point solution of the $\Omega_{\hat{k}}$ field theory is exact. Consequently, the saddle point value may be used to determine $[P_{\{kk'\}}(q)]$, and its symmetry properties will determine the nature of the overlap distribution. Although we have only discussed the two-state overlap, a sequence of higher overlaps between more than two states is also determined by the saddle point value of $\Omega_{\hat{k}}$.

6. Conclusions

Finding the stable stationary point of the effective Hamiltonian $\mathcal{S}[\Omega]$ is a non-trivial task. For the infinite range vector spin glass there are $\frac{1}{2}n(n-1)m^2$ independent components to the order parameter $q_{\mu\nu}^{\alpha\beta}$ with respect to which the effective Hamiltonian must be made stationary. In contrast, the vulcanization problem has an effective Hamiltonian $\mathcal{S}[\Omega]$ whose argument is a complex replicated-space-dependent field; rather a large space of functions must be searched.

We can identify a continuous transition as the mean number of cross-links is increased. The precise location of this transition depends on the interaction between macromolecules. If the ficticious fluid is non-interacting then rigidity first occurs when the mean number of cross-links per chain exceeds 1/2.

Ideally, one could locate a stable solution in the rigid phase. Less ideally, a longitudinally stable replica symmetric solution should be found, which we presume would be unstable with respect to transverse replica symmetry-breaking fluctuations. We have not yet succeeded with either of these tasks. Some information about the solid phase may be obtained by working extremely close to the transition and assuming that $\Omega_{\hat{k}} = 0$ unless two d-vectors in $\hat{k}$ are the smallest possible non-zero vectors and the remainder vanish. In this case $\mathcal{S}$ takes the same form (with the same signs) as the effective Hamiltonian of the S-K model [14] and thus translational and replica symmetry are broken [15].

A reasonable replica symmetric *ansatz* for the vulcanization problem, motivated by the model amorphous solid introduced in Section 3 to illustrate the self-overlap, is

$$\Omega_{\hat{k}} = e^{-\frac{1}{2}\xi^2 \hat{k}^2} \delta_{\sum_\alpha k^\alpha, 0}$$

with stationarity being enforced with respect to the localization length, ξ. This form has the virtue of being invariant under common translations and rotations of the replicas.

What avenues should be pursued in the future? Apart from solving the aforementioned problems it would be useful to have results for

- the shear modulus, as a function of cross-link density;

- the distribution of localization lengths, analogous to the spin glass local field distribution;

- the non-linear response, or large deviation behavior;

- whether or not there are non-symmetry related states within a topology;

- the effect of incorporating correlations between cross-links;

- dynamical correlation functions.

Recent experiments [16] have indicated a host of interesting phenomena associated with the related problem of gelation, notably the existence of dynamical scaling at the rigidity transition, and glassy behavior near the onset of the rigid phase. It is possible that these phenomena may be explained within the present framework, as consequences of a continuous phase transition to an amorphous solid with striking similarities to spin glasses.

Acknowledgements: This work was partially funded by the National Science Foundation through grant number DMR-87-01393, and through the UIUC Materials Research Laboratory under National Science Foundation grant number DMR-86-12860. NDG gratefully acknowledges the support of the Alfred P. Sloan Foundation. PMG gratefully acknowledges the receipt of an Arnold O. Beckman Research Award from the University of Illinois Campus Research Board. The authors acknowledge the support of the UIUC Polymer Group.

REFERENCES

1. C. Goodyear, *Gum-Elastic and its Varieties, with a Detailed Account of its Applications and Uses, and of the Discovery of Vulcanization* (New Haven, 1855).
2. P.W. Anderson, *Basic Notions of Condensed Matter Physics* (Benjamin, Reading, 1984); D. Forster, *Hydrodynamic Fluctuations, Broken Symmetry and Correlation Functions* (Benjamin, Reading, 1975).
3. For a discussion, see P.G. de Gennes, *Scaling Concepts in Polymer Physics* (Cornell University Press, 1979), especially Chapter 5.
4. Statistical mechanical formulations have been introduced by Edwards and collaborators: see, e.g., R.T. Deam and S.F. Edwards, *Philos. Trans. Roy. Soc. (London), Ser. A* **280**, 317 (1976)and R.C. Ball and S.F. Edwards, *Macromol.* **13**, 748 (1980). These works do not attempt to describe the transition to the solid state, but should be viewed as replica-symmetric *ansätze*.
5. P.M. Goldbart and N.D. Goldenfeld, *Phys. Rev. Lett.* **58**, 2676 (1987).

6. L.R.G. Treloar, *The Physics of Rubber Elasticity* (Clarendon Press, Oxford, 1975).

7. For an introduction to knot theory, see L.H. Kauffman, *On Knots* (Princeton University Press, 1987).

8. F.W. Wiegel, *Introduction to Path-Integral Methods in Physics and Polymer Science*, Chapter 4 (World Scientific, Singapore, 1986).

9. For an introduction to equilibrium states, see J. Glimm and A. Jaffe, *Quantum Physics* (Springer-Verlag, New York, 1981); the introduction by A.S. Wightman to *Convexity and the Theory of Lattice Gases* by R.B. Israel (Princeton: Princeton University Press, 1978); G. Parisi *Statistical Field Theory* (Benjamin, Reading, MA, 1988); J.L. van Hemmen, in J.L. van Hemmen and I. Morgenstern (Eds.) *Proceedings of the Heidelberg Colloquium on Spin Glasses*, Lecture Notes in Physics Vol. 192 (Springer, Berlin, 1983).

10. S.F. Edwards and P.W. Anderson, *J. Phys. F: Metal Physics* **5**, 965 (1975); G. Parisi, *Phys. Rev. Lett.* **50**, 1946 (1983).

11. G. Kotliar, H. Sompolinsky and A. Zippelius, *Phys. Rev. B* **35**, 311 (1987); P.M. Goldbart, *Thesis*, University of London (1985), unpublished.

12. P.M. Goldbart and N.D. Goldenfeld, *A Microscopic Theory for Cross-linked Macromolecules I: Broken Symmetry, Rigidity and Topology*, submitted for publication.

13. P.M. Goldbart and N.D. Goldenfeld, *A Microscopic Theory for Cross-linked Macromolecules II: Replica Theory of the Transition to the Solid State*, submitted for publication.

14. D. Sherrington and S. Kirkpatrick, *Phys. Rev. Lett.* **35**, 1972 (1975).

15. G. Parisi *Phys. Rev. Lett.* **43**, 1754 (1979).

16. H.H. Winter and F. Chambon, *J. Rheol.* **30**, 367 (1986); M. Adam, M. Delsanti, J.P. Munch and D. Durand, *Phys. Rev. Lett.* **61**, 706 (1988).

Superlocalization of Fractons:
Direct Observation by Supercomputer

K. Yakubo and T. Nakayama

Department of Applied Physics, Hokkaido University, Sapporo 060, Japan

Evidence is given for the superlocalization of fracton wave-functions excited on two-dimensional percolating networks. It is shown that fractons are superlocalized as $\exp[-(r/\Lambda(\omega))^{d_\phi}]$, where the exponent d_ϕ takes the value of 2.3 ± 0.1. Our result provides new insight into the underlying properties of fractons.

1. Introduction

There has been a great deal of interest in the dynamical properties of fractal systems. Particular attention has been paid to the localized nature of fracton excitations [1]. ALEXANDER et al. [2] have supposed that fractons on percolating networks are strongly localized in the form of $\exp[-(r/\Lambda(\omega))^{d_\phi}]$, where $\Lambda(\omega)$ is the localization length depending on frequency. The exponent d_ϕ indicates the strength of the localization. Excitations with d_ϕ larger than unity are called "superlocalized" modes. Theoretical predictions on the exponent d_ϕ, however, are very confusing at the present stage [3-5].

We present a numerical evidence for the superlocalization of fractons excited on two-dimensional site-percolating networks, paying attention to the numerical value of the exponent d_ϕ. The fractons were excited on very large percolating networks (network size $>10^5$). We have obtained the value of $d_\phi=2.3\pm0.1$, which is large compared with the existing theoretical predictions. Our numerical result indicates that the strong localization is a consequence of extreme localization within weakly connected segments.

2. Numerical Method and Results

Consider site-percolating networks consisting of N atoms with unit mass, where two nearest-neighbor atoms are connected by linear springs. Usually, one diagonalizes numerically the dynamical matrix of the equation of motion to obtain the eigenmodes of the system by the well-known scheme such as the Householder method or the recursive technique. These methods, however, require a large amount of computer memory space. The method employed here, introduced by WILLIAMS and MARIS [6], enables us to obtain the eigenmodes of very large system, because the algorithm requires less memory space and is suitable for a vectorizable program. The method is based on the idea that the eigenmode of the system satisfies the resonance condition when applying the periodic external force. The detail of this algorithm has been described in Ref. [7].

We have applied this numerical technique to obtain the wave-function of fractons excited on percolating networks. The networks were formed on 700x700 square lattices. We excited 129 fractons with frequencies close to $\omega=0.01$ on nine percolating networks at the percolation threshold $p_c(=0.593)$. Figure 1 shows the shape of fractons in a log-log scale (filled circles). The abscissa

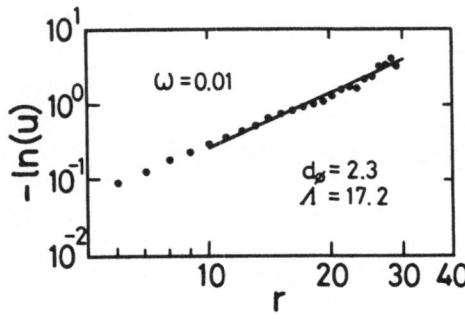

Fig. 1 The shape of fracton wave-function averaged over 129 fractons. Filled circles show our numerical results, and the straight line is from the least squares fitting in the range 10≤r≤30.

represents the distance r from the center of fracton and the ordinate the averaged amplitude. The straight line through filled circles denotes the least squares fitting in the range 10≤r≤30. The gradient of this line indicates the exponent d_ϕ of the fracton wave-function. From this figure, we can find that the exponent becomes d_ϕ=2.3±0.1 and the localization length Λ(ω=0.01)=17.2. We have also calculated d_ϕ and $\Lambda(\omega)$ for different five frequencies smaller than ω=0.01. The results show the exponent d_ϕ takes the same value independent of frequencies within our numerical accuracy. The localization length $\Lambda(\omega)$ is found to be proportional to $\omega^{-\bar{d}/D}$, which is in good agreement with the scaling argument on the frequency dependence of $\Lambda(\omega)$ [1]. This means these are surely fractons. These numerical results provide evidence of the strong localization of fractons on percolating networks.

The value of d_ϕ is larger than those of the theoretical predictions [3-5]. The theories assume that the localization of fractons is caused by the interference on topologically disordered lattices. Our results indicate that this view is not appropriate for the fracton superlocalization. Our recent results give the evidence that the fracton is confined within weakly coupled segments, and cannot diffuse through such weakly coupled segments [8]. We emphasize that the conventional theories [3-5] based on the interference of waves passing through sinuous paths pay no regard to the fact that the weak segments play a crucial role for the superlocalization.

The Hokkaido University Computing Center is acknowledged for the use of the supercomputer facilities. This work was supported by the Suhara Memorial Foundation.

References

1. S. Alexander, R. Orbach: J. Phys. (Paris) Lett. 43, L625(1982)
2. S. Alexander, O. Entin-Wohlman, R. Orbach: J. Phys. (Paris) Lett. 46, L549(1985); Phys. Rev. B32, 6447(1985)
3. Y. E. Lévy, B. Souillard: Europhys. Lett. 4, 233(1987)
4. A. B. Harris, A. Aharony: Europhys. Lett. 4, 1355(1987)
5. A. Aharony, O. Entin-Wohlman, R. Orbach: Phys. Rev. Lett. 58, 132(1987)
6. M. L. Williams, H. J. Maris: Phys. Rev. B31, 4508(1985)
7. K. Yakubo, T. Nakayama: Phys. Rev. B36, 8933(1987)
8. K. Yakubo, T. Nakayama: to be published

A Possibility of Spin-Glass Ordering in the Oxide Superconductor $(La_{1-x}Sr_x)_2CuO_{4-\delta}$

K. Katsumata and H. Kitazawa

The Institute of Physical and Chemical Research (RIKEN),
Wako, Saitama 351-01, Japan

Recently, we have succeeded in growing sizable single crystals of the oxide superconductor $(La_{1-x}Sr_x)_2CuO_{4-\delta}$ (LSCO). We have performed positive muon spin relaxation(μSR) experiments on these single crystals and found that a magnetic ordering coexists with the superconductivity/1/. The concentration(x) vs temperature(T) phase diagram determined from the μSR experiments is shown in Fig.1. Qualitatively, this phase diagram is the same as that of the similar compound $(La_{1-x}Ba_x)_2CuO_{4-\delta}$ /2,3/ except the coexistence. A theoretical phase diagram of LSCO has been proposed by AHARONY et al./4/ in which the spin-glass(SG) phase appears in the concentrations between the antiferromagnetic(AF) and superconducting(SC) phases. Experimentally, LSCO is AF at low temperatures up to at least x=0.025/5/. Then, SG is limited, if present, to the concentrations x>0.025.

In order to elucidate the nature of the magnetic phase found in the x=0.04 sample, we have made a magnetization measurement. Figure 2 shows the external magnetic field(H) dependences of absolute values of the magnetizations($|M|$) observed in the x=0.04 and 0.08 single crystals. The $|M|$-H curve obtained in the Sr-rich sample shows the behavior typical of a type II superconductor: $|M|$ in-

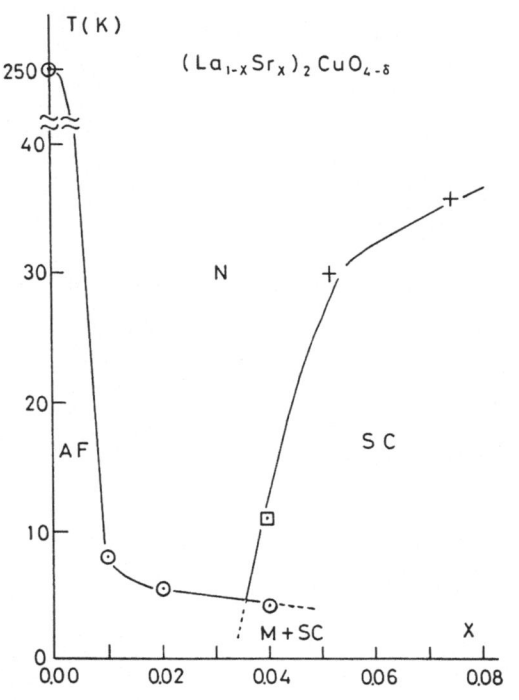

Fig.1. The concentration(x) vs temperature(T) phase diagram of $(La_{1-x}Sr_x)_2CuO_{4-\delta}$ determined from the μSR experiment/1/. AF:antiferromagnetic phase, SC:superconducting phase, M:magnetic phase, N:normal (paramagnetic) phase

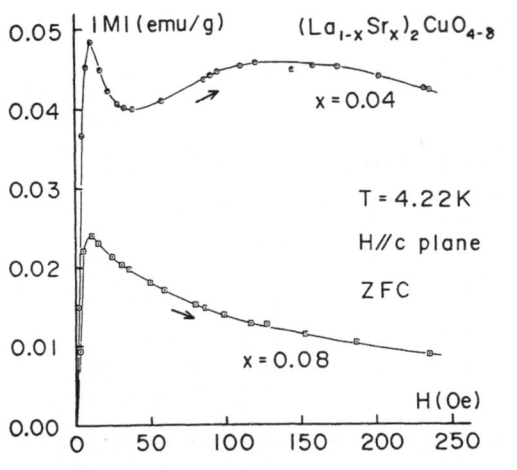

Fig.2. The external magnetic field(H) dependences of absolute values of the magnetizations($|M|$) obtained in the x=0.04 and x=0.08 single crystals. The external field is applied in the c-plane. The samples have been cooled from room temperature to 4.22K in zero field and the magnetizations are measured in an increasing field

creases with H until the lower critical field(H_{C1}) is reached and then decreases when H is increased further. On the other hand, the $|M|$-H curve observed in the x=0.04 sample is largely different from that of the x=0.08 sample: there is another peak in the $|M|$-H curve above H_{C1}.

The unusual behavior of the magnetization found in the x=0.04 sample can be understood if we assume that the low temperature magnetic phase is SG. In the T vs H phase diagram of a spin-glass, one has a line(L) separating SG from the paramagnetic phase. We consider the superconductor with which SG coexists at the temperatures below the spin-glass transition temperature(T_{SG}). When H(>H_{C1}) is increased above L at a given temperature, SG is destroyed. A superconducting phase coexistent with a magnetic ordering usually costs extra energy compared to that without the coexistence. From these arguments we see that SC shows up again at a field above H_{C1} when SG is destroyed. Consequently, some of the magnetic flux which have been present in the sample are excluded out of the sample. This corresponds to the broad maximum in the $|M|$-H curve of Fig.2.

Acknowledgement
This work was partially supported by a Research Fund from RIKEN.

References
1. H. Kitazawa, K. Katsumata, E. Torikai and K. Nagamine, Solid State Commun. (in press).
2. K. Kumagai, I. Watanabe, H. Aoki, Y. Nakamura, T. Kimura, Y. Nakamichi and H. Nakajima, Physica 148B, 480 (1987).
3. Y. Kitaoka, K. Ishida, S. Hiramatsu and K. Asayama, J. Phys. Soc. Jpn. 57, 734 (1988).
4. A. Aharony, R. J. Birgeneau, A. Coniglio, M. A. Kastner and H. E. Stanley, Phys. Rev. Lett. 60, 1330 (1988).
5. Ch. Niedermayer, J. I. Budnick, B. Chamberland, A. Golnik, E. Recknagel, M. Rossmanith, A. Weidinger and D. P. Yang, preprint.

Percolation Treatment of Frustrated Ising Lattices

Y. Kasai and K. Ohnaka

Department of Applied Physics, Osaka University, Suita, Osaka 565, Japan

In a previous paper [1], we proposed an exact transformation from frustrated Ising lattices into percolation systems. The percolation cluster is considered to express the spin cluster which is able to make up the spin-glass type order. As an example, we applied the above consideration to Bhatt and Young's 3D spin glass simulation data and obtained the critical concentration of the corresponding frustrated percolation system [2]. With the use of dimensional invariant relation for the critical concentration, we can obtain a value of effective dimension for this system, which is far lower than that of lattice dimension. This result can be explained by the coexistence of low dimensional infinite clusters. The avoiding tendency between clusters by frustration is consistent with the dimensional lowering. However, the structure of the coexisting state is not clear. To obtain more detailed information, we intend to perform the direct simulation of frustrated percolation system and to show the snap shot of the clustering state. Since the display of a 3D lattice is difficult, we treat the 2D frustrated lattice in this work. As a result, we observe at a sufficiently high concentration that the macroscopic cluster has a low dimensional structure although the magnitude of macroscopic cluster is the order of site number. In other words, the lower dimensional branches fill up nearly the whole lattice. It is probable that this tendency occurs similarly in 3D frustrated lattices. Our simulation has been done as follows.

1. Generating Function

We write down the transformation from the partition function $Z(L)$ of an Ising system with mixed $\pm J$ to the generating function $\Xi(\xi)$ of the frustrated percolation system:

$$Z(L) = \sum_{\{S(i)\}} \exp(L \sum_{\langle ij \rangle} S(i)S(j)\sigma(i,j))$$

$$= e^{-BL} \sum_{\{g\}n.f.} \xi^{\ell(g)} 2^{C(g)} = e^{-BL} \Xi(\xi) , \qquad (1.1)$$

where $L=J/kT$, $\xi = e^{2L}-1$, B denotes the number of whole bonds on the lattice, $S(i)$ the i Ising spin variable, $\sigma(i,j)$ the sign of interaction between the i and j sites, g an on-bond graph, $\ell(g)$ the number of on-bonds on g, $C(g)$ the number of clusters on g and $\{g\}n.f.$ means the summation only over the non-frustrated on-bond graphs. This restriction on graphs produces the avoiding property between the large clusters [2]. The concentration p of on-bonds is given as

$$p = B^{-1} \xi \, \partial \ln \Xi / \partial \xi . \qquad (1.2)$$

2. Master Equation

The generating function gives us the equilibrium probability $Pe(g)$ of state g:

$$Pe(g) = \xi^{\ell(g)} 2^{C(g)} \qquad \text{for an allowed g}. \qquad (2.1)$$

The time dependent probability $P(g)$ of state g satisfies the master equation

$$\partial P(g)/\partial t = -\sum_{\{g\}} \Gamma(g \rightarrow g')P(g) + \sum_{\{g'\}} \Gamma(g' \rightarrow g)P(g') . \qquad (2.2)$$

Springer Series in Synergetics, Vol. 43   **Cooperative Dynamics in Complex Physical Systems**   221
Editor: H. Takayama   © Springer-Verlag Berlin, Heidelberg 1989

If $\Gamma(g \rightarrow g')$ is confined within a one-bond alternating process, the detailed balance between allowed g and g' at equilibrium is given as

$$\Gamma(g \rightarrow g')/\Gamma(g' \rightarrow g) = Pe(g')/Pe(g)$$

$$= \left\{ \begin{array}{ll} \xi/2^\Delta & \text{for } g'=g + \text{'on-bond'} \\ 2^\Delta/\xi & \text{for } g'=g - \text{'on-bond'} , \end{array} \right. \tag{2.3}$$

where Δ takes value 0 if both terminal sites of the 'on-bond' has a path connecting them on g other than 'on-bond' itself and 1 otherwise.

3. Simulation

Making use of the relation (2.3), we formulate the simulation scheme. The graph theoretical approach is used to calculate the value of Δ for an allowed graph. As a model case, we simulate the full-frustrated square lattice (8×8 cyclic). From the simulation data, we observe the following facts:

(ⅰ) At the maximum concentration 0.75 (0° K for the Ising system), all the sites belong to the macroscopic cluster each elementary face of which is made of six on-bonds. This clustering state is kept below but near concentration 0.75.

(ⅱ) As the concentration decreases, the branching of the macroscopic cluster (low dimensional structure) proceeds, which is enhanced by the avoiding property between large clusters (see Fig. 1).

(ⅲ) As concentration decreases above ∼0.5, the macroscopic cluster vanishes.

The magnitude of macroscopic cluster is essentially the order parameter. In this sense, the full-frustrated square lattice has an ordered state (spin-glass type) at low temperatures (high concentration). At the transition concentration (∼0.5), the thermodynamic function has no analytic anomaly [3]. Note that branching enhanced by frustration occurs from sufficiently lower concentration than the transition concentration.

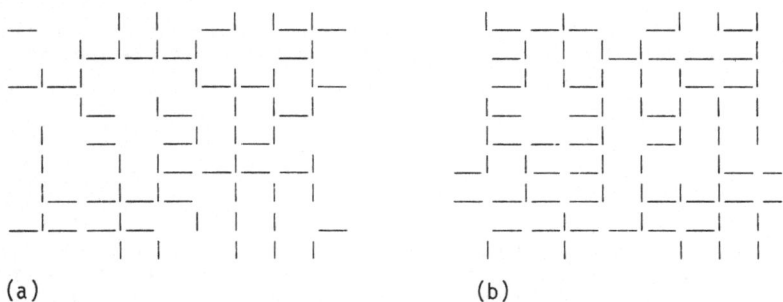

(a)   (b)

Fig. 1. A snap shot of the clustering state for the full-frustrated square lattice (8×8 cyclic) (a) p=0.55, kT/J=1.24, (b) p=0.62, kT/J=0.868

References

1. Y. Kasai and A. Okiji, Prog. Theor. Phys. 75, 1076 (1986).
2. Y. Kasai and A. Okiji, Prog. Theor. Phys. 79, 1080 (1988).
3. G. André et al., J. Phys. (Paris) 40, 479 (1979).

Critical Properties of the Fuzzy Model

T. Kawasaki

Department of Physics, College of Liberal Arts, Kyoto University, Kyoto 606, Japan

When the random spin system is quenched in the computer simulation experiment from the infinite to a temperature below the critical temperature T_C, it should be inevitably trapped at one of the metastable states, depending on the random sequence in the Monte Carlo simulation. Then the random spin system shows similar behavior in some respects as the spin glass due to the fact that both have highly degenerate metastable states in the energy./1/, /2/ The randomness in the random spin system is usually introduced either in the coupling strength or in the site occupation. As far as we are concerned with the quadratic spin Hamiltonian, the more general introduction of the randomness may be achieved by the Fuzzy Spin Model(FSM)/3/, defined by

$$H = -\sum_{ij} J_{ij} S_i S_j$$

where spins S_j are Ising ones and uniformly random between 0 and 1 in their magnitudes site by site. The coupling J_{ij} is set ferromagnetic ($J_{ij} = J$) for nearest neighbor pairs only. The model is different from the Mattis model in that spin magnitude varies continuously between 0 and 1. We call static when spin magnitude once chosen is fixed forever, and time-dependent when spins fluctuate themselves in their magnitudes independently of the Hamiltonian.

1. Static Fuzzy Spin Model

When the system is quenched from a completely random phase to the ordered phase (the ferromagnetic phase by definition), the total magnetisation M in this system is found zero. One of the typical spin pattern at 0 K with M=0 is shown in Fig.1 where black and white areas indicate up and down spins respectively. Since the model is essentially ferromagnetic, all spins should align in the same direction at the true ground state. However due to the existence of degenerate metastable states one of which the system is trapped in, the spin clusters are not compact but rather ramified: The total perimeter estimated by counting the total number of broken bonds defined as bonds with antiparallel spin pairs is found proportional to the system size N. Furthermore most of the metastable states lie in a narrow region centred at 0.78×(energy of the true ground state). As the temperature goes up but still below the critical temperature, the clusters becomes rather compact due to thermal fluctuations which favor the system to overcome barriers between metastable states. These features are common to random systems having highly degenerate metastable states.

While the magnetisation keeps vanishing even below T_C in the case of quenching, the spin glass order parameter defined by q($= \langle\langle S\rangle^2_{time}\rangle_{system}$ grows sharply below a temperature, denoted by T_G, as is seen in Fig.2. Moreover the values q are found almost the same, except near T_G, with those obtained by the annealing treatment, gradual cooling or heating of the system in the simulation. (The magnetisation of course becomes finite in this treatment.) Susceptibility and specific heat are also plotted in Fig.3. Since the system is not yet found at equilibrium after more than 10^5 Monte Carlo Steps near the critical temperature, whether or not there is a cusp at T_C in the specific heat is not certain now. Also within the present accuracy of the simulation we cannot discriminate T_G from T_C.

Springer Series in Synergetics, Vol. 43   **Cooperative Dynamics in Complex Physical Systems**
Editor: H. Takayama   © Springer-Verlag Berlin, Heidelberg 1989

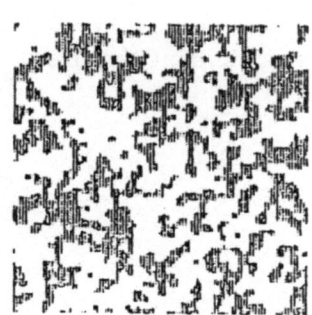

Fig.1 Spin pattern at one of the meta-stable states (sample size $N = 100 \times 100$)

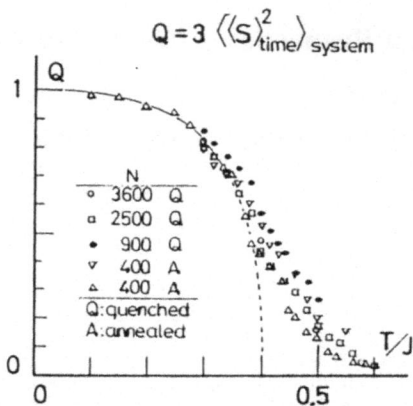

$$Q = 3 \left\langle \langle S \rangle^2_{time} \right\rangle_{system}$$

N		
○	3600	Q
□	2500	Q
●	900	Q
▽	400	A
△	400	A

Q:quenched
A:annealed

Fig.2 Temperature dependence of the spin glass order parameter.

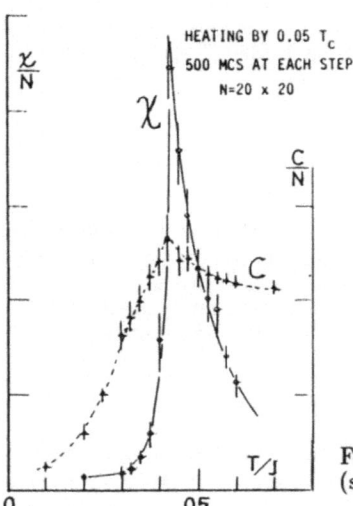

HEATING BY 0.05 T_c
500 MCS AT EACH STEP
N=20 x 20

Fig.3 Susceptibility and specific heat. (sample size = 20×20)

2.Time-dependent Fuzzy Spin Model

As mentioned above, the model is defined as the one in which magnitude of each spin makes self-sustaining, random fluctuation, independent of the Hamiltonian, while keeping the initial random distribution of the spin magnitude. When the fluctuating period is shorter than that of thermal fluctuations and the relaxation time of the system, the system will have equilibrium states. Due to the self-oscillation of the magnitude, each spin can take any energy between -zJ and +zJ and then the system has a possibility to behave like a continuous spin model. In this model there is no difficulty in the simulation as was met in the Static Fuzzy Model.

Finally we mention that these models are proposed to study phase separation of particles with different sizes or collective motion of the self-sustaining species.

1. A.E.Jacobs and C.M.Coram, Phys.Rev.**B36**(1987),3844.
   M.Cieplak and T.R.Gawron, J.Phys.**A20**(1987),5657.
2. T.Kawasaki, Prog.Theor.Phys.**80**(1988) in press.
3. The model DTRM named in ref.2 is renamed as the Fuzzy Spin Model.

Monte Carlo Simulations on a Controlled-Frustration Model of $\pm J$ Ising Systems

I. Ono and K. Ishikawa

Department of Physics, Tokyo Institute of Technology,
Oh-Okayama, Meguro-ku, Tokyo 152, Japan

Frustration is believed to be essential for the occurrence of the spin-glass phase[1]. Nevertheless, quantitative relations between spin-glass transitions and frustration have not yet been established. For the regularly frustrated Ising lattices the phase transition has been exactly solved in some two-dimensional models[2,3].

Here Monte Carlo simulations have been performed to investigate how the phase transition is influenced by the concentration of frustration or the spatial correlation of frustrated plaquettes. In order to extract the pure effect of frustration, we propose, under the condition of a fixed 50-50 concentration of $\pm J$ bonds, a special random bond-distribution which allows an arbitrary concentration of frustration. We start with the following Ising Hamiltonian of Mattis type consisting of random nearest-neighbor interaction:

$$\mathcal{H} = - \sum_{<i,j>} J_{ij}\, \sigma_i \sigma_j \; , \qquad \sigma_i = \pm 1, \qquad (1)$$

where $J_{ij} = J\epsilon_i \epsilon_j$ and ϵ_i is a random parameter on the i-th site which may take the value $+1$ or -1. One half of the lattice sites are randomly selected and are set to $\epsilon_i = 1$ and the rest to $\epsilon_i = -1$. This model clearly contains no frustration. Then picking out any bond and changing its sign of interaction, we have a wrong bond[4] and thus a pair of frustration plaquettes is created. By repeating this process the concentration of frustration increases linearly as long as the positions of frustration are selected so as not to overlap. The above-mentioned process never results in any change in the ratio of ferro- and antiferromagnetic bonds on the average. The ground state energy increases proportional to the concentration of wrong bonds C_W as

$$E_g(C_W) \, / \, |E_g(0)| = -1 + 2C_W, \qquad (2)$$

where $E_g(0) = - NzJ/2$. For the square (SQ) and simple cubic (SC) lattices the concentration of frustrated plaquettes C_F is connected to C_W by the relation $C_F = 4C_W$ when frustration pairs do not overlap.

Monte Carlo simulations have been performed on dimensions up to 64×64 for the SQ lattice or $20 \times 20 \times 20$ for the SC lattice using a complete periodic boundary condition. After the first 5000 steps had been discarded, the following 5000 steps were utilized to evaluate thermal averages. The order parameter similar to that of the Mattis model is defined by

$$\langle O \rangle = (1/N) \, \langle \, \sum_i \epsilon_i \sigma_i \, \rangle \; , \qquad (3)$$

where N is the total number of spins. Its temperature dependence

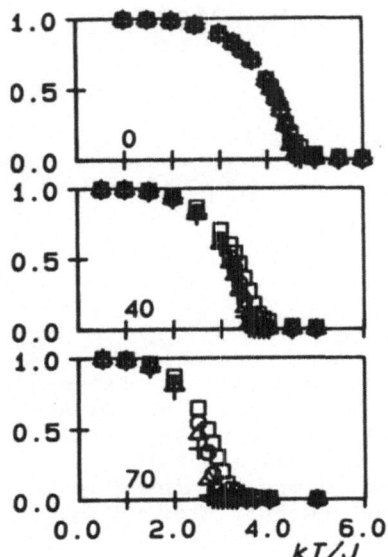

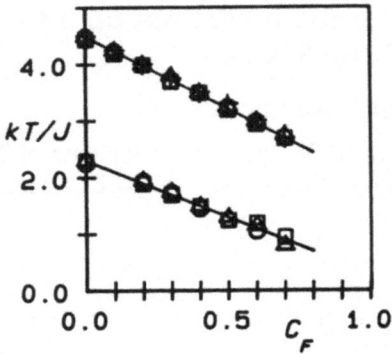

Fig. 2. Transition temperatures versus the concentration of frustration C_F . The upper and lower parts stand for SC and SQ lattices, respectively. Marks □ or Δ are determined from the peak of the specific heat or the staggered susceptibility, and o or +, from the maximum gradient of the order parameter or the uniform susceptibility, respectively.

Fig. 1. Temperature dependences of squared order parameters for 0, 40 and 70% of C_F. Marks □, o, Δ, and + represent 8^3, 12^3, 16^3, and 20^3 SC lattices.

is plotted in Fig. 1, where the order parameter at T = 0K is shown to be equal to the value for perfect order even when C_F = 0.7. Both the specific heat and the staggered susceptibility conjugate to this order parameter, not shown here, also exhibit a sharp peak at the transition temperature. The transition temperatures T_c estimated in several different ways are satisfactorily coincident with each other, as shown in Fig.2. They have been found to decrease proportional to C_F, according to an empirical rule as $T_c(C_F) / T_c(0)$ = 1 - 0.57 C_F for the SC lattice or 1 - 0.87 C_F for the SQ lattice . As long as the wrong bonds are isolated from each other, in other words the pairs of frustrated plaquettes do not overlap each other, the phase transition seems not to be spin-glass-like but to be 2nd-order-like until nearly C_F = 0.7 for the SC lattice. This should be compared with the fact that the random ±J Ising model, which is known to exhibit the spin-glass transition on the SC lattice, contains a 50% concentration of frustrations.

Finally it should be emphasized that the occurrence of a spin-glass phase is attributed not only to the concentration of frustration but also to its spatial correlation. It should probably be related to the degeneracy of spin arrangements in the ground state.

References

1. G. Toulouse: Commun. Phys. 2 115 (1977).
2. J. Villain: J. Phys. C10 1717 (1977).
3. G. Andre, R. Bidaux, J.-P. Carton, R. Conte, and L. de Seze: J. de Physique 40 479 (1979).
4. I. Ono: J. Phys. Soc. Japan 41 345 (1976).

Simulation of Spin Dynamics for FMR in Compounds with Competing Anisotropies

Y. Natsume

Department of Physics, Faculty of Science, Chiba University,
Yayoi-cho, Chiba 260, Japan

The direct simulation is adapted for the first time to the study of ferromagnetic resonance in magnetic compounds by the method of spin dynamic simulation, which has been developed by the present author and his co-workers. This simulation is, in particular, useful for investigating the behavior of spins in systems including the competition between magnetic interactions. Here, the calculation is made and discussed in connection with experiments of FMR in Co-doped K_2CuF_4.

1. Introduction

Pure K_2CuF_4 is a typical two-dimensional ferromagnetic compound [1] with an easy-plane type anisotropy $+DS_z^2$ and confirmed to be the system including the co-operative Jahn-Teller distortion of an octahedron for F^- ions surrounding a Cu^{2+} ion [2,3]. In order to make clear the effect of magnetic impurity with a large single-ion type anisotropy $-D'S_x^2$ or $-D'S_Y^2$ in this system, ANDE et al.[4] have recently made FMR measurements in spherical samples of $K_2Cu_xCo_{1-x}F_4$ at 1.3K. As a result of x= 0.995, they have obtained the following remarkable facts in the case where the external field $\vec{H}_o$ lies in the c-plane; though the resonance frequency v remains the value for K_2CuF_4 (1.67v_o)[5] with the narrow linewidth (Δv=0.008v_o)(v_o is the Larmor frequency $\gamma H_o/2\pi$ for a free electron) for $\vec{H}_o$//[110], v becomes 1.20v_o with quite a large width such as Δv=0.43v_o for $\vec{H}_o$//[100]. Namely, the lineshape is quite disturbed as the direction of $\vec{H}_o$ gets near to [100].

2. The Method of Simulation

In order to explain the remarkable features mentioned in 1, we carry out the computer simulation for spin dynamics which has been proposed [6,7] and adapted to the study of spin glass [7-9] by the present author and his co-workers. In short, we investigate the behavior of spins by the following equation for a magnetic moment at site i:

$$d\vec{M}_i/dt = \gamma \vec{M}_i \times H_i^{\textbf{eff}} + (\vec{M}_i^{\textbf{eff}} - \vec{M}_i)/T_1 \quad , \tag{1}$$

where

$$\vec{H}_i^{\textbf{eff}} = \sum_{n(i)} J_{i,n(i)} \cdot \vec{M}_{n(i)}/(\hbar \gamma)^2 + H_0 + 2D_i \cdot \vec{M}_i/(\hbar \gamma)^2 \quad , \tag{2}$$

$$\vec{M}_i^{\textbf{eff}} = \vec{H}_i^{\textbf{eff}} \cdot |\vec{M}_i|/|\vec{H}_i^{\textbf{eff}}| \quad . \tag{3}$$

In (2), n(i) means the nearest neighbor of i-site and summation is made for n(i). Here, $J_{i,n(i)}$ is an exchange interaction between i-site and n(i) site. In this paper, we adopt the model schematically shown in Fig.1. On the basis of this model, we make the calculation; in the first place, we make the simulation using (1)-(3) from the initial state where directions of spins are given randomly. Here, the value of T_1 is determined to be sufficiently large ($T_1 v_o \gg 1$) under the condition that it never affects the obtained state. As such a state corresponds to an equilibrium one for sufficiently low-temperature ($J \gg k_B T \gg \mu_B H_o$), it can be adopted as the initial state for the subsequent operation for FMR. Namely, we make, furthermore, simulation from such a state considering only the first term (torque term) in (1) so as to obtain information on the behavior of spins in FMR.

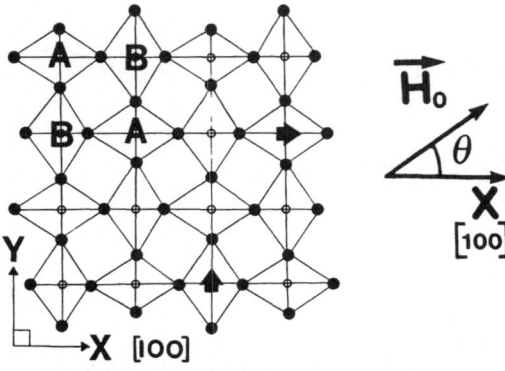

Fig.1. The present model: Open (closed) circles represent Cu^{2+}(F^-) ions in the c-plane. By the co-operative Jahn-Teller distortion, sites for Cu^{2+} are classified into A and B sites. A Cu^{2+} ion has a weak easy-plane type anisotropy ($+DS_z^2$) for both sites. If a Co^{2+} ion is placed at an A (B) site, it has a strong single-ion type anisotropy expressed as $-D'S_x^2$ ($-D'S_Y^2$) as schematically shown by black arrows. The angle between the field $\vec{H_o}$ (lying in the c-plane) and [100] axis is θ.

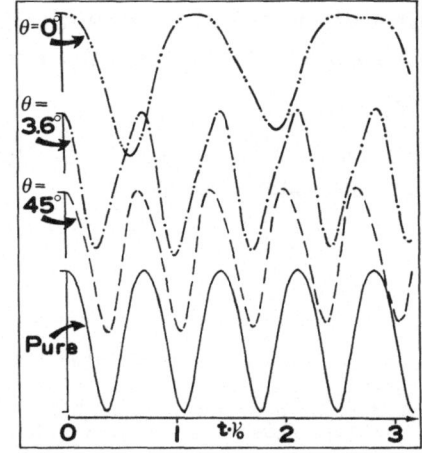

Fig.2. The calculated time correlation $C(t)$ of total magnetization is shown for pure K_2CuF_4 (the lowest curve) and $K_2Cu_{0.99}Co_{0.01}F_4$ (upper three curves for θ=45°,3.6°and 0°). Resonance frequency decreases effectively with decreasing θ. Such a decrease reflects the complex motion of spins.

3. Results of Simulation and Discussion

Here, we discuss the time correlation $C(t)$ of the total magnetization. The values of parameters are determined on the basis of pure K_2CuF_4 and K_2CoF_4: D=0.05J, D'=J. It is supposed that the exchange interaction between Cu^{2+} and Co^{2+} ions is antiferromagnetic and its absolute value is the same as that between Cu^{2+} ions.

The typical example for $C(t)$ in 40x40 square lattice is shown in Fig.2. We see that a curve is nearly a cosine with ν=1.41ν_o for any θ in K_2CuF_4. On the other hand, the curve has quite strong dependence on θ; with decreasing θ ν decreases significantly, while it is near to that in K_2CuF_4 at θ=45°. This dependence of ν on θ agrees well qualitatively with the experimental results [4]. Furthermore, we can discuss the feature of each waveform; it is noticeably disturbed for $\theta\sim$0°. This pronounced feature corresponds to an experimental fact that the resonance for $\theta\sim$0° has quite a large linewidth.

The present author would like to thank Mr. K. Fujimoto, Mr. Y. Yoshihara and Mr. A. Yamamoto for their collaboration with this spin dynamic simulation.

References

1. I. Yamada: J. Phys. Soc. Japan 33, 979 (1972)
2. I. Khomskii and K. I. Kugel: Solid State Commun. 13, 763 (1973)
3. Y. Ito and J. Akimitsu: J.Phys.Soc.Japan 50, 1333 (1976)
4. T. Ande, Y. Yamaguchi, M. Itoh and I. Yamada: to be submitted.
5. H. Yamazaki, Y. Morishige and M. Chikamatsu: J.Phys.Soc.Japan 9, 2872 (1981)
6. K. Fujimoto, T.Yoshihara and Y. Natsume: Theoretical and Applied Mechanics, 36, (University of Tokyo Press, Tokyo 1988) p.413
7. Y. Natsume, K. Fujimoto and Y. Yoshihara: Proceedings of International Symposium on Physics of Magnetic Materials, Sendai 1987 (World Scientific,1987) p.318
8. T. Yoshihara, K.Fujimoto and Y. Natsume: Theoretical and Applied Mechanics, 36, (University of Tokyo Press, Tokyo 1988) p.421
9. Y.Natsume, T.Yoshihara and K. Fujimoto: IEEE Trans.Magnetics MAG-23, 2239 (1987)

Part III

Optimization Problems and Neural Network Models

Optimization Problems at Thermal Equilibrium

M. Mézard

Laboratoire de Physique Théorique de l'Ecole Normale Supérieure*,
24 rue Lhomond, F-75231 Paris Cedex 05, France

Abstract : We review the present status of the analytic study of some
optimization problems using concepts and techniques from spin glass
theory. Particular emphasis is put on the matching and travelling salesman problems. These have been worked out in detail in the case of
random independent distances. The replica symmetric solution to the
matching problem has been shown to be stable and in good agreement
with numerical simulations. The travelling salesman problem is still
waiting for a stability analysis and some good numerical results. The
extension of the analysis to Euclidean finite dimensional problems is
explained (in the case of the matching problem) ; it suggests new approaches to the analytic study of spin glasses in finite dimensions.

In recent years the spin glass family has grown considerably. As
strange "materials" as neural networks or travelling salesmen have
been added to the original family of metallic (CuMn...) or insulating
($Eu_xSr_{1-x}S...$) spin glasses. Furthermore, some of the most interesting
representatives of these newborn spin glasses turn out to be under-
standable by mean field techniques. As we shall see in the case of
optimization problems such as the travelling salesman problem (TSP)
the use of concepts and methods from spin glass theory in these new
fields is doubly interesting : first of all it allows one to derive
new results in these fields, secondly it greatly enriches our understanding of spin glasses, providing some information on the generality of the results obtained in mean field theory, and suggesting new
approaches to the theoretical analysis of finite dimensional spin
glasses.

We shall concentrate on matching and TSP. In both cases one is
given N points (N must be even for matching) i = 1...N, and the distances l_{ij} between them. The allowed configurations are: a set of N/2
links such that each point has one link only for matching, and a clo-

---

*Laboratoire Propre du Centre National de la Recherche Scientifique,
associé à l'Ecole Normale Supérieure et à l'Université de Paris-Sud.

sed tour of N links going through all the points for the TSP. In both cases the problem is, given a sample (i.e. a set of l_{ij}), to find the configuration of shortest length.

These are typical examples of combinatorial optimization problems : in every such problem there is a discrete set of allowed configurations, each configuration C has a cost E_C and one must find the configuration of lowest cost. What will interest us here is not so much the algorithmic problem of finding the optimal configuration, but rather the possibility of predicting some properties of this configuration (beginning with its cost) for some sets of samples. For the two problems above, the most studied such sets are: the random link problems in which the l_{ij} (=l_{ij}) are independent random variables with a probability distribution $\rho(l)$, and the Euclidean problems in which the N points are chosen randomly inside a hypercube of side 1 in a D-dimensional space, and l_{ij} is the Euclidean distance between i and j.

Statistical physics deals with the asymptotic behaviour of optimization problems in the limit $N \to \infty$ (thermodynamic limit). A temperature parameter $T = 1/\beta$ is introduced and one assigns a Boltzmann weight $\exp(-\beta E_C)$ to each configuration. The properties of the optimal configuration (the ground state) are then recovered in the limit $T \to 0$ [1]. After some thought one realizes that the interesting optimization problems are disordered and frustrated systems, so that they share many properties of spin glasses.

This analogy has led to the analytic study of several optimization problems using the methods of spin glass theory, mainly the replica method and the cavity method [1]. For matching and TSP, the random link models are naturally mean-field-like models so that they are presumably exactly solvable by the above methods. In contrast, the Euclidean problems bear some resemblance to the finite dimensional spin glasses and their analysis is more complicated. Hereafter we shall review what is known about these problems today.

Let us begin with the simplest problem, matching. From the algorithmic point of view it is a polynomial problem: there exist algorithms which solve it in computer time growing like N^3 [2]. The replica symmetric solution of the random link matching problem was found in [3]. For the much studied case of flat distances (uniform distribution of the length of each link on [0,1]), the total length of the optimal tour is predicted to tend towards $\pi^2/12$ when $N \to \infty$. (The reason why it does not increase with N is that, the larger N, the higher the probability of there being small links in each sample). This result, together with the similar result for bipartite matching (a variant where there are N/2 red points and N/2 blue points, and there can be links only between points of different colours) are by now rather well established. The replica symmetric solution has been shown to be locally stable with respect to replica symmetry breaking [4], and the length of the optimal matching coincides with the results of numerical simulations (extrapolated to $N \to \infty$) within error bars (due to statistical fluctuations and extrapolation to $N \to \infty$) of the order of 2 or 3 per 1000.

As for its impact on spin glass theory, the study of the matching problem has prompted us to analyse mean field models in which each spin interacts effectively with a few other spins only. Most of the effort on spin glass theory had concentrated on the infinite range SK model [5]. However, what is important for a spin glass model to be solvable by mean field techniques is not that the interaction is of infinite range, but rather that a given spin has the same probability of interacting with all the others. If this probability is very small ($\sim 1/N$), the spin will interact offectively with only a few neighbours (but the total number of neighbours still increases exponentially with the "distance"). Such a strongly diluted spin glass had been considered a few years ago by Viana and Bray [6], and has seen renewed interest recently [7, 8, 9]. It is typically what happens in the random link matching, where the distances between all pairs of points are random and independent, but a given point will be matched with only one of its "near neighbours" - for flat distances, those points are at distance $\sim 1/N$ - and there is a finite number of them. Matching is the first such problem in which it was realized that a whole order parameter function is needed even at the replica symmetric level [3, 10]. The reason is easily understood in the cavity method [11]: in the SK model the local field is the sum of uncorrelated contributions from all the N-1 other sites, therefore its distribution tends towards a gaussian, and only one order parameter is needed, namely q, the width of the gaussian. In strongly diluted models the local field is the sum of a finite number of terms, and its distribution is no longer gaussian : the whole distribution has to be kept as the order parameter. This raises new problems, and it is not completely settled yet which properties of the SK model will be carried along to this diluted case [9].

Let us now turn to the TSP. On the algorithmic side this is an NP complete problem [12]. There is of course no known polynomial algorithm for this problem, but many clever heuristics have been worked out [13]. Application of spin glass ideas to this famous problem was initiated in [14] and has been tried in several works since then [15-

232

19]. The study of the random link TSP with the replica method was undertaken in [17]. Like in matching, and for the same reasons, the order parameter is a function, even at the replica symmetric level. However, even within this replica symmetric approximation, the saddle point equations turned out to be extremely difficult to solve. An approximate solution was found in [17] for the case of flat distances, but the procedure was too complicated to be generalizable to other distances, or to get more accurate predictions. On the other hand, the cavity equations have been written for this problem some time ago [18], and recently we have been able to obtain a tractable solution of the random link TSP by solving these cavity equations directly at zero temperature [19]. This is an example which can be solved relatively easily by the cavity method (always at the replica symmetric level) and which is extremely difficult to solve with replicas.(Although it should be possible to map the saddle point equations of the replica method onto the self-consistency equations of the cavity method through some sophisticated changes of variables). The reason is that the cavity method deals directly with physical quantities like the distribution of local magnetizations. The resulting prediction for the length of the optimal tour is $L_0 = 2.0415...$ for flat distances, $L_0 = 1.8175... \sqrt{N}$ when the distribution of distances behaves as $\rho(1) \sim 1$ at small distances. The results for flat distances, as well as the distribution of lengths of the occupied links in the optimal tour, have been compared with numerical simulation of the TSP obtained with efficient heuristics and seem to be in agreement with it. However, the stability analysis of the replica symmetric solution in the TSP has not yet been done, so that these analytic results are still not as well established as the corresponding ones of the matching problem.

To conclude I would like to mention some recent progress on the Euclidean matching problem, which is generalizable to the Euclidean TSP and to finite dimensional metallic spin glasses [20]. In the Euclidean matching problem the points i = 1...N are located at random positions and $l_{ij} = |\vec{x}_i - \vec{x}_j|$ is the Euclidean distance. So the l_{ij} are now correlated and it is not so clear how to use mean field techniques. A zero-order approximation to this problem would be to neglect these correlations. For instance for the 2D Euclidean problem one could consider as a zero-order approximation the random link problem with distribution $\rho(1) \sim 2\pi 1$ when $1 \to 0$. Surprisingly enough the result of this approximation is off by only 5%. What is interesting is that this approximation can be improved on systematically. The idea is to take the correlations at increasing orders into account.

For instance one can exactly take account of the three link correlations (including triangular inequalities), while one continues to neglect all the connected correlation functions involving at least four links. At a given order of this approximation the model is still solvable, though the complexity of the solution improves fast.

This kind of approach can be used also in metallic spin glasses [20]: the coupling constant J_{ij} between spins i and j is a function $j(\,|\,\vec{x}_i - \vec{x}_j\,|\,)$, where the $\vec{x}_i$ are random positions. The zero-order approximation, neglecting the correlations, corresponds to some kind of strongly diluted spin glass model as described above. It will be quite interesting to study these various approximations to the finite dimensional spin glass problem. I think that this is a good example of how the study of new spin glasses like matching, TSP, etc., can bring new ideas into the field of old spin glasses like CuMn...

Acknowledgements

The work presented here has been done mostly in collaboration with G. Parisi, and also with W. Krauth and J. Vannimenus. It is a pleasure to thank them for these collaborations.

References

[1] For a review see : M. Mézard, G. Parisi and M.A. Virasoro, in Spin glass theory and beyond (World Scientific, Singapore 1987)

[2] R.E. Burkardt and U. Derigs eds, Lecture Notes in Economics and Math Systems, vol. 184 (Springer-Verlag, Berlin, Heidelberg 1980)

[3] M. Mézard and G. Parisi, J. Physique Lett. 46 (1985) L 771

[4] M. Mézard and G. Parisi, J. Physique 48 (1987) 1451

[5] D. Sherrington and S. Kirkpatrick, Phys. Rev. Lett. 35 (1975) 1792

[6] L. Viana and A.J. Bray, J. Phys. C18 (1985) 3037

[7] M. Mézard and G. Parisi, Europhys. Lett. 3 (1987) 1067

[8] I. Kanter and H. Sompolinsky, Phys. Rev. Lett. 58 (1987) 164

[9] P. Mottishaw and C. De Dominicis, J. Phys. A20 (1987) L375
     C. De Dominicis and P. Mottishaw, Saclay preprint SPhT/87-139

[10] H. Orland, J. Physique Lett. 46 (1985) L763

[11] M. Mézard, in Heidelberg colloquium on glassy dynamics, Lecture Notes in Physics 275 (1987) 354

[12] M.R. Garey and D.S. Johnson, Computers and intractability (Freeman, New York 1979)

[13] See for instance : E.L. Lawler, J.K. Lenstra, A.H.G. Rinnooy Kan and D.B. Shmoys, eds, The Traveling salesman problem Wiley, Chichester 1985)

[14] S. Kirkpatrick, C.D. Gelatt Jr., M.P. Vecchi, Science 220 (1983) 671

[15] S. Kirkpatrick and G. Toulouse, J. Physique 46 (1985) 1277

[16] J. Vannimenus and M. Mézard, J. Physique Lett. 45 (1984) L1145

[17] M. Mézard and G. Parisi, J. Physique 47 (1986) 1285

[18] M. Mézard and G. Parisi, Europhys. Lett. 2 (1986) 913

[19] W. Krauth and M. Mézard, in preparation

[20] M. Mézard and G. Parisi, LPTENS 88/21, and in preparation

Bayesian Statistics and Statistical Mechanics

Y. Iba

Institute of Physics, College of Arts and Sciences,
University of Tokyo, Tokyo 153, Japan

In this note we give a brief exposition of a fruitful interaction
between statistical mechanics and modern statistical methods which
might be rather unfamiliar to most physicists. Details of the
author's contribution will appear in forthcoming papers [1,2].

1.Bayesian Statistics and Statistical Mechanics

Bayesian statistics provides a convenient framework to study inverse
problems. In these problems prior knowledge is known to play an
important role. By the prior knowledge we mean information such as
'smoothness of the answer',' smallness of a certain parameter', etc.
In Bayesian language, they are expressed in terms of 'prior
distribution' $\pi(\{x\})$ and 'likelihood function' $L(\{y\},\{x\})$. The
former represents the prior knowledge on parameters $\{x\}$ and the
latter represents the relation of parameters with data $\{y\}$. For
given data $\{y\}$, 'posterior distribution' $P(\{x\})$ is defined by

$$P(\{x\}) = \frac{L(\{y\},\{x\})\,\pi(\{x\})}{\int L(\{y\},\{x\})\,\pi(\{x\})\,d\{x\}} \qquad (1)$$

Calculation of (1) is not easy if the number of parameters
becomes large. For example, in image processing problems we should
deal with hundreds (or thousands) of 'microscopic' parameters. In
these cases the situation becomes similar to that in the equilibrium
statistical mechanics where cooperative behavior among large number
of degrees of freedom are crucial. In this context increasing
attention has been directed to the analogy between statistics and
statistical mechanics [3,4,5,6].
Monte-Carlo approach to the image processing problem [3,7] is the
most promising one. 'Connectionist's networks' [5] can also be
considered as large scale statistical models. Another related
topic is 'simulated annealing' approach to combinatorial
optimization problems. We can use this method in the optimization
problems in statistics, i.e., search for the parameters $\{x\}$
maximizing (1) ('mode'). However, generation of the distribution
(1) with the Monte-Carlo method ('simulation without annealing') is
also important in itself as shown in latter sections.

2.Estimation of Macroscopic Parameters

One of the most important problems in Bayesian statistics is how to
control the prior knowledge. In most cases the prior distribution
(and the likelihood function) contains several 'macroscopic'
parameters $\{\alpha\}$, e.g., threshold values in the image processing. A
powerful method to estimate these macroscopic parameters is the
maximization of

$$(-1/2)ABIC(\{\alpha\}) = \ln \int L(\{y\},\{x\}) \pi (\{x\},\{\alpha\}) d\{x\} \qquad (2)$$

[8,9,10,6], where the integration (or summation) is taken over all the 'microscopic' parameters $\{x\}$.

Methods developed in statistical mechanics work effectively in evaluating (2),an analogous quantity to free energy (or entropy). Monte-Carlo realization of the distribution (1) [10,11,12] is the most interesting one. In this case maximization of (2) by the gradient-ascent method gives generalized 'Bolzmann machine-type' equation [5] for the macroscopic parameters. Approximation techniques in statistical mechanics (e.g., mean field approximation) to (2) are also applicable [1,6]. In [6] some theoretical aspects of these topics have been discussed.

3.Numerical Classification

Applications of these analogies are not necessarily restricted to the image processing problems. An interesting example is the 'unsupervised' classification problem based on an explicit statistical model. In classifying objects into categories, the possible number of partition diverges exponentially as the number of the objects increases. Here we can use the simulated annealing method to obtain the optimal solution (or one of the 'near optimal' solutions). It is, however, not the whole story. Calculating (1) or (2) with the Monte-Carlo method, we can deduce more information such as the reliability or statistical significance of the optimal solution. Detailed treatment of a specific classification problem (classification based on pair comparison) will be published in the forthcoming paper[2]. In [2] roles of concepts in statistical mechanics (e.g., spontaneous symmetry breaking, frustration) will also be discussed.

4.Applications of Bayesian Statistics to Physics

In preceding sections, we have mainly discussed the applications of the analogy between statistical mechanics and statistics. On the other hand, modern statistical methods will be effective in analyzing the numerical (or real) experiments in statistical mechanics. Actually, there are many 'improper' inverse problems in physics; a typical example is inverse Laplace transformation appearing in the study of relaxation phenomena. Research in this direction is now in progress.

Reference
1,2. Y.Iba: in preparation
3. S.Geman, D.Geman: IEEE PAMI-6 p.721 (1984)
4. Y.Ogata, M.Tanemura: J.Roy.Stat.Soc. B46 No.3 p.496 (1984)
5. Parallel Distributed Processing, Vol.1 Chap.6-8
   ed. by D.E.Rumelhart and J.L.McClelland(MIT Press,Cambridge 1986)
6. W.Bialek, A.Zee: Phys.Rev.Lett 58 (1987) p.741; preprints
7. T.Poggio, V.Torre, C.Koch: Nature 317 p.314 (1985)
8. I.J.Good: The Estimation of Probabilities
   (MIT Press,Cambridge 1965)
9. H.Akaike: In Bayesian Statistics ed. by Bernard et al.
   (Univ. Press, Valencia 1980)
10.D.Geman, S.Geman: In Disordered Systems and Biological
   Organization ed. by E.Bienenstock et al.
   (Springer, Berlin, Heidelberg, 1986)
11.Y.Ogata: Inst.Stat.Math. Res.Memo. No.347 (1988)
12.Y.Iba: unpublished

Graph Partitioning and Spin Glasses

S. Katsura and M. Sasaki

Faculty of Science and Engineering, Tokyo Denki University,
Hatoyama, Saitama 350-03, Japan

Recently Sherrington and his group [1] related the theory
of graph partitioning to that of spin glasses. They treated their
problem in terms of an integral equation for the distribution
function of the effective field which our group [2] used to treat
the spin glass. The $\pm J$ model for which $P(J) = p\delta(J-1) + (1-p)\delta(J+1)$
is considered. The integral equation for the distribution
function $g(h)$ of the effective field h is given by [2,3]

$$g(h) = \int \delta(h - (1/\beta) \operatorname{th}^{-1}(\operatorname{th} \beta J \operatorname{th} \beta J^{(2)})) P(J) dJ G(H^{(2)}) dH^{(2)} \qquad (1)$$

$$G(H^{(2)}) = \int \delta(H^{(2)} - \Sigma(1/\beta) \operatorname{th} \beta J_j \operatorname{th} \beta H_j) \Pi P(J_j) dJ_j G(H^{(2)}) dH^{(2)}. \qquad (2)$$

Let $S(x)$ be the Fourier transform of the $g(h)$. The integral
equation for $S(x)$ reads

$$S(x) = (1/2\pi) \int dy \, K(x,y) [S(y)]^{z-1}. \qquad (3)$$

The kernel $K(x,y)$ in the case of general p is derived to be

$$K(x,y) = 2\pi\delta(y)\cos x - 2p\frac{\sin(y-x)}{y} - 2(1-p)\frac{\sin(y+x)}{y}$$
$$+ 2p\frac{\sin(y-x)}{y-x} + 2(1-p)\frac{\sin(y+x)}{y+x}. \qquad (4)$$

We put

$$S(x) = a + b\cos x - ic \sin x + \Sigma(-1)^n d_n \, j_n(x). \qquad (5)$$

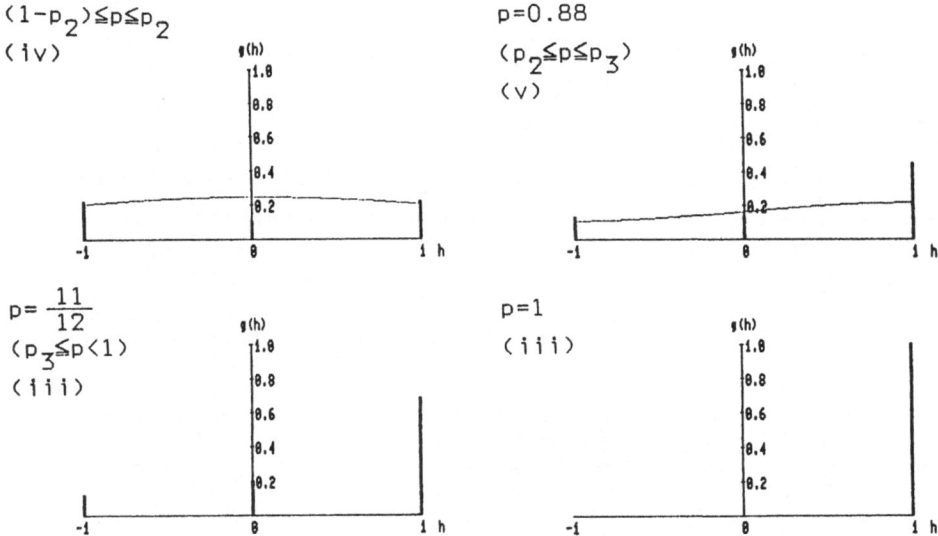

Fig. 1   Distribution of effective field of z=3 for
several values of the concentration p.

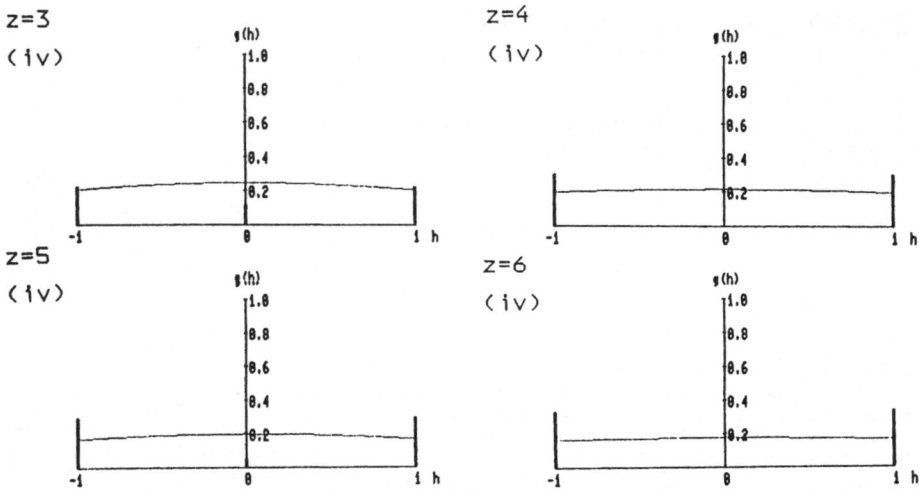

Fig. 2   Distribution of effective field of
          continuous spin glass solution for z=3, 4,
          5, 6.

Substitutiing (4) and (5) into (3), we get a system of algebraic
equations for unknowns a, b, c, and d_ℓ (ℓ=0, 1, 2, 3).

     I. Phase boundaries and several solutions are examined for
z=3. The system of the algebraic equations has the following
solutions (i) a=1, b=c=d_0=d_1=d_2 =d_3=0 (0$\leq$p$\leq$1), paramagnetic
solution, (ii) a=1/3, b=2/3, c=d_0=d_1=d_2=d_3=0 (0$\leq$p$\leq$1), discrete
spin glass solution, (iii) a=1/(2p-1)-1, b=2 -1/(2p-1),
c=$\sqrt{\{(8p-7)(4p-3)\}}$/(2p-1), d_0=d_1=d_2=d_3=0 (7/8$\leq$p$\leq$1), ferromagnetic
solution, (iv) a=0.10683, b=0.43686, d_0=0.45631, d_2=0.05759
c=d_1=d_3=0 (0$\leq$p$\leq$1), continuous spin glass solution, (v) asymmetric
continuous solution as shown in Fig. 1. Solution (v) is composed
of asymmetric three delta functions and an asymmetric continuous
part exists for $p_2$$\leq$p$\leq$$p_3$. The p_2 is the value of p at which the
asymmetric part disappears and connects to the symmetric
continuous distribution (iv). We put a, d_0, d_2, to be the same
as (iv), and d_1/c$\neq$0, $d_3$$\neq$0. Then we got p_2=0.869427. The p_3 is the
value of p at which the continuous part of (v) disappears and
connects to the solution (iii). By substituting d_0=d_1=d_2=d_3=0
(d_1/$d_0$$\neq$0), we got p_3=11/12 (which agrees with Kwon and Thouless
[4]). It is expected that the solution (iv) is realized for
(1-p_2)$\leq$p$\leq$$p_2$, that the solution (v) for $p_2$$\leq$p$\leq$$p_3$ and that solution
(iii) for $p_3$$\leq$p$\leq$1.

     II. The continuous spin glass solution (iv) is calculated
for z=3, 4, 5, 6. They are shown in Fig. 2, The result for z =4,
5, 6 confirmed the results of Wong et al [1]. Solutions other
than (ii) and (iv) are shown to be unstable.

1) K. Y. M. Wong et al, J. Phys. A: Math. Gen. 21 L99.
2) S. Katsura and S. Fujiki, J. Phys. C 12 (1979) 1087.
3) S. Katsura, Prog. Theor. Phys. Suppl. 87 (1986) 139.
4) C. Kwon and D. J. Thouless, Phys. Rev. B: 37 (1988) 7649.

Statistical Neurodynamics –
Associative Memory and Self-Organization

S. Amari

University of Tokyo, Bunkyo-ku, Tokyo 113, Japan

Some results of statistical neurodynamics are presented, and some theoretical difficulties underlying its method are pointed out. The statistical neurodynamical method is applied to the analysis of associative memory models, leading to the dynamical (non-equilibrium) equation of recalling processes.

1. Introduction

The brain is a complex system consisting of a huge number of mutually connected neurons. Since the connections are very complicated and the behavior of elements is highly non-linear, it is difficult to analyze its dynamical behavior in detail from the microscopic point of view. Statistical neurodynamics is one method to overcome this difficulty. Here, we assume that connections are randomly determined subject to a prescribed probability distribution, and we then search for its macroscopic dynamical behavior, which is assumed to be common to almost all the networks generated by the same probability law. In other words, we search for such properties that are common in the ensemble of networks thus generated. It is somewhat related to the field theory of self-organizing neural systems proposed and analyzed mathematically in detail (by Amari [1, 2, 3], Takeuchi and Amari [4]).

The present paper reviews some results on statistical neurodynamics, and applies the method to associative memory models. Macroscopic dynamical behavior of randomly connected neural networks has been studied by many researchers (e.g., Amari [5, 6], Harth et al. [7]). It was pointed out by Rozonoer [8] that the derivation of macroscopic state equations involves some difficulty, which is of the same nature that we encounter in deriving the Boltzmann equation (the H-theorem). The related problem was treated by Amari [9], which gave some interesting results on the microscopic state transition diagram from the macroscopic point of view. The problem was partially solved by Amari et al. [10], but is still unsolved. We introduce these results in section 1 briefly. There are many interesting problems to be solved by the statistical neurodynamical method.

It is fruitful to apply the statistical neurodynamical method to associative memory models. The associative memory models were introduced by Nakano

[11], Anderson [12] and Kohonen [13], and its dynamical behavior and stability were analyzed mathematically by Amari [14, 15]. Recently, Hopfield [16] suggested the spin-glass analogy, and pointed out the important role of the "energy function". Since then, there have appeared many important papers on this subject (e.g., Amit et al. [17], Meir and Domany [18], etc.). It seems interesting to compare our results (Amari and Maginu [19], Amari [20]) with those from the spin-glass analogy method (e.g., Sherrington and Kirkpatrick [21]). The statistical neurodynamical method treats the non-equilibrium dynamical processes with zero temperature, while the spin-glass method mainly treats equilibrium states with a finite temperature. These two methods possess many common aspects, and will be unified to give a powerful statistical method of computational neuroscience. It is my hope that the present short paper will be useful for such a purpose.

2. Statistical Neurodynamics and Macroscopic Equations

Let us consider a very simple neuron model, which receives n input signals x_1, $\cdots x_n$ and emits one output signal z. The input-output relation of the neuron is

$$z = sgn\left(\sum_i w_i x_i\right),$$

where sgn is the signature function, w_i is the connection weight of the i-th input x_i, and all the signals x_i and z take on the binary values $+1$ and -1. Let us consider a network consisting of n neurons, which receives a common bundle of input signals x_1, x_2, $\cdots x_n$ (Fig. 1). Let w_{ij} be the connection weight of the j-th input x_j to the i-th neuron. The output z_i of the i-th neuron is then written as

$$z_i = sgn\left(\sum w_{ij} x_j\right)$$

or

$$z = T_W x = sgn(Wx)$$

in the vector notation. Here, T_W is the nonlinear operator determined by the connection matrix $W = (w_{ij})$. It is a mapping from the input signal space $X = \{1, -1\}^n$ to the output signal space $Z = \{1, -1\}^n$.

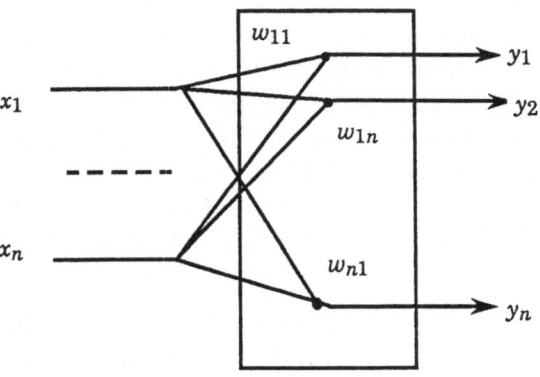

Fig. 1 Neural network

We next consider a network in which the output is fed back to the input (Fig. 2). This is a network in which neurons are mutually connected. When the neurons emit signals at discrete times $t = 1, 2, \cdots$, synchronously, its dynamical behavior is described by the dynamical.equation

$$\mathbf{x}_{t+1} = T_W \mathbf{x}_t , \qquad (2.1)$$

where $\mathbf{x}_t$ is the output vector of the net at time t. We may regard this vector as the state of the net. The non-linear operator T_W is a mapping from the state space $X = \{1, -1\}^n$ to itself. The mapping is represented by the associated state transition graph, in which every $\mathbf{x}$ is connected to its next state $T_W \mathbf{x}$.

It is important to study properties of the operator T_W of a network. However, it strongly depends on the connection W. When w_{ij} are determined randomly subject to a probability law, we study an ensemble E of networks which are generated by the same probability law. Statistical neurodynamics searches for those properties that hold for almost all networks in the ensemble E, provided the number n of the neurons is sufficiently large.

Let us show a simple example. We assume that all w_{ij} are independently determined subject to a common probability distribution $p_n(w)$, where the expectation of w_{ij} satisfies

$$\lim_{n \to \infty} E[nw_{ij}] = \overline{w} ,$$

and the variance of w_{ij} satisfies

$$\lim_{n \to \infty} V[nw_{ij}] = \sigma^2_w ,$$

E and V denoting the expectation and the variance, respectively. The non-linear operator T_W strongly depends on W, so that every net in the ensemble E

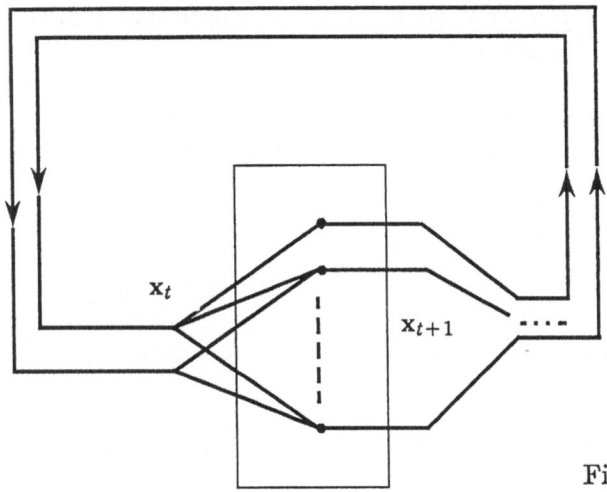

Fig. 2 Dynamical neural network

has different microscopic properties. We therefore look for properties related to macroscopic quantities such as the activity or mean firing rate of a signal x,

$$A(\mathbf{x}) = \frac{1}{n} \sum_{i=1}^{n} x_i . \qquad (2.2)$$

It is easy to show that, when the input activity of $\mathbf{x}$ is A, the activity $A' = A(\mathbf{z})$ of the output $\mathbf{z} = T_W \mathbf{x}$ is obtained from a relation

$$A^{'} = \Phi(A)$$

which is common for almost all the networks in E as n tends to infinity. Here, the function Φ is given by $\Phi(A) = err(wA/\sigma_w)$, where err is the error integral

$$err(A) = \int_{-A}^{A} (1/\sqrt{2\pi}) exp\{-u^2/2\} du, \qquad (2.3)$$

provided the central limit theorem holds for Σw_{ij} (see Amari [5, 6]).

A more interesting example is as follows : Let $D(\mathbf{x}, \mathbf{x}')$ be the normalized Hamming distance between two signals $\mathbf{x}$ and $\mathbf{x}'$,

$$D(\mathbf{x}, \mathbf{x}') = \frac{1}{2n} \sum_{i=1}^{n} |x_i - x'_i|. \qquad (2.4)$$

Then, it is interesting to know the distance $D' = D(T_W\mathbf{x}, T_W\mathbf{x}')$ between the outputs $\mathbf{z} = T_W\mathbf{x}$ and $\mathbf{z}' = T_W\mathbf{x}'$ when the distance between $\mathbf{x}$ and $\mathbf{x}'$ is D. We also have the following relation which holds for almost all the nets,

$$D' = \psi(D). \qquad (2.5)$$

Here, ψ is explicitly given, when $w = 0$, by

$$\psi(D) = \frac{2}{\pi} arcsin(\sqrt{D}/\sigma_w).$$

See Amari [9]. This shows that the map T_W is continuous, but its derivative $\psi'(D)$ diverges to ∞ at $D = 0$. We can show such a microscopic property of the map by the macroscopic method of statistical neurodynamics.

Now let us consider a network of Fig.2 with mutual connections. Let $A_t = A(\mathbf{x}_t) = A(T_w{}^t \mathbf{x}_0)$ be the activity of the state $\mathbf{x}_t$ at time t. It is easy to show

$$A_1 = \Phi(A_0).$$

However, it is not certain whether or not

$$A_{t+1} = \Phi(A_t) \qquad (2.6)$$

holds for $t = 1, 2, \cdots$. This is because, when calculating

$$A_{t+1} = A(T_W \mathbf{x}_t),$$

we must not neglect the correlation between T_W and $\mathbf{x}_t$, which latter depends also on the same W. In other words, the dynamics (2.1) can be expanded as the transmission of a signal through a cascaded network of Fig. 3, where all the networks have the same connection matrix W. The correlation between the same T_W's may not be able to be neglected here. If the connections W of the component networks are independent, (2.6) holds obviously.

Let

$$X_t = X(\mathbf{x}_t) = X(T_W^t \mathbf{x}_0)$$

be any (vector-valued) macroscopic quantity which is a function of the microscopic state $\mathbf{x}_t$. Or more generally, it is a function of two states $\mathbf{x}_t$ and $\mathbf{x}_t'$, $\mathbf{x}_t = T_W{}^t \mathbf{x}_0$ and $\mathbf{x}_t' = T_W{}^t \mathbf{x}_0'$. The problem is to study when the macroscopic equation of the type

$$X_{t+1} = F(X_t)$$

holds, provided it holds for $t = 0$. Let $\widetilde{X}_t$ be the solution of the above equation. Then, we have two propositions concerning the validity of the macroscopic equations.

Weak proposition. For any t_0 and $\varepsilon > 0$,

$$\lim_{n \to \infty} Prob\{|X(\mathbf{x}_t) - \widetilde{X}_t| < \varepsilon\} = 1 \tag{2.7}$$

holds for $t < t_0$.

Strong proposition. For any $\varepsilon > 0$,

$$\lim_{n \to \infty} \sup Prob\{|X(\mathbf{x}_t) - \widetilde{X}_t| < \varepsilon\} = 1. \tag{2.8}$$

These two are fundamental problems for statistical neurodynamics, having the same structure as the Boltzmann equation in statistical dynamics. We can prove the weak proposition under a certain probability distribution of W for the dynamics of activity A_t, and for the dynamics of distance

$$D_t = D(T_W^t \mathbf{x}_0, T_W^t \mathbf{x}_0').$$

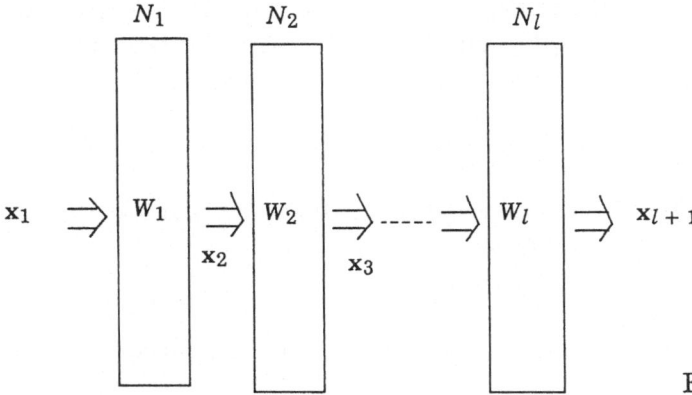

Fig. 3 Cascaded network

The equation

$$D_{t+1} = \Psi(D_t) \qquad (2.9)$$

gives some information about the microscopic structure of the state transition diagram of randomly generated neural networks. For example, let q_i be the probability that a state x has i predecessors,

$$q_i = Prob\{|T_W^{-1}x| = i\}, \qquad (2.10)$$

where $T_W^{-1}x$ is the set of the states y such that $T_W y = x$ and $|\quad|$ denotes the cardinality, i.e., the size of the set. It is obvious that

$$\sum q_i = 1 \; , \quad \sum i q_i = 1 \, .$$

However, it is rather surprising (Amari, [9]) that

$$\lim_{n \to \infty} \sum i^2 q_i = \infty \, . \qquad (2.11)$$

This implies that the variance of the number of the precedent states of a state diverges to infinity as n tends to infinity, while its expectation remains 1.

We have so far studied the case where all the w_{ij} are independently and identically distributed. When $w_{ij} = w_{ji}$ holds, the state transition diagram is quite different. There exist no limit cycles in it. The expected number of the equilibrium states is about $2^{0.3n}$, while it is equal to 1 in the non-symmetric case. However, (2.6) and (2.9) hold in this case, too.

We show that the associative memory model can be treated also in the framework of statistical neurodynamics, but the weak proposition does not hold in this case. There remain many problems in statistical neurodynamics to be studied further.

3. Statistical neurodynamics of Associative Memory

Let us consider a neural network of Fig. 2, whose state transition is given by (2.1). Given m vectors $x^1, \cdots, x^m$, the autocorrelation associative memory is defined by the connection matrix W

$$W = \frac{1}{n} \sum_{\mu = 1}^{m} x^\mu (x^\mu)', \qquad (3.1)$$

where x'is the transposition of a column vector x. If all the m vectors x^μ ($\mu = 1$, $2, \cdots, m$) are fixed points of the dynamics,

$$T_W x^\mu = x^\mu, \quad \mu = 1, \cdots, m, \qquad (3.2)$$

we may say that m vectors x^μ are memorized in the network as its equilibrium states of dynamics.

Moreover, when x_0 belongs to the basin of attraction of x^μ, starting with initial state x_0, the network falls in the state x^μ after a finite number of state transitions,

$$(T_W)^k x_0 = x^\mu$$

for some k. This is interpreted as the memorized pattern x^μ being recalled from an incomplete key pattern x_0.

The associative memory model was proposed by Nakano [11], Anderson [12], and Kohonen [13], and its dynamical behavior was analyzed mathematically by Amari [14, 15]. Later, Hopfield [16] suggested the spin glass analogy, and a large number of theoretical studies appeared (see, e.g., Amit et al., [17], Meir and Domany [18], etc.).

When the signals x^μ to be memorized are randomly chosen, one can apply the method of statistical neurodynamics to the analysis of its dynamical recalling processes. Here, we assume that the components $x_i{}^\mu$ of each signal are independently assigned to be equal to $+1$ or -1 with probability 1/2. The connection matrix W is then randomly determined depending on them, but their components w_{ij} are no longer independent.

Let us consider the dynamical behavior of recalling one of x^μ, say x^1, by assuming that the initial state x_0 satisfies

$$x^1 \cdot x_0 > x^\mu \cdot x_0 \ , \quad \mu = 2, \cdots, m .$$

We evaluate the similarity of a vector x to x^1 by the direction cosine,

$$a (x) = \frac{1}{n} x^1 \cdot x . \tag{3.3}$$

Let

$$a_t = a (x_t) \tag{3.4}$$

where

$$x_t = T_W x_{t-1} = (T_W)^t x_0 .$$

We first search for a_1 as a function of a_0,

$$a_1 = G (a_0) . \tag{3.5}$$

We have

$$x_1 = sgn (Wx) = sgn (a_0 x^1 + N),$$

where the components of N are

$$N_i = \frac{1}{n} \sum_{\mu = 2}^{m} \sum_{j = 1}^{n} x_i^\mu x_j^\mu x_{0j} , \tag{3.6}$$

which are regarded as noise or cross-talk terms. When the number m of patterns is

$$m = r n$$

and n is large, N_i is asymptotically normally distributed with mean 0 and variance

$$\sigma_0^2 = r.$$

Therefore, we have, by using the law of large numbers,

$$a_1 = G(a_0) = err(a_0 / \sqrt{r}).$$

This shows that, even when $a_0 = 1$, $a_1 \neq 1$, implying that

$$T_W x^\mu = x^\mu$$

does not hold when the number m of patterns is proportional to n. It is not difficult to show that (3.2) holds when

$$\frac{m}{n} \leqq \frac{1}{2 \log n} + smaller\, order\, terms.$$

When $m / n = r$, we cannot recall the precise x^1 but some others close to it. Equation (3.5) shows how x_0 approaches the memorized one by a one-step state transition.

The problem is to see if

$$a_{t+1} = G(a_t)$$

holds for the recalling process (2.1) of the dynamics of state transitions. Unfortunately, this does not hold in the present case. This is because the crosstalk noise at time t,

$$N_t = \frac{1}{n} \sum x^\mu (x^\mu \cdot x_t), \tag{3.7}$$

is no longer a sum of independent random variables, because $x_t = T_W x_{t-1}$ also depends on the random variables W or x^μ's. A careful calculation suggests that N_i^t is subject to a normal distribution

$$N_i^t \sim N(x_i^t b_t, \sigma_t^2), \tag{3.8}$$

where the bias term b_t and the variance σ_t^2 are to be determined recursively. By using these quantities, the direction cosine is written as

$$a_{t+1} = <err(\overline{a_t})> \tag{3.9}$$

$$b_{t+1} = (2 r / \sigma_t) < p(\overline{a_t})> \tag{3.10}$$

$$\sigma_{t+1}^2 = r + 4 < p(\overline{a_t})>^2 + 4 r \overline{a_t} a_{t+1} < p(\overline{a_t})>, \tag{3.11}$$

where

$$\overline{a_t} = a_t / \sigma_t,$$

$$<f(\overline{a_t})> = \frac{1 + a_{t-1}}{2} f((a_t + b_t) / \sigma_t) + \frac{1 - a_{t-1}}{2} f((a_t - b_t) / \sigma_t),$$

$$p(u) = (1 / \sqrt{2\pi}) exp\{ - u^2 / 2\}.$$

showed that $r \doteq 0.156$ is the capacity of associative memory net by using a simplified version of the above equation.

The network of Fig. 2 can be used as an associative memory of a sequence of states $\{x^1, x^2, x^3, \cdots, x^m\}$, when the connection matrix W is given by

$$W = \frac{1}{n} \sum_{\mu=1}^{m} x^{\mu+1} (x^\mu)' \ , \tag{3.12}$$

Amari [14]. It is expected that

$$T_W x^\mu = x^{\mu+1}$$

holds approximately. When an initial state x_0 is close to some of x^μ, the sequence of states

$$x_{t+1} = T_W x_t \tag{3.13}$$

produced by the network approaches the memorized sequence. Let

$$a_t = \frac{1}{n} x^{\mu+t} \cdot x_t,$$

where x_0 is close to x^μ. The dynamics of a_t is then given, in this case, by

$$a_{t+1} = err(a_t / \sigma_t), \tag{3.14}$$

$$\sigma_{t+1}^2 = r + 2\{p(a_t / \sigma_t)\}^2 . \tag{3.15}$$

It is interesting that the memory capacity of the network is about $r_c = 0.27$, which is larger than that of autoassociative memory.

References

1. S. Amari : Topographic organization of nerve fields, Bull. Math. Biol., 42, 339-364, 1980

2. S. Amari : Field theory of self-organizing neural nets, IEEE Trans. SMC-13, 741-748, 1983

3. S. Amari : Competitive and Cooperative Aspects in Dynamics of Neural Excitation and Self-Organization, ed. by S. Amari and M.A. Arbib, Competition and Cooperation in Neural Nets, Springer Lecture Notes in Biomathematics, Vol.45, pp.1-28, 1982

4. A. Takeuchi and S. Amari : Formation of topographic maps and columnar microstructures, Biol. Cybern., 35, 63-72, 1979

5. S. Amari : Characteristics of randomly connected threshold element networks and network systems, Proc. IEEE, 59, 35-47, 1971

6. S. Amari : Characteristics of random nets of analog neuron-like elements, IEEE Trans., SMC-2, 643-657, 1972

7. E. M. Harth, J. T. Csermely, B. Beek and R. D. Lindsay : Brain functions and neural dynamics. Journal of Theoretical Biology, 26, 93-120, 1970

8. L. I. Rozonoer : Random logical nets : I. Automatika Telemekhanika, 5, 137-147, 1969

9. S. Amari : A method of statistical neurodynamics, Kybernetik, 14, 201-215, 1974

10. S. Amari, K.Yoshida and K. Kanatani : A mathematical foundation for statistical neurodynamics, SIAM J. App. Math. 33, 95 - 126, 1977

11. K. Nakano : Associatron--- a model of associative memory, IEEE Trans., SMC-2, 381-388, 1972

12. J. A. Anderson : A simple neural network generating interactive memory, Math. Biosciences, 14, 197-220.

13. T. Kohonen : Correlation matrix memories, IEEE Trans., C-21, 353-359, 1972

14. S. Amari : Learning patterns and pattern sequences by self-organizing nets of threshold elements, IEEE Trans., C-21, 1197-1206, 1972

15. S. Amari : Neural theory of association and concept-formation, Biol. Cybernetics, 26, 175-185, 1977

16. J. J. Hopfield : Neural networks and physical systems with emergent collective computational abilities, Proc. Nat. Acad. Sci. U.S.A., vol. 79, 2445-2458, 1982

17. D. J. Amit , H. Gutfreund and H. Sompolinsky : Spin-glass models of neural networks, Phys. Rev., A2, 1007-1018, 1986

18. R. Meir and E. Domany : Exact solution of a layered neural network memory, Phy. Rev. Lett., 59, 359-362, 1987

19. S. Amari and K.Maginu : Statistical neurodynamics of associative memory, Neural Networks, 1, 63-73, 1988

20. S. Amari : Associative memory and its neurodynamical analysis, in Neural and Synergetic Computers, ed. H. Haken, Springer Series in Synergetics, Vol. 42 (Springer, Berlin, Heidelberg 1988)

21. D. Sherrington and S. Kirkpatrick : Solvable Model of a Spin-Glass, Phy. Rev. Lett. 35, 1792, 1975

Reconstruction of Images in the Visual Cortex

S. Shinomoto

Department of Physics, Kyoto University, Kyoto 606, Japan

1. Introduction

Receptive properties of neurons in visual pathways of mammalian brains have become clarified as a result of physiological and anatomical research. There are a number of neurons with rather simple receptive functions, which respond to different types of spatial differentiation of the lightness field. The significance of each function, however, is not yet understood in terms of visual perception. For instance, although the neurons for early stage visual processing participate in spatial differentiation, we are not particularly conscious of the spatial derivatives of images, but are able to recover the general images.

I propose here an inverse problem in which local information due to the differentiation is integrated to recover the original images. Reasonable constraints on the transformation hereafter called 'differentiation' will be introduced on the basis of physiological and psychological knowledge. Under these conditions the differentiation is not ideal. However, the constraints result in several information processes which help to comprehend images. They may be classified within linear and non-linear frameworks. It will be seen that most of fundamental processing tasks such as the regularization and the emphasis on edges are resolved within the linear framework. The thresholding effect which will be introduced in the non-linear framework provides the system with the capacity for removing the slowly varying part of images. The principle in this framework is similar to the one employed by LAND and McCANN [1], and HORN [2]. However, the present model can regularize more general images which may include noise, and does not require iteration procedures such as simulated annealing employed by GEMAN and GEMAN [3], POGGIO, TORRE and KOCH [4], and KOCH, MARROQUIN and YUILLE [5] to regularize degraded images. The time scale of neuronic response is of $O(10^{-2})$ sec., and a leisurely procedure like the simulated annealing would not be appropriate for the image processing in the real brain.

2. Actual Visual Pathways

Let us briefly summarize the knowledge of the receptive functions of neurons in mammalian visual pathways.

The lightness field on the retina is first transformed into a potential field by means of receptors. The information in the potential field is then converted into a set of output pulses from the on-center and the off-center neurons. Their receptive function is approximately to take the second order differentiation $\Delta = \partial^2/\partial x^2 + \partial^2/\partial y^2$ of the two-dimensional intensity field $u(x,y)$ (see Fig.1a). The on-center and off-center cells are found at the retina and the lateral geniculate nucleus which is a relay station from retina to visual cortex. In the visual cortex, there are a number of neurons which are called the simple cells. The simple cells are classified into several types according to their receptive functions. A typical one is what is called the line (gap) detector discovered by Hubel and Wiesel. Figure 1b shows its function. Another type of the simple cells are the edge detectors. Each neuron responds to the distinct intensity edge in a particular direction (see Fig.1c). It can convey the amount of local intensity gradient by means of the frequency of output pulses. Thus the terminology appears

Springer Series in Synergetics, Vol. 43   **Cooperative Dynamics in Complex Physical Systems**
Editor: H. Takayama   © Springer-Verlag Berlin, Heidelberg 1989

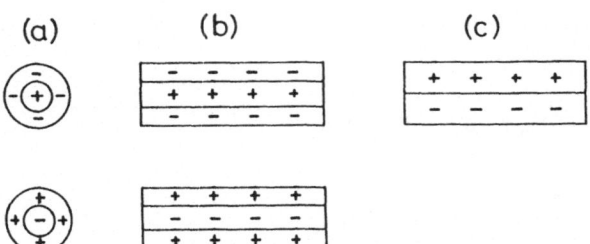

Fig.1. Schematic representation of the receptive fields of (a): on-center and off-center cells, (b): line and gap detectors, and (c): gradient (edge) detector. Each neuron is particularly responsive if (+) region is illuminated   while (−) region is not.

inappropriate, and I will call the neurons the gradient detectors. In addition to this, additional integrated operations are found in other neurons called complex cells. The 'features' to which these neurons respond are abundant. The features can be the 'tokens' (MARR [6]) which will be the fundamental factors for the further visual perception. I will not go deeply into the problem of the visual perception. In the present paper, I will simply discuss the reconstruction of images from signals transformed by the gradient detectors. For simplicity, I will restrict the present consideration to the monocular and monochromatic problem. Thus the receptive field $u(x,y)=u(\underset{\sim}{r})$ can be assumed to be unique and real-valued.

3. Inverse Problem

I consider an inverse problem with respect to the transformation by the gradient detectors. If the operation of the gradient detectors is simply to get the true gradient field of the intensity field $u(\underset{\sim}{r})$,

$$\underset{\sim}{g}(\underset{\sim}{r}) = \underset{\sim}{\nabla} u(\underset{\sim}{r}) \quad , \tag{1}$$

then its left inverse operation recovers the original image, i.e.,

$$v(\underset{\sim}{r}) = \underset{\sim}{\nabla}^{-1} \underset{\sim}{g}(\underset{\sim}{r}) = \int ds\, \underset{\sim}{K}(\underset{\sim}{s}) \cdot \underset{\sim}{g}(\underset{\sim}{r}-\underset{\sim}{s}) = u(\underset{\sim}{r}) \quad , \tag{2}$$

where $v(\underset{\sim}{r})$ represents an integrated field and $\underset{\sim}{K}$ is

$$\underset{\sim}{K}(\underset{\sim}{s}) = \underset{\sim}{s} / 2\pi s^2 \quad . \tag{3}$$

In practice, however, the gradient detector does not function as an ideal gradient operation. I introduce physiologically plausible constraints on the gradient detectors. Their effect on the inverse problem will be investigated, assuming that the 'inverse' transformation (2,3) remains as above. The constraints can be classified within linear and non-linear frameworks which can be discussed separately.

(I) Linear Problems
(i) non-locality
The actual gradient detector is particularly responsive to a straight intensity edge in a specific direction (see KUFFLER, NICHOLLS and MARTIN [7]). Thus the operation is not strictly local. To explain the physiological data, it would be natural to introduce an integration of the gradient values along a line orthogonal to the direction ℓ to which the detector is responsive. The output of the gradient detector for the direction ℓ is not $\partial_\ell u(\underset{\sim}{r})$, but

$$g_\ell(\underset{\sim}{r}) = \int ds\, \delta(\underset{\sim}{s}\cdot\underset{\sim}{\ell})\, \rho(\underset{\sim}{s})\, \partial_\ell u(\underset{\sim}{r}-\underset{\sim}{s}) \quad , \tag{4}$$

where $\delta(x)$ is the Dirac delta function, $\rho(s)$ is a weighting function of the integration which is localized on a scale σ, and ∂_ℓ is the spatial derivative with respect to the direction ℓ. Then, the gradient field will be approximated by the sum of outputs of detectors for all directions:

$$g(r) = \sum_\ell \ell \, g_\ell(r) \, / \, \sum_\ell 1 \quad . \tag{5}$$

The above equation can be rewritten as a linear integral transformation of the true gradient field $\nabla u(r)$. The transformation introduces a kind of smoothing of the vector field $\nabla u(r)$. However, the integral transformation generally involves a tensor operation. This point will be discussed elsewhere.

(ii) lateral inhibition

It is rather general that the neighboring neurons inhibit their response to each other via inhibitory neurons. I will point out that a kind of lateral inhibition between gradient detectors can simulate some well-known illusions such as the Mach bands and the Hermann-Hering grid (see JUNG [8], and LINDSEY and NORMAN [9]). There are a number of explanations based on the property of on-center and off-center cells. Most explanations, however, suppose implicitly that the output signals of these neurons are recognized directly. There is no ground for this.

I will introduce the 'lateral inhibition' which is of the form

$$\tilde{g}(r) = g(r) + \int ds \, i(s)g(r-s) = g + i*g \tag{6}$$

for feed-forward type interaction. Here, $i(s)$ is assumed to be a negative-valued scalar function with its range $\gamma(\gg\sigma)$. The 'inhibition' for the vector field is not so simple as it looks, because each mutual inhibition between detectors depends on the difference between their receptive angles. For the case of feedback inhibition, we have to rewrite eq.(6) as

$$\tilde{g} = g + i*\tilde{g} = \frac{1}{1-i} * g \quad . \tag{7}$$

We are able to translate eq.(7) in the same form as (6) if we introduce an operation $j\equiv i+i*i+i*i*i+ \cdots$. Then the lateral interaction $j(s)$ involves oscillation. Let us consider the feed-forward inhibition (6) and explain the Mach bands (see Fig.2). This type of lateral inhibition induces a weak gradient field in the opposite direction to the original gradient signal. This would result in overshooting in the integrated field. The Hermann-Hering grid which is essentially two-dimensional can also be interpreted in terms of this mechanism. The lateral inhibition thus plays a role in emphasizing the edges. I have not found other advantages of such inhibition.

These linear operations can restore the image blurred by noise (see Fig.3a and 3b). The restoration is not complete, however, and nonlinear effects are helpful for further processing.

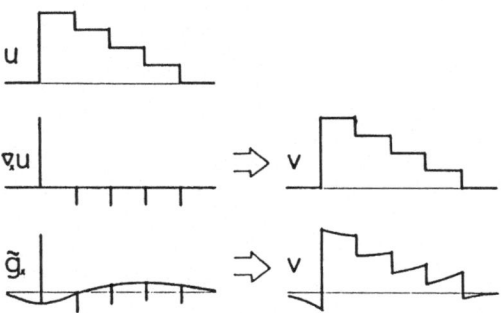

Fig.2. Effect of lateral inhibition on the integrated field. See text for further details.

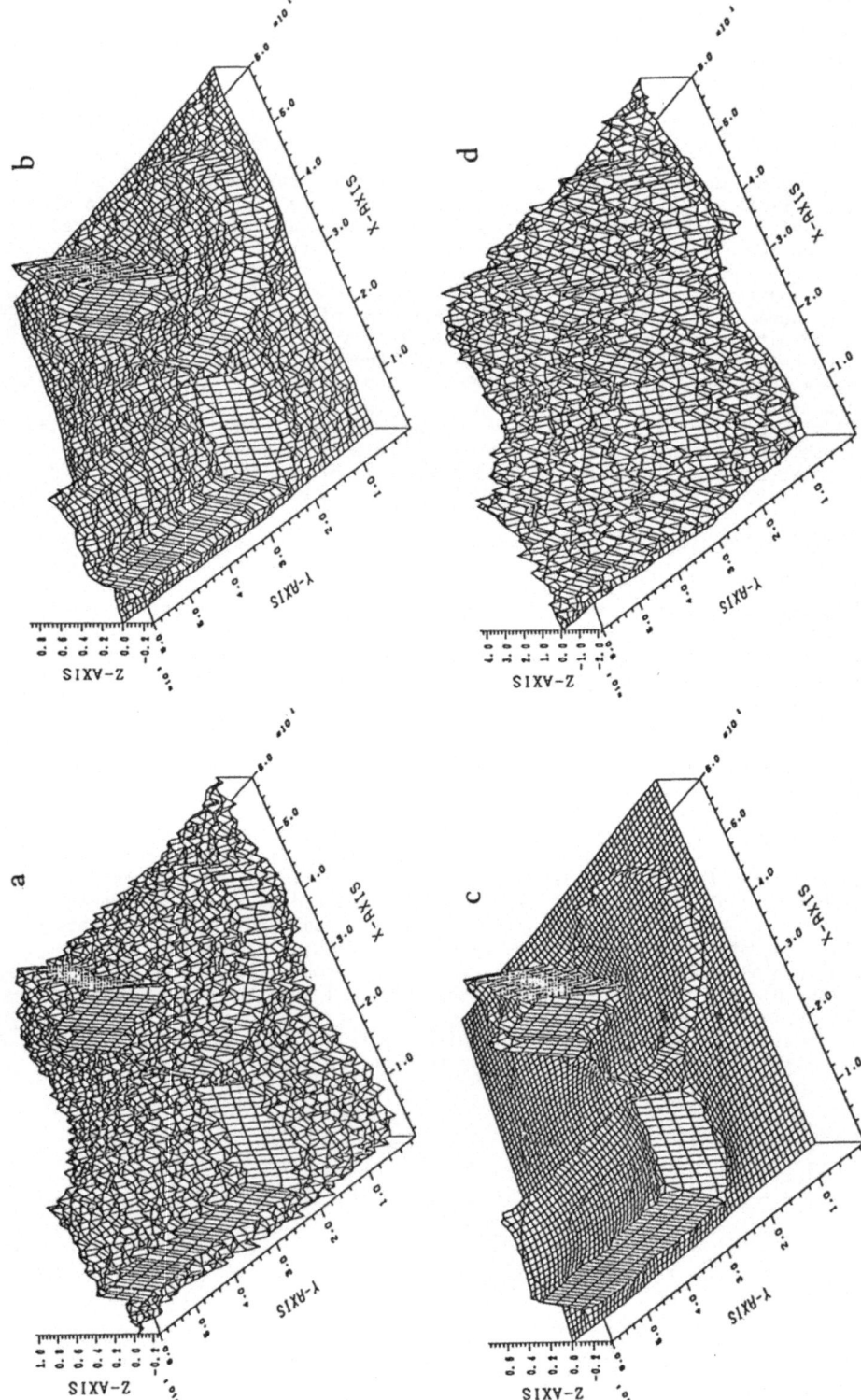

Fig.3. Results of numerical experiments; (a): original intensity field u(χ); (b): integrated field v(χ) obtained by the linear transformation; (c) and (d): v(χ)'s obtained by the non-linear transformation. See text for further details.

(II) Non-linear Problems

The neuron is a non-linear transducer of the signals from input to output. The non-linearity is characterized by the thresholding and saturation effects, which specify features at which each cortical site of the integration is looking. For instance, the slowly varying part of the input pattern can be eliminated by the thresholding effect. This was pointed out and shown by LAND and McCANN [1] for a one-dimensional first-order derivative, and by HORN [2] for a two-dimensional second-order derivative Δ. On the contrary, rapidly changing part of the input images would effectively be eliminated by the saturation effect.

Results of numerical experiments on this proposition are shown in Fig.3. I prepared a degraded image in which sharp patterns, white noise, and long-range variation are buried. The smoothing of the white noise and the emphasis on edges are already seen in the linear framework. The smoothing effect also helps the non-linear operation to extract sharp patterns. This is an advantage of the present model compared with those of LAND and McCANN [1] and HORN [2], which are unable to regularize degraded images. Fig.3c is a typical example in which white noise and long-range variation are eliminated; Fig.3d is an example obtained by emphasizing the saturation effect. In this integrated image, sharp patterns are effectively eliminated and smooth variation is relatively emphasized.

4. The 'Inverse' Operation

Throughout the present scenario, the presence of the 'inverse gradient operation' ∇^{-1} is assumed. The operation ∇^{-1} is a global integral operation as shown in eq. (2). The integral kernel (3) is a kind of a dipole connection to the integrated field $v(\underline{r})$. Otherwise, it can be given by a linear sum of the pair of monopoles, $\pm 2\pi \ln|\underline{s}|$. Even if the presence of the intact 'integrated field' cannot be assumed, there should be a connection of this kind to the site of image comprehension. If the present scenario is more or less true and the information from gradient detectors is used for the recovery of images, the polarization should be conveyed to the sites of integrated perception. The polarization of the post neuronic connections such as eq.(3) or the asymmetry of axons may be worth investigating.

5. Conclusion

A scenario for the reconstruction of images is presented by studying the inverse problem for the gradient operation ∇. It is found that the fundamental information processing such as the regularization and the emphasis on edges can be achieved within the present framework. The present transformation does not require the leisurely iteration procedure such as used in the simulated annealing approach to regularization [3, 4, 5]. Thus the present scenario would be a candidate for the processing mechanism of real mammalian brains.

Acknowledgements

I would like to thank   Yoshiki Kuramoto, Shun-ichi Amari, Yukito Iba, Tetsuya Hattori, Hiroyuki Hata, Toshio Inui, Mitsuo Kawato, Sei Miyake, and Edgar Knobloch for helpful discussions.

References

[1] E. Land and J.J. McCann : J. Opt. Soc. Am. 61, 1 (1971).
[2] B.K.P. Horn : Computer Graphics and Image Processing 3, 277 (1974).
[3] S. Geman and D. Geman : IEEE Trans. PAMI 6, 721 (1984).
[4] T. Poggio, V. Torre and C. Koch : Nature 317, 314 (1985).
[5] C. Koch, J. Marroquin and A. Yuille : Proc. Natl. Acad. Sci. USA 83, 4263 (1986).
[6] D. Marr : 'Vision' (W.H. Freeman and Company, NY 1982).

[7] S.W. Kuffler, J.G. Nicholls and A.R. Martin : 'From Neuron to Brain' 2nd ed. (Sinauer, Massachusetts 1984).
[8] R. Jung : in 'Handbook of Sensory Physiology Vol. VII/3' (Springer-Verlag, Berlin 1973).
[9] P.H. Lindsey and D.A. Norman : 'Human Information Processing' 2nd ed. (Academic Press, Orlando 1977)

Modelling Brain Lesions in Neural Networks

M.A. Virasoro

Dipartimento de Fisica, "La Sapienza", Ple. A. Moro 2, I-00185 Roma, Italy

Abstract

The observation of patients with lesions in the brain and specific cognitive functions impaired has led during the last 130 years to stress modular organization and physical localizations of different functions. If this point of view is pushed to the extreme, one would have to conclude as a consequence that there is no place left for disordered models as the ones that derive from neural networks. We argue that theoretical estimates of the burden so imposed on the organization capabilities of neural development and/or natural evolution make this extreme position untenable. We then show in one particular example (the syndrome called "prosopagnosia") that an alternative explanation is possible. Assuming a completely amorphous neural model that stores categorized ultrametric patterns we use the probability measure on synaptic interactions introduced by Gardner to analyze the effect of a random destruction of synapses. At an intermediate level of destruction the retrieval of the pattern is impaired particularly in those items that permit the distinction among exemplars in a class.

1. Introduction

Faced with the task of modeling the brain, two extreme, opposite, but in a certain sense complementary, attitudes are possible.

One of them is the natural point of view for a "neuroanatomist" that discovers inside the brain an extremely complicated variety of neurons connected in a way which looks like the result of a careful design. He has experienced many times in the past that a detail of the organization whose existence seemed to be at first fortuitous turned out to play a crucial role in the brain's performance. He is led to conclude that **every** detail has been optimized.

The other point of view, the "physicist" point of view, estimates the number of degrees of freedom that defines a possible brain configuration.

Counting 10^{10} neurons with 10^4 synapses, each of which can be positive or negative, he guesses 10^{14} degrees of freedom. Experience with optimization problems tells him that the time required to optimize such a system would be exponential in 10^{14}. He concludes that neither natural evolution nor neural development can deal with such complexity and assumes in the first approximation the system to be amorphous.

Of course, the truth lies somewhere in between but where exactly it is difficult to guess. For the time being it is reasonable to choose one of the two extreme points and work out all its possible consequences. In this way some theoretical understanding will be gained of the range of possibilities. A realistic model remains for the future.

2 Brain Damage and Localization

In the last 127 years starting from the fundamental work by Broca on language impairment as a consequence of a localized lesion in the cortex, an impressive number of clinical cases have accumulated demonstrating a rather tight correlation between the area of the lesion and the type of impairment[1]. If two patients A and B suffer from lesions in two separated areas of the brain and as a consequence patient A has function a impaired and function b normal while patient B has function b impaired while a is normal then functions a and b are said to be anatomically dissociated. Many examples of this type have led to the hypothesis of **modularity** of the brain organization: different regions of the brain would work almost independently of each other and would perform specific functions. Undoubtedly this hypothesis reinforces the "neuronatomist" point of view.

At a certain point however an alternative explanation has to be found. For instance, there is in the literature a description of a patient who, as a result of a car accident, is incapable of retrieving the names of objects belonging to the categories of fruits and vegetables[2]. If a peach or an orange is presented visually to him he will not be able to find the name. On the other hand he has no problem with other objects belonging to other categories. Should one conclude that a specific area of the brain is dedicated to store fruits and vegetables? This is perhaps one exceptional case, but in the next section we will describe the prosopagnosia syndrome and face a similar question.

2.1 Lesions to the Visual Cortex. Agnosia and Prosopagnosia.

The path of visual data in the cortex starts in Area 17 in the occipital lobe and from there progresses towards the temporal lobe. It is known that a serious lesion in Area 17 may produce *cortical blindness*[3]: the patient simply complains of total blindness though the eyes respond to changes of intensity. Lesions in Area 18 or ahead produce different types of agnosia:

Apperceptive Agnosia: the patient discriminates changes of intensity and/or color. However, he does not perceive the shape of an object. He typically fails when trying to copy the shape of an object or match object to sample through the identification of the same shape.

Associative Agnosia: In this case shape discrimination tests demonstrate that shape perception is normal. However, the object cannot be identified on the basis of visual presentation. Presented with an ashtray, or a picture of it, the patient can copy it but does not know what it is.

Prosopagnosia : Originally the defining symptoms were thought to be restricted to an impairment in face recognition. The patient could recognize common objects but in front of a face he was unable to recognize the individual person behind it. Recently[4], however, a more careful analysis has shown that a similar difficulty applies to categories other than faces. The syndrome can be better defined as a generalized impairment in identifying one individual among those that are visually similar to it and that therefore can be said to belong to the same class. In this perspective *prosopagnosia is a malfunctioning in the categorization process* . The patient, in front of the stimulus, correctly recognizes the category to which it belongs but is unable to recognize the individual. H. Ellis[5] has suggested that the explanation for the existence of this syndrome lies in a two-stage recognition process: the first stage recognizes the class while the second distinguishes the individual; the lesion destroys the second but not the first. This is in the direction of further localization. We will present in the next section an alternative explanation based on a maximally disordered neural network.

For completeness we mention *visual anomia* where semantic recognition as tested by silent use demonstration or semantic matching is normal but naming is impaired. It is assumed to correspond to a lesion in the link between the visual and language regions.

3 Categorization in Neural Networks and Prosopagnosia

Categorization is a simple basic way of processing information and as such a good candidate for a computational capability that could emerge from neural networks[6]. Many authors have studied how to construct such models[7-12] but the situation is not completely satisfactory because of the large degree of arbitrariness in model building. On the other hand prosopagnosia is an impairment of the categorization process in the real brain. Is it possible to simulate it in a neural network, for instance as a consequence of random synapse destruction?

First of all it is important to notice that the apparent fact that the patient loses simultaneously the ability to recognize individuals belonging to different classes, for example faces and automobiles, suggests that the same module is coding for all those individuals and as such weighs the balance in favor of distributed storage of the information[13].

But then, as stressed by H. Ellis, the same logic suggests that the two operations, recognition of the class and recognition of the individual, would be performed by different (physically separated) families of neurons. This possibility is consistent with a neural network[10,12] and could eventually be proved, or disproved, by more neuroanatomical data (see ref[14-15] for a recent discussion of the relevant evidence).

Still, as discussed in the introduction, we prefer to explore the possibility that the relevant neural network associative memory is as amorphous as possible and therefore we consider a simple neural network with two-state neurons and fully connected asymmetric couplings(see ref[16] and references therein).

The similarity among the patterns is modeled through an *ultrametric* organization. We generate Q classes of P_α patterns coded as N-bit words $\{s_i^{\alpha\beta}=\pm1;\ i=1,...N;\ \alpha=1,...Q;\ \beta=1,...P_\alpha\}$ by the following two-step stochastic process[7]:

1) Q N-bit words are generated by random *uncorrelated* choice of each components: $s_i^\alpha=+1$ or -1 with equal probability.
2) Inside each class, at fixed α, P_α N-bit words are generated by a *correlated* choice of the components: $s_i^{\alpha\beta}=\pm s_i^\alpha$ with probability $(1\pm m)/2$ for $\beta=1,...P_\alpha$. P_α is chosen proportional to N for N->∞ while Q is finite.

The fixed point equations of such a simple network are

$$s_i=\text{sign}(\textstyle\sum_j J_{ij}s_j) \tag{1}$$

where the threshold has been chosen arbitrarily equal to zero.

The fact that the patterns { $s_i^{\alpha\beta}$} have been stored is reflected in the following N·P (with P=$\Sigma_\alpha P_\alpha$) inequalities:

$$s_i^{\alpha\beta}\sum_j J_{ij}s_j^{\alpha\beta}\geq 0 \qquad i=1,...N;\ \alpha=1,...Q;\ \beta=1,...P_\alpha \quad . \tag{2}$$

For one particular pattern the sites where $s_i^{\alpha\beta}=s_i^\alpha$ (defined to be the plus sites) can be said to confirm the class while sites where $s_i^{\alpha\beta}=-s_i^\alpha$ (the minus sites) characterize the individual.

A generalized destruction of synapses will obviously produce the symptoms of *agnosia* . We now show that a limited destruction, **in spite of the fact that it acts randomly on synapses,** will affect the minus sites more than the plus sites.

The net effect of the death of synapses will be a noise term added to the local molecular field h in (2)

$$h_i^{\alpha\beta}=\sum_j J_{ij}s_j^{\alpha\beta}\text{------->}h_i^{\alpha\beta}+\delta_i^{\alpha\beta} \quad . \tag{3}$$
$$\text{lesion}$$

The noise may lead to a violation of (2) only at those sites and for those patterns where $|h_i|$ is smaller than $|\delta_i^{\alpha\beta}|$. If the molecular field h is normally, before the lesion, larger at the plus sites than at the minus sites then the different degree of robustness will immediately follow.

At this point we would have to introduce the learning rule. This is inconvenient because it introduces a large degree of arbitrariness with little biological relevance. Fortunately, Gardner[17] has shown how to analyze these systems in a way which is largely independent of the learning rule. One introduces a probability distribution in the volume defined by the inequalities (2):

$$P (\{J_{ij}\})\Pi_{ij}dJ_{ij}$$

$$= \frac{(\Pi_{ij}dJ_{ij})(\Pi_i\delta(\Sigma_j J_{ij}^2-1))(\Pi_{i\alpha\beta}\theta(s_i^{\alpha\beta}\Sigma_j J_{ij}s_j^{\alpha\beta}))}{\int(\Pi_{ij}dJ_{ij})(\Pi_i\delta(\Sigma_j J_{ij}^2-1))(\Pi_{i\alpha\beta}\theta(s_i^{\alpha\beta}\Sigma_j J_{ij}s_j^{\alpha\beta}))} \tag{4}$$

and in that measure we calculate the average probability distribution of the local molecular field h=$\Sigma_j J_{ij}s_j^{\alpha\beta}$ on the plus sites and on the minus sites separately[18]:

$$F_\pm(h;\{J_{ij}\},\{s_i^{\alpha\beta}\})$$

$$= (2/NP(1+m))\sum_{\alpha\beta i}\delta(s_i^{\alpha\beta}\Sigma_j J_{ij}s_j^{\alpha\beta}-h)\theta(\pm s_i^\alpha s_i^{\alpha\beta}) \ . \tag{5}$$

In other words these functions count the fraction of the plus (minus) sites such that the local field falls in the range (h,h+Δh).

The calculation of

$$\overline{F_\pm(h)} = < F_\pm(h;\{J_{ij}\},\{s_i^{\alpha\beta}\}) >_J \qquad , \tag{6}$$

where $< \ >_J$ indicates the average over the measure (4) and the bar indicates the quenched average over the patterns, is done in the limit N->∞ using mean field approximation and the replica method. The saddle point that contributes is the one derived in ref[17]. The result is

$$F_\pm(h) = -\int \frac{dt\,e^{-\frac{t^2}{2}}}{\sqrt{2\pi}}\ \frac{\dfrac{d}{dh}\,\mathrm{erfc}\!\left(\left(\dfrac{h-mM}{\sqrt{1-m^2}}+\sqrt{q}\,t\right)/\sqrt{2(1-q)}\right)}{\mathrm{erfc}\!\left(\left(\dfrac{\pm mM}{\sqrt{1-m^2}}+\sqrt{q}\,t\right)/\sqrt{2(1-q)}\right)} \tag{7}$$

where M and q are determined by Gardner's saddle point equations as functions of m and P, and erfc is the complementary error function.

If the number of patterns stored in the network is near its maximum possible value then q≈1 and (7) simplifies to

$$F_\pm(h) = \delta(h)\mathrm{erfc}((\pm mM)/\sqrt{2(1-m^2)}\,)+\frac{\theta(h)\exp(-(-h\pm mM)^2/2(1-m^2))}{\sqrt{2\pi(1-m^2)}} \ . \tag{8}$$

There is a clear asymmetry between F_+ and F_- . The latter corresponds to a distribution more concentrated at small values of h. The asymmetry increases with m (the ambiguity of ref[4]). It follows that category-specific information is more robust than individual information. This is in essence the meaning of prosopagnosia.

Similar ideas can be applied to other syndromes. As an example we mention the work on *bitemporal amnesia* by McClelland and Rummelhart[13].

Acknowledgements: A fellowship of the John Simon Guggenheim Memorial Foundation is gratefully acknowledged.

References

1. M. L. Barr and J. A. Kiernan "The Human Nervous System" Harper International Ed. Harper and Row, Philadelphia 1988.
2. J. Hart, R.S. Berndt and A. Caramazza, Nature $\underline{316}$, 439-440 (1985).
3. See for instance A. Kertesz "The Clinical Spectrum and Localization of Visual Agnosia" in "Visual Object Processing: a cognitive neuropsychological approach" ed G. W. Humphreys and M. J. Riddoch, Lawrence Erlbaum Associates, Hove and London 1987.
4. A.R.Damasio, H.Damasio and G.W. Van Hessen Neurology (NY)$\underline{32}$ 331-341 (1982).
5. H. Ellis "Theoretical Aspects of Face Recognition" in "Perceiving and Remembering Faces" ed. G. Davies, H. Ellis and J. Shepherd, Academic Press, London 1981.
6. J. J. Hopfield Proc. of Nat. Acad. Sci. USA $\underline{79}$ 2554-2558 (1982).
7. M.A. Virasoro in "Disordered Systems and Biological Organization", ed. by E. Bienenstock, F. Fogelman-Soulie, G.Weisbuch. Springer Verlag, Berlin (1986). N. Parga and M.A.Virasoro J. Physique $\underline{47}$ 1857-1864 (1986).
8. G. Toulouse, S. Dehaene and J.P.Changeux Proc. of Nat. Acad. Sci. USA $\underline{83}$ 1695-1698 (1986).
9. D. J. Amit, H. Gutfreund and H. Sompolinsky Phys. Rev. A $\underline{35}$ 2293-2298 (1987).
10. V. Dotsenko Physica $\underline{140A}$ 410-415 (1986)
11. M. V. Feigelman and L.B. Ioffe International Journal of Mod. Phys. $\underline{B1}$, 51 (1987).
12. H. Gutfreund Phys. Rev. A $\underline{37}$ 570-577 (1988)
13. D.E.Rummelhart and J.L. McClelland eds. "Parallel Distributed Processing: Explorations in the Microstructure of Cognition" Vol I and II, MIT Press, Cambridge Mass. 1986.
14. D.I. Perrett, P.A.J. Smith, D.D.Potter, A.J. Mistlin, A.J.Head, A.D. Milner and M.A. Jeeves, Human Neurobiology $\underline{3}$ 197-208 (1984).
15. E. T. Rolls in "Models of Visual Perception: from Natural to Artificial", ed. M. Imbert. Oxford University Press. Oxford 1988.
16. M. Mezard, G. Parisi, and M.A. Virasoro "The Spin Glass Theory and Beyond" Chapter X-XIII, World Scientific, Singapore 1987.
17. E. Gardner J. of Physics A $\underline{21}$ 257 (1988).
18. W. Krauth, M. Mezard and J.P. Nadal, LPTENS preprint 88-8 have done an independent calculation of the h distribution though without distinguishing plus and minus sites.

Neural Network Formation and the Activity of an Aggregate of Dissociated Hydra Cells

T. Itayama and Y. Sawada

Research Institute of Electrical Communication, Tohoku University,
Sendai 980, Japan

Introduction

The characterization of a living system is an important problem. The development of a living system in natural conditions has attracted the attention of many scientists throughout the years. It is, however, too complicated to understand the whole mechanism of these processes of development from a physical viewpoint.

Hydra is well known for its strong regenerative capacity. Hydra tissue can be dissociated into single cells. When those cells are reaggregated by centrifugation, the reaggregated cell mass can regenerate a complete hydra within a few days. This process of hydra regeneration from reaggregated cells involves a transition from a state of random cell mass to a state of an organic system.

The nervous system of hydra is a scattered nervous system without a central nervous system, but with two concentrated regions in the head and foot parts. It is suspected that some trigger cells which excite epitheriomuscular cells exist in the network at the head part. The nerve cells are known to have several kinds of neuropeptides.

Methods

The hydra cells were dissociated by pipetting them into a dissociation medium, and were gathered by centrifugation to make aggregates. The neural network can be clearly visualized by an immuno-histochemical method using antibodies which specifically stain nerve cells containing particular neuropeptides. In this present work anti-RFamide and anti-Vasopressin were used as antibodies. The contractive motion of aggregates was monitored by a time-lapse video recorder. At the same time the motion was recorded by measuring the change of scattering light intensity from the aggregates.

Results

(1) The temporal variation of the average density of the nerve cells and the average number of connections of a nerve cell with others are shown Fig.1 and Fig.2 respectively. At about 40 hrs from the starting time of regeneration, the average density and average number of connections begin to increase rapidly.

(2) The frequency of the contractive motion of the aggregates was measured by the light scattering method (Fig.3). The frequency starts increasing at about 40 hrs.

Springer Series in Synergetics, Vol. 43   **Cooperative Dynamics in Complex Physical Systems**
Editor: H. Takayama   © Springer-Verlag Berlin, Heidelberg 1989

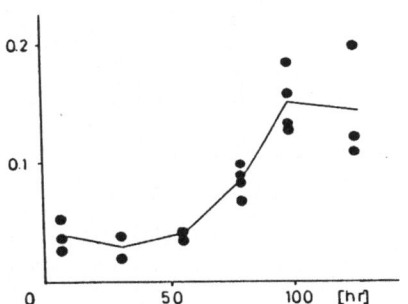

Fig.1 The average density of nerve cells

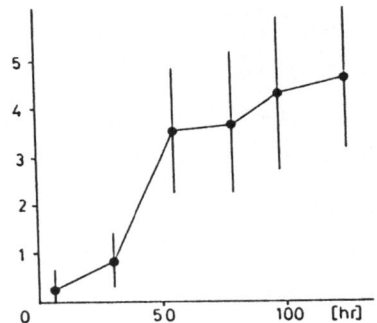

Fig.2 The average number of connections

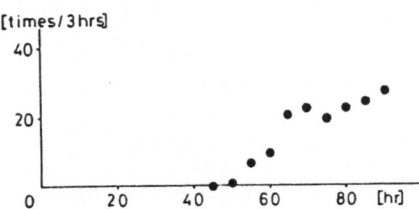

Fig.3 The frequency of contractive motion

Discussion

The results show that the average number of nerve cells and contractions as well as activity of the aggregates increases rapidly at about 40 hrs, suggesting some correlation between them. Formation of clusters of nerve cells is reminiscent of a percolation model. According to the percolation theory the cluster size becomes infinite at the percolation threshold corresponding to the particular density of nerve cells. The observed rapid increase at 40 hrs of activity might correspond to this transition.

References

1. K.M.Flick,H.R.Bode: HYDRA, Research Methods. Chap 18
2. C.J.P.Grimmelikhuzen: Cell Tissue Res. 241:171-182,(1985)

Glauber Dynamics of Neural Network Models

M. Schreckenberg and H. Rieger

Institut für Theoretische Physik, Universität zu Köln, Zülpicherstr. 77,
D-5000 Köln 41, Fed. Rep. of Germany

1. Introduction

The dynamical behaviour of spin-glass like systems of neural networks, e.g. the LITTLE–HOPFIELD model [1,2], is an active research field. Due to the complexity of the investigated models, analytic solutions are available only in some mean-field approaches, which give insight into the basic mechanisms of the systems. The asymmetric SK model belongs to this class of models and even some of its extensions, e.g. models with additional local random fields or hierarchically arranged couplings can be treated analytically.

Among various available methods to describe the dynamics we have chosen the path-integral description of the Glauber dynamics due to SOMMERS [3]. That is, we are dealing with master equations for the probability distribution of spin configurations. In the case of 1–spin–flip processes the transition rates are given by

$$w(\sigma_i \rightarrow -\sigma_i) = \frac{\Gamma}{2}\left[1 - \sigma_i \tanh \beta(h_i)\right] \quad , \tag{1}$$

where Γ is the spin–flip rate, β the inverse temperature and

$$h_i = b_i + \sum_{i(\neq j)}^{N} J_{ij}\sigma_j \tag{2}$$

the local field of the i–th neuron with external field b_i and synaptic couplings J_{ij}.

The advantage of the dynamical approach via path integrals is that it avoids the unphysical replica trick, since all quenched random variables, appearing linearly in the exponent of the functional, can be averaged out.

2. Little–Hopfield Model

In the Little–Hopfield model the synaptic couplings are specified as

$$J_{ij} = \frac{1}{N}\sum_{\mu=1}^{p}\xi_i^\mu\xi_j^\mu \quad , \quad J_{ii} = 0 \, , \tag{3}$$

where the ξ_i^μ are independently distributed, quenched random variables, representing the stored patterns of the network. The case of an extensive number of stored patterns is of special interest and can be studied analytically with the help of the path-integral formulation without using the replica trick.

As a result it is possible to extract the equilibrium properties of the model [4,5] and to confirm the replica symmetric solution of the Little–Hopfield model found

Springer Series in Synergetics, Vol. 43   **Cooperative Dynamics in Complex Physical Systems**
Editor: H. Takayama   © Springer-Verlag Berlin, Heidelberg 1989

by AMIT et al.[6]. Within a high–temperature expansion of the response function (or equivalently, an expansion for small concentrations of stored patterns), it is also possible to identify the generalized AT line. Furthermore the time-dependent evolution of the macroscopic overlaps with the stored patterns for a finite number of them can be investigated.

3. Asymmetric SK Model

The asymmetric neural network models take into account the asymmetry of real biological networks. These models cannot be treated by means of equilibrium statistical mechanics, because no energy function exists. The investigation of asymmetric networks therefore requires the dynamical definition of neural networks.

In the asymmetric SK model pairs of synaptic couplings (J_{ij}, J_{ji}) are independent and randomly distributed according to a bivariate Gaussian distribution with

$$\langle J_{ij} \rangle = \langle J_{ji} \rangle = 0 \quad , \quad N \langle J_{ij}^2 \rangle = N \langle J_{ji}^2 \rangle = J^2 \quad , \quad N \langle J_{ij} J_{ji} \rangle = \lambda J^2 . \quad (4)$$

The correlation parameter λ varies between -1 and $+1$. The case $\lambda = 1$ yields the SK model with symmetric couplings and $\lambda = 0$ the fully asymmetric SK model with completely uncorrelated couplings.

The case $\lambda = 0$ is of special biological interest and can be solved exactly [7]. The local response function can be determined directly and for the averaged autocorrelation function at zero temperature a self-consistency equation can be derived. This equation was also found by CRISANTI and SOMPOLINSKY [8] with a different method. Both functions show exponential decay in vanishing external field.

The nonequilibrium dynamics of the autocorrelation function is governed by an increasing asymptotic relaxation time starting with Γ^{-1} at the initial time. For the equilibrium dynamics one gets a second-order differential equation:

$$C'' = \Gamma^2 \cdot \left[C - \frac{2}{\pi} \arcsin (C) \right] . \quad (5)$$

The asymptotic solution of this equation yields an equilibrium relaxation time $\Gamma^{-1} \cdot \sqrt{\pi/(\pi - 2)}$. The physical meaning of the different relaxation times is that the system starting in a nonequilibrium state relaxes faster than starting in the vicinity of the equilibrium. It is also possible to extend the calculations to the case of small finite λ. A first-order expansion yields an increasing relaxation time of the response function and a vanishing long–time limit of the autocorrelation function. For small enough λ, the spin–glass state seems to be excluded even at zero temperature.

4. Extensions

In real biological networks one would expect some local noise which acts on the postsynaptic potentials. This effect can be included by introducing random local fields. The calculation yields a nonvanishing long-time limit of the autocorrelation function, but all relaxation times remain finite, as one would expect [7].

The Gaussian distributions in the fully asymmetric SK model are the same for all couplings. One simple way to change this situation is to arrange the distributions in a hierarchical way. This change has no influence on the zero-temperature dynamics, but would be seen at finite temperatures [9].

If one tries to store a finite number of patterns in a system with fully asymmetric couplings, one finds that the width of the Gaussian distribution of the couplings provides some kind of temperature in the network, even at zero temperature [7].

References

1. W. A. Little, *Math. Biosci.* **19**, 101 (1974)
2. J. J. Hopfield, *Proc. Natl. Acad. Sci. USA* **79**, 2554 (1982)
3. H. J. Sommers, *Phys. Rev. Lett.* **58**, 1268 (1987)
4. H. Rieger, M. Schreckenberg, J. Zittartz *J. Phys.* **A21**, L263 (1988)
5. H. Rieger, M. Schreckenberg, J. Zittartz, *Z. Phys.* **B72**, 523 (1988)
6. D. J. Amit, H. Gutfreund, H. Sompolinsky, *Ann. Phys. (NY)* **173**, 30 (1987)
7. H. Rieger, M. Schreckenberg, J. Zittartz, *Z. Phys.* **B** in press
8. A. Crisanti, H. Sompolinsky, *Phys. Rev.* **A37**, 4865 (1988)
9. M. Schreckenberg, H. Rieger, in preparation

Stochastic Dynamics of an Analog Neural Network

Jong-Hoon Oh

Department of Physics, University of Edinburgh,
The King's Buildings, Edinburgh, EH9 3JZ, UK, and
Department of Physics, Pohang Institute of Science and Technology,
PO Box 125, Pohang, Kyongbuk 790-600, Korea

Many optimizing problems can be interpreted as the problem of minimizing the cost functions in the form of the spin-glass Hamiltonian, i.e.

$$H = -\sum_{i,j} J_{ij} s_i s_j.$$
(1)

Kirkpatrick et al. /1/ used a simulated annealing method to minimize this cost function. The analog neural network given by Hopfield and Tank (HT) is is an alternative approach. The equation of motion of the HT network is defined as

$$\frac{du_i}{dt} = -u_i + \beta h_i,$$

$$h_i = -\frac{\partial H}{\partial s_i} = \sum_j J_{ij} s_j,$$
(2)

$$s_i = \tanh u_i,$$

where h_i is the derivative of the Hamiltonian with respect to s_i, and a local field to s_i at the same time. At its fixed point, $s_i = \tanh(\beta h_i) = <s_i>$. We can see that under this dynamics the system approaches the point where spin values become the same as mean field expectation values. This method has been utilized in various applications, and in many cases it is faster than simulated annealing. However, as it is a fully deterministic algorithm, its solutions are those corresponding to the local minima of the cost function. If it falls into a local minimum, it has no chance of escaping from it. Also, the dynamics is very sensitive with respect to initial conditions, and very often goes to irrelevant fixed points. Recent work by Wilson and Pawley/3/ shows this method is unsatisfactory for the Traveling Salesman Problem (TSP). Usually it is desirable to update spins asynchronously and to select good initial parameters/4/. In order to avoid these problems, I designed and tested several stochastic algorithms. I introduce two of them. Detailed numerical simulation results will be published elsewhere.

The simplest possibility is a HT network with noise, i.e. we can add a Gaussian noise term to the first equation of (2). The modified form is

$$\frac{du_i}{dt} = -u_i + \beta h_i + \Gamma_i(t),$$
(3)

where Γ is Gaussian noise with the property $< \Gamma_i(t)\Gamma_j(t') > = \Gamma_0\delta_{ij}\delta(t - t')$. In the long time limit

$$u_i = \beta h_i + \Gamma_i(\infty),$$
$$s_i = \tanh \beta(h_i + T\Gamma_i(\infty)), \tag{4}$$

where $T = 1/\beta$. If β is constant, the long time limit is no longer fixed because of the noise term. Usually we lower the value of T during the iteration a to get fixed final configuration. This net behaves similarly to the HT network at steep slope of the cost function, but when it approaches local minima, the noise term dominates, and provides a mechanism of escape. I tested the algorithm against the HT network and simulated annealing for SK spin glasses. It is almost as fast as the HT network and also approaches the global minimum value steadily.

The second network is based on the Langevin equation. It has three different variables. It is written as

$$\frac{dv_i}{dt} = -v_i - k_i(s_i - m_i) + \frac{1}{\beta}\Gamma_i(t),$$
$$\frac{ds_i}{dt} = v_i, \tag{5}$$
$$m_i = \tanh(\beta h_i),$$

where the definitions of $\Gamma_i(t)$ and h_i are the same as before. It may thought of as a multilayer version of the optimizing analog neural network. As we can see the first two equations have the form of the Langevin equation. In the long time limit, $< v_i^2(\infty) > = 1/\beta$ and $< k_i(s_i - m_i)^2 > = 1/\beta$. As the change of s_i is controlled by the first constraint, this dynamics becomes more stable and insensitive to parameters. Also parallel update of spin values is possible. An interesting application of the equation of motion is optimization problems with varying constraints. This type of problem is found in restoration of dynamic image. Furthermore, we may extend this network to have more than one layer of m_i. In that case, it is possible to define different J_{ij} for different optimization constraints in each layer.

References

1. S. Kirkpatrick, C. D. Gelatt, M. P. Vecchi: Science 220, 671 (1983)
2. J. J. Hopfield, D. W. Tank: Biol. Cybern. 52, 142 (1985)
3. G. V. Wilson, G. S. Pawley: Biol. Cybern. 58, 63 (1988)
4. G. C. Fox, W. Furumanski: preprint

Rule Dynamics and the Fuzzy Attractor: New Approach to EEG

Y. Aizawa[1] *and Y. Nagai*[2]

[1]Department of Applied Physics, Waseda University,
 Shinjuku 160, Tokyo, Japan
[2]Department of General Education, Azabu University,
 Fuchinobe, Sagamihara 229, Japan

1. Introduction

The EEG is believed to be generated from the flow of neural impulses to the Pyramidal cell in cortex. The detailed pathway of the impulse flow is controlled by many rules which process the signals in a brain. Therefore, the essence of EEG should be explained in terms of the rule-dynamics which unifies the global activity of neurons, if the real mechanisms are extremely complex. The purposes of this article are to propose a simple rule-dynamical model of cellular automata and to discuss the possibility of the rule-dynamical approach to EEG.

2. Fuzzy Attractor of Rule Dynamics -Intermittency-

The 1-D C.A. with nearest-neighbor interaction is uniquely represented by

$$S_i(t+1) = \sum_{k=1}^{5} \varepsilon_k f_k , \qquad (1)$$

and $f_1 = S_{i-1} + S_i + S_{i+1}$, $f_2 = S_{i-1}S_i + S_iS_{i-1} + S_{i-1}S_{i+1}$, $f_3 = S_{i-1}S_iS_{i+1}$,

$f_4 = (S_i - S_{i-1})(S_i - S_{i+1})$, $f_5 = f_2 f_4$ (mod.2),

where the spin variable S and the switching parameter ε take the values 0 or 1 depending on the site i and time t [1]. Dynamics in rule space $\varepsilon = (\varepsilon_1, \varepsilon_2, \varepsilon_3, \varepsilon_4, \varepsilon_5)$ can be introduced under several constraints. This article treats the autonomous case described by

$$\varepsilon_k(i,t) = \theta [\eta_k \{S^*(i,t) - C_k\}] \qquad (2)$$

and

$$S^*(i,t) = \Sigma e^{-\gamma |i-j|} S_j(t) / \Sigma e^{-\gamma |i-j|} ,$$

where $\theta [\ ]$ is a step function, η $\pm$ sign, C the threshold and $1/\gamma$ the effective range of local feedback. In what follows, $\gamma \rightarrow 0$ is assumed and the one-dimensional rule-space R is defined by the ordering used in ref.[1] ($1 \leq R \leq 32$). Fix the values of parameters; C = (0.6,0,0,0.48,0) and $\eta = (-,-,+,+,-)$. Changing the parameters C the system reveals intermittency among three rules (R = 4, 7, and 22), and the return map of the firing rate T: $r(t) \rightarrow r(t+1)$ ($r(t) \equiv <S(t)>$) shows a fuzzy attractor (Fig.1). The correlation dimension D(m) of the attractor is shown in Fig.2, where m stands for the embedding dimension [2]. The dimension is less than the case of a single fixed rule for R=4.

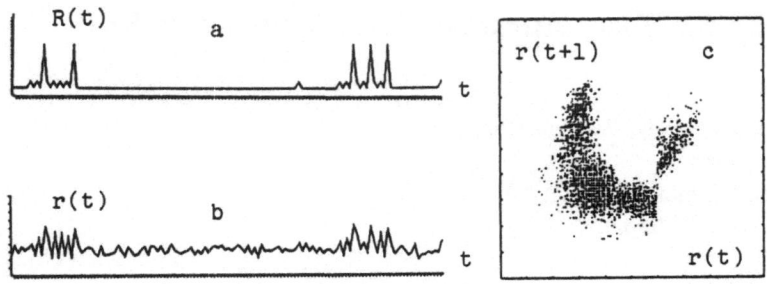

Fig.1:(a)(b)Intermittent time courses of R(t) and r(t), (c)Fuzzy attractor

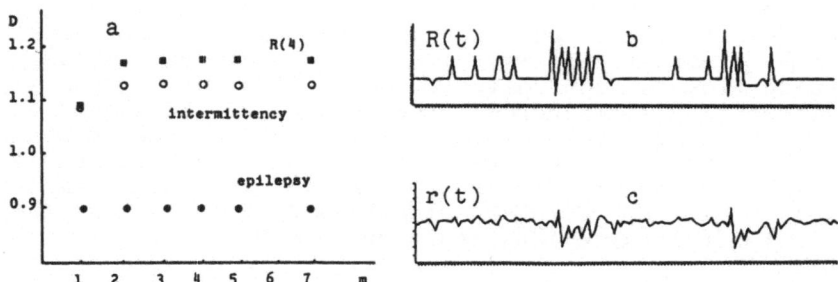

Fig.2:(a)Correlation dimension of attractor, (b)(c)Epileptic time courses of R(t) and r(t).

3. Rule Dynamical View Points of EEG -Epilepsy-

The time course of r(t) has several similarities with EEG, e.g., figure 1-(b) shows the slow alternation of the small amplitude rhythm and large fluctuation which are usually observed in normal EEG as the alpha rhythm and the bursts of slow variant. Though the frequency of the fluctuation is quite different with each other, the structures of both have a strong similarity. Changing the constraints of eq.(2), the epileptic time series can also be generated as is shown in Fig.2-(c) where the spike and wave rhythms inherent to epilepsy are observed [3].   A striking result of Fig.2-(a) is that the dimension of the epileptic case is smaller than that of the intermittency though the number of the contributing rule in the former case is larger than the latter. This implies that the strong time-ordering is created in the epileptic rule dynamics.   The interesting results in ref.[4] should be revisited from the rule dynamical viewpoints.

References

1. Y. Aizawa, I. Nishikawa; In Dynamical Systems and Nonlinear Oscillators, ed. G.Ikegami (World Scientific, Singapore, 1986) p.210
2. P. Grassberger, I. Procaccia; Phys. Rev. Lett. 50 (1983),346
3. Y. Aizawa, Y. Nagai; submitted to Prog. Theor. Phys. 1988
4. A. Babloyantz, C. Nicolis, M. Salazer; Phys. Lett. 111A (1985),152

Phase Selection of the Immune Network

T. Ikegami

Department of Physics, Faculty of Science, University of Tokyo,
Hongo 7-3-1, Bunkyo-ku, Tokyo 113, Japan

1. Introduction

The immune system performs cognitive functions like pattern recognition, learning, memory, etc., typical of the central nervous system. Both systems consist of complex adaptive networks, but underlying dynamics and interactive means are considered to be quite different.

A model presented here is aimed at understanding the cognitive function of the immune system. In a system composed of antibodies, each antibody element(Ab) is known to connect with other Ab's as well as with a foreign body called an antigen (Ag). For this, one can connect Ab_1 with Ab_2 provided that an 'idiotope'(acceptor) of Ab_1 contacts with a 'paratope' (donor) of Ab_2. The idiotope and paratope form a complementary junction with each other.

When the immune system is invaded by Ag which has an idiotope, we assume that Ab having a matching paratope against Ag is generated. The Ag is linked with Ab and only the resultant Ag-Ab complex is assumed to be removed from the system as a foreign element. Any open idiotope of the Ab will induce another matching Ab and a cascade reaction takes place among Ab's. Dynamical behavior of the immune system is considered to originate from such an immune response among Ab elements in the internal network.

This model of idiotype network for the immune system is examined by a computer simulation on the analogy of automata, since in the absence of foreign Ag, such a network exhibits paratope/idiotope characteristics as an autonomous behavior.

2. Model

Labelings are made for the idiotope-paratope pair(X_i, Y_i) of Ab_i (i=1,2,...,p), where X_i and Y_i are integers. We assume a cyclic network of size p, that is,

$$X_i = i,$$

$$Y_i = X_i + 1, \tag{2.1}$$

with a cyclic boundary condition,

$$X_p = p,$$

$$Y_p = 1. \tag{2.1'}$$

There can be many Ab_k's as a system element of k-type. We assume that Ab_k can interact only with Ab_{k-1} or Ab_{K+1}, generating linked complexes; (X_{k-1}, Y_k) or (X_k, Y_{k+1}). Thus generating complexes can further react with other elements, but the reaction rate between the complexes is assumed to be limited below a certain value.

If the number ratio of a certain idiotope Y_k to the total elements exceeds a given birth threshold T_B, we assume that there is injected an amount Q_B of elementary Ab's having a complementary paratope, that is, X_{k+1}. In addition to this, one element will be injected for every Ab_k at each step. As to the removal of the resultant complexes, a similar assumption is made. If the number ratio of a certain complex idiotope to the total elements exceeds a given death threshold T_D, we remove those complexes by an amount up to Q_D.

In the above handling of source and sink for system elements, the threshold values T_B, T_D which satisfy the condition $T_B > T_D$ and the birth/death numbers Q_B, Q_D are adjustable parameters. It must be noted that a removal of antigen Ag is always in a form of a complex containing Ag.

3. Discussion and Conclusion

Present model of the immune system is computer simulated under the initial condition where the dominant element is Ab_1. The system is observed to have multi-stable phases, which correspond to three distinctive phases called S, B, and N phase. In S-phase, one specific idiotope is rich in numbers and others are suppressed. B-phase is characterized by many successive bursts of various Ab_k quantities. N-phase has an equal distribution of every idiotopes. The rich idiotopes in S-phase(Fig.1) are observed to conserve their numbers during the phase, because the supply of elementary Ab's to the abundant idiotope is just compensated by generating complexes. The immune function will be the best in S-phase, because the system is most reactive against a specific antigen or idiotope.

As the network size p is increased, the irrelevant N-phase appears more frequently. The critical size of the system to stay in S-phase can be determined to be $p_c = 10$, if we assume a hierachical structure among the number of idiotopes in S-phase. Beyond the critical size, the network stays only in N-phase and no S-phase appears.

In conclusion, the critical size of the immune network is obtained to be about 10, if we take a model of antigen removal to be only in the form of a complex containing Ag. It is considered crucial in the present model to retain the complexes until they reach a given threshold T_D. More details are given in IKEGAMI[1].

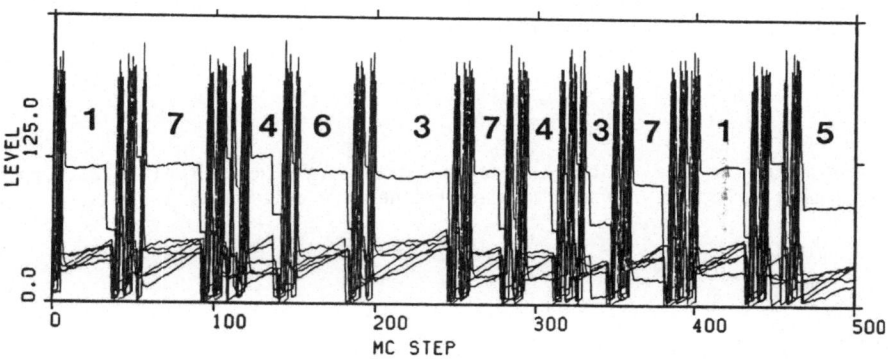

Fig.1 Time evolution of number levels for each of the seven idiotopes(p=7) is simultaneously recorded, where the numbers denote the idiotope of a maximum level in each S-phase which is almost periodically switched to B-phase associated with a bunch of spikes, for the case of T_B=0.5, Q_B=200, T_D=0.14 and Q_D=80.

Reference

1. T.Ikegami: "Dynamical Behaviors of the Immune Network",(preprint,1988)
        submitted to Prog.Theor.Phys.

Part IV

Nonlinear Dynamics in Fluids, Chemical and Biological Systems, etc.

Are Earthquakes, Fractals, and $1/f$ Noise Self-organized Critical Phenomena?

P. Bak, Chao Tang, and K. Wiesenfeld

Brookhaven National Laboratory, Upton, NY 11973, USA

Spatially extended dynamical systems can evolve towards a "self-organized" critical state. We suggest that several phenomena which are known to have temporal or spatial power law correlations are manifestations of this state. The Gutenberg-Richter law for the distribution of earthquakes indicates that the occurrence of earthquakes is a critical phenomenon.

1. INTRODUCTION

There exists a couple of phenomena in nature with long spatial or temporal correlations. One is the "$1/f$" noise[1] which occurs in resistors, sunspots, quasar-light etc[1]. Another is the occurrence of self-similar structures, called fractals, in mountain landscapes, clouds, river-branching, growth phenomena, and so on[2]. Despite the fact that $1/f$ noise is always found in spatially extended systems, little effort has been put into the investigation of the spatial organization of these systems. Nor has there been any substantial attempts to understand the dynamics responsible for the formation of fractal structures. We believe that these phenomena are often two sides of the same coin: they are fingerprints of a self-organized critical state[3].

Critical states are known from condensed matter systems undergoing equilibrium second order phase transitions. To arrive at the critical point, one has to tune a parameter such as temperature. As the temperature of a ferromagnetic system is lowered, there will be larger and larger clusters of aligned spins: the correlation length increases. Eventually, precisely at the critical point, there are fractal clusters of all sizes with no characteristic length scale. If one measures the magnetization at a given position, it will show power-law "$1/f$" fluctuations. Thus, there are some striking similarities between the systems above and critical equilibrium systems. But how can non-equilibrium dynamical systems be critical since there is no need to turn any knobs to fix any parameters to arrive at the critical state? Indeed, attempts to explain $1/f$ noise as a dynamic intermittency phenomenon fail because a parameter has to be adjusted in order to obtain the desired behavior[4].

We argue that spatially extended non-equilibrium systems naturally evolve into a statistically stationary state, which is also critical. The

Springer Series in Synergetics, Vol. 43 **Cooperative Dynamics in Complex Physical Systems**
Editor: H. Takayama © Springer-Verlag Berlin, Heidelberg 1989

criticality is self-organized since it does not require an outside agent to fine-tune the system. Our conjecture is supported by numerical simulations, scaling considerations, mean field theories, and hand-waving arguments. Of course, it is impossible (and probably not even desirable) to construct models which are both general and realistic. The models that we study are extremely simple cellular automatons, which may serve as "Ising" models of the self-organized critical state. Despite their simplicity, we believe that the models contain the essential features of many real dynamical systems, and since they operate at a critical point, universality may apply in the sense that real systems may have the same critical exponents. On the other hand, the models cannot be used to predict precisely which systems in nature will actually approach the critical point.

2. A SIMPLE MODEL

Figure 1 shows an interacting dynamical system, represented by an array of N particles connected by springs. The springs represent some interaction which does not necessarily have to be harmonic. The particles experience "friction" with a rough surface, which causes the particle to be pinned. The number of metastable pinned states increases exponentially with the size of the system. The system can be driven either by pushing particles into the array at one end, or by adding a uniform driving force to all the particles, i. e. by tilting the system. The model may represent for instance the motion of earth or "tectonic plates" along a fault, the motion of a dislocation line in a resistor, a sliding superconducting vortex lattice, or a sliding charge-density-wave.

An individual particle, i, will experience the elastic force from the springs connecting it with its neighbors plus an external driving force, which can be visualized as a tilting of the chain. When this force overcomes the random pinning force, the particle will jump forward. During the jump, potential energy is first converted to kinetic energy, and then dissipated. The elastic force on the particle decreases, and the elastic forces on the two neighbors increase. Let us for simplicity (and

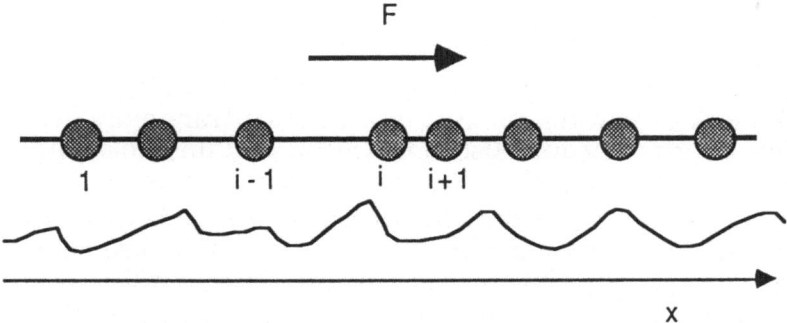

Figure 1. The earthquake model. Individual particles move to the right when the spring force plus the constant driving force overcome the pinning potential.

with little loss of generality) assume that the jump distance will be an integer so that we can specify the positions of the pinning well, and the particle, by an integer, X_i. We shall monitor only the elastic force on the individual particles. The dynamics is very simple: if the elastic force on some particle, i, exceeds a critical value Z_c determined by the pinning force plus the driving force, the particle moves one unit to the right. Without loss of generality, we can set the critical value equal to zero. The system has exponentially many metastable states where all Z_i's are non-positive. Mathematically, the dynamics can be expressed by the simple rule

$$Z_i \rightarrow Z_i - 2$$
$$Z_{i\pm1} \rightarrow Z_{i\pm1} + 1, \qquad\qquad \text{if } Z_i > 0. \qquad\qquad (1)$$

Let us start, for instance, with a situation where the forces on the particles are small, so that all the Z_i's assume negative numbers. We feed the system by increasing the force on one individual particle, which could for instance be pushing the extreme left particle,

$$Z_i \rightarrow Z_i + 1. \qquad\qquad (2)$$

As this process is repeated, then somewhere Z must necessarily exceed the critical value zero, and the particle will be depinned following the rule (1). This process may or may not cause the neighbors to become unstable. If it does, then the operation (1) is performed on that (or those) neighbors. If it does not then the process (2) is performed again.

It is not difficult to see that the system will evolve until it reaches the point where *all* the Z_i's assume the critical value, $Z_i = 0$. This is the *least* stable of all the metastable states. Any further increase of Z's will propagate through the chain, leaving it in the least stable state. In other words, the effect of a small local perturbation is communicated throughout the system, but the system is robust with respect to noise insofar as it returns to the least stable state. If the chain is perturbed randomly by the rule (2), the resulting flow is also random white noise, i.e. with power spectrum $1/f^0$. As we shall see in the next section, the robustness of the least stable state is lost in two and higher dimensions.

The dynamical selection principle leading to the least stable state is insensitive to the way the system is built up. One can randomly (with respect to i) apply (2); or one could also start with a very unstable state, $Z_i > Z_c$ for all i, and let the system relax. In all these cases the least stable state will be reached. In one dimension, the least stable state is critical in the restricted sense that any small perturbation can just propagate infinitely through the system, while any lowering of the Z_i's will prevent this. This is analogous to some other 1D critical phenomena, such as percolation where at the percolation threshold particles can just percolate to infinity. Also, like other 1D systems the critical state has no spatial structure, and correlation functions are trivial. In the next section we shall see that in higher dimensions the critical states and their dynamics are dramatically different.

3.SELF-ORGANIZED CRITICALITY

The rules (1) and (2) for the one-dimensional model can easily be generalized to higher dimensions. When the particles move left, the force increases on all its nearest neighbors. In two dimensions,

$$Z(i,j) \; \rightarrow \; Z(i,j) \; - \; 4$$
$$Z(i \pm 1, j) \; \rightarrow \; Z(i \pm 1, j) \; + \; 1$$
$$Z(i,j \pm 1) \; \rightarrow \; Z(i,j \pm 1) \; + \; 1, \qquad \qquad \text{if } Z(i,j) > Z_c. \qquad (3)$$

where we have the square array (i, j), for $1 \leq i,j \leq N$. Naively, one might expect that the situation is the same as in one dimension, namely that the system will build up (or collapse) to the least stable state where the forces $Z(i,j)$ all assume the critical value. A moment's reflection will convince us that it cannot be so. Suppose we perturb the least stable state on one site. This will render the surrounding sites unstable $(Z > Z_c)$, and the noise will spread to the neighbors, then *their* neighbors, by a "domino" effect, *ever amplifying* since the sites are generally connected with more than two least stable sites. The perturbation eventually propagates throughout the entire lattice. The least stable state is thus unstable with respect to small fluctuations and cannot represent an attracting fixed point for the dynamics. As the system further evolves, more and more more-than-least stable states will be generated, and these states will impede the motion of the "noise". <u>The system will become stable precisely at the point when the network of least stable clusters has been broken down to the level where the noise signal cannot be communicated through infinite distances. At this point there will be no length scale, and consequently no time scale in the problem.</u>

Hence one might expect that the slowly perturbed system approaches, through a self-organizing process, a critical state with power law correlation function for physically observable quantities, including the power spectrum of the particle current. In analogy with the discussion for the one dimensional case, the the average force on the particles, $<Z>$ will build up to the point where statistical stationarity is obtained: *this is assured by the self-organized critical state* , but not by the least stable state.

Now the reader can guess what will happen if we perturb the critical state locally via Eq. (2). The perturbation will grow over all length scales. That is, a given perturbation can lead to anything from a shift of a single unit to a large avalanche. The lack of a characteristic length scale leads directly to a lack of a characteristic time scale for the fluctuations. As is well known, a random superposition of pulses of a physical quantity with a distribution of lifetimes $D(t) \approx t^{-b}$ (weighed by the average value of the quantity during the pulse) leads to a power frequency spectrum, $S(f) \approx f^{-2+b}$, so we also expect a $1/f$ like power spectrum for the system. Extensive numerical simulations in two and three dimensions have been carried out to testify the above argument. Here we show only two of the results. Plotted in Fig. 2 is the frequency spectrum of the slowly perturbed system, that is the spectrum of the flow along a fault in an earthquake region. Figure 3 shows the distribution of cluster (or "avalanche") size at the stationary critical state. A cluster is defined as those sites being affected by a single perturbation before it dies out. Hence the clusters are defined dynamically. The distribution function fits a power law $D(s) \approx s^{-\tau+1}$ with $\tau \approx 2$. The fall off at large s is due to the finite size effect. Fig. 3 shows that there is no macroscopic length scale except the system size.

The cluster size is a measure of the energy dissipated during the "earthquake". Indeed, data on the occurrence of earthquakes have been interpreted in terms of a power law distribution, known as the Gutenberg-Richter law[5]. *The Gutenberg-Richter law can thus be interpreted as a manifestation of the self-organized critical behavior of the earth dynamics.*

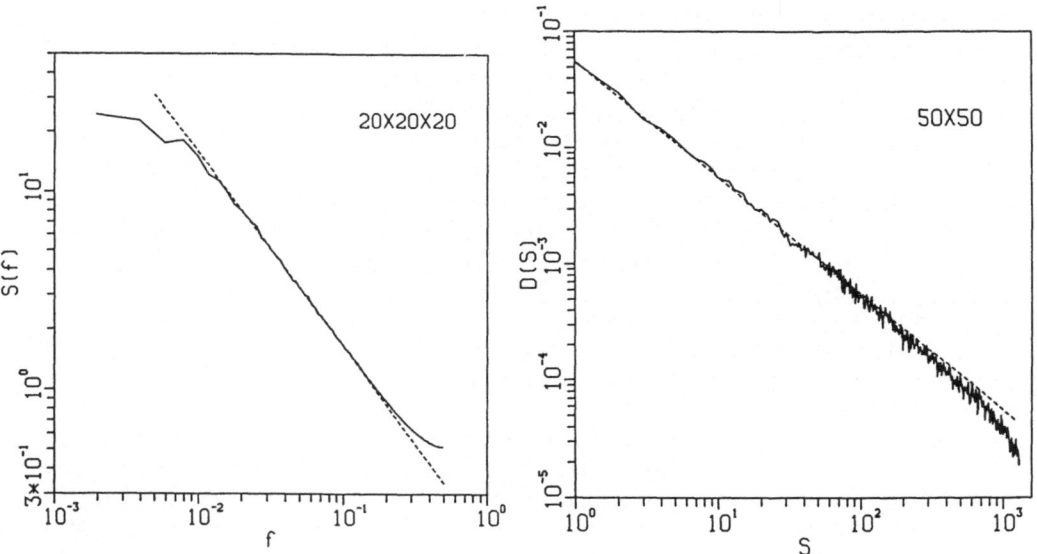

Figure 2. Power spectrum of the particle current in the self-organized critical state. The dashed straight line has a slope of -0.98.

Figure 3. Cluster size distribution for a two-dimensional system of size 50×50. The measurement was taken by applying rule (2) 100,000 times after the system reaches a stationary state. The dashed straight line has a slope of -1.

4. SUMMARY AND DISCUSSION

Our general arguments and numerical simulations show that dissipative dynamical systems with extended degrees of freedom can evolve towards a self-organized critical state, with scale invariant spatial and temporal fluctuations. Although this new concept is demonstrated via simple models, we believe that it can be taken much further to explain a wide range of observed spatial and temporal scaling phenomena. We now discuss a few applications.

The widespread phenomenon of "$1/f$" noise has long been a myth. There are probably many different mechanisms generating the noise, but the picture presented here might be one of the most general mechanisms. The $1/f$ noise is not "noise", but the temporal signal of the self-organized critical state. As the readers can see, the "$1/f$" like spectrum comes naturally out of the dynamics and no further assumptions and parameter tunings are needed. We have constructed a mean field theory[3] which gives a value 1 for the noise exponent. This gives us a hint on why "$1/f$" noise is usually so close to $1/f$. The deviations of the exponent from unity in many real systems can thus be understood as deviations from mean-field theory. Obukov has developed a renormalization group theory to first order in 4-d, 4 being the upper critical dimension above which mean field exponents are exact[6].

When a liquid is rapidly quenched into a glass state, the dynamical process involved seems to have much in common with our model, when started far from equilibrium (all $Z>Zc$) and let it relax. At the beginning all "particles" can move and then the system gets stuck at some metastable state after getting rid of excess energy. But this speculation has to be developed further.

This work was supported in part by the Division of Materials Science, US Department of Energy, under contract DE-AC02-76CH00016.

REFERENCES

1. For a review on $1/f$ spectrum, see W. H. Press, Commun. Mod. Phys. C **7**, 103 (1978); P. Dutta and P. M. Horn, Rev. Mod. Phys. **53**, 497 (1981)
2. B. Mandelbrot, *The Fractal Geometry of Nature* (W. H. Freeman, San Fransisco, 1982).
3. P. Bak, C. Tang, and K. Wiesenfeld, Phys. Rev. Lett. **59**, 381 (1987); Phys. Rev. A **38**, 364 (1988); C. Tang and P. Bak, Phys. Rev. Lett. **60**, 2347 (1988); C. Tang and P. Bak, J. Stat. Phys. **51**, 797 (1988).
4. P. Manneville, Journal de Physique **41**, 1235 (1980).
5. B. Gutenberg and C. F. Richter, Ann. di Geofis, **9**, 1 (1956).
6. S. A. Obukov, private communications.

Marginal Stability, Memory and Nonlinear Dynamics of Charge Density Waves

P.B. Littlewood and S.N. Coppersmith

AT&T Bell Laboratories, Murray Hill, NJ 07974, USA

1. INTRODUCTION

There are now some half-dozen compounds known in which collective transport by sliding of a charge density wave (CDW) has been observed. The most familiar manifestation of CDW sliding is of course a non- linear current-voltage characteristic, where the CDW motion contributes to the current above a threshold electric field E_T. Below this field, the CDW is pinned to the underlying lattice. CDW's also possess complex dynamical properties which have been studied on time scales ranging from 10^{-10} to 10^4 seconds.[1] The pinning of the CDW is spatially distributed and non-uniform, but locally periodic (if the CDW is displaced by precisely its wavelength, the configuration is identical). Hence the response to a time-dependent driving is complex, arising from interplay between the internal "washboard" frequency ω_0 ($\equiv v/\lambda$, the ratio of the uniform CDW velocity to its wavelength) and the frequency of the driving field.

The elucidation of the complicated behavior of CDW's under periodic forcing will be the primary object of this paper. We use a simple classical model for CDW motion, first written down by Fukuyama, Lee and Rice (FLR)[2], that has been very successful in interpreting a large body of experimental data. This model treats the CDW as an elastic continuum which is deformed by interaction with randomly-distributed impurities; it can be idealised as a model of coupled nonlinear relaxational oscillators, driven by an applied field. The feature of paramount importance is that the behavior is intrinsically described by many effective degrees of freedom (EDF), and cannot be reduced to the dynamics of a small number of collective coordinates. Nevertheless, there is indeed unusual collective behavior of this system, despite the absence of a low-EDF description. We find that this type of dynamic behavior is generic to a number of models, including certain types of coupled maps, and some cellular automata; under some conditions, coupled maps can be derived rigorously from the FLR equations of motion.[3]

Before studying the behavior under time-dependent driving fields, it is necessary to review briefly the properties of the FLR model in purely dc electric fields.

2. PROPERTIES OF THE FLR MODEL

2.1 Equations of motion

A CDW can be regarded as a weak periodic modulation of the conduction electron charge density, of the form

$$\rho(r) = \rho_c + \rho_o \cos(\mathbf{Q}\cdot\mathbf{r}+\phi(\mathbf{r})) , \qquad (1)$$

or, equivalently, as a small periodic lattice distortion. For simplicity, we have chosen a purely sinusoidal modulation with wavelength $\lambda=2\pi/|\mathbf{Q}|$, and we shall take $\mathbf{Q}=Q\hat{z}$, with $\hat{z}$ the direction of sliding. Only the "soft" phase fluctuations are relevant at long wavelengths, long-range order being destroyed by pinning of the CDW phase $\phi(\mathbf{r})$ to impurities (that long wavelengths dominate can be checked self-consistently). Including an elastic energy for CDW deformations, as well as a local interaction with impurities at sites $\mathbf{R}_i$, leads to the energy[2]

Springer Series in Synergetics, Vol. 43 **Cooperative Dynamics in Complex Physical Systems**
Editor: H. Takayama   © Springer-Verlag Berlin, Heidelberg 1989

$$H(\phi) = \int d^d r \left[\tfrac{1}{2} K |\nabla\phi|^2 + \sum_i \rho_o V(r - R_i)\cos(Q \cdot r + \phi) \right], \tag{2}$$

together with an equation of motion

$$\gamma\dot\phi = -\frac{\delta H}{\delta\phi} + \rho_c E(t)/Q . \tag{3}$$

The distortion of the CDW is determined by competition between the elastic energy and impurity pinning, while the driven dynamics are purely relaxational, controlled by a large viscous damping (at the low frequencies of interest here ($\leq$10MHz) the inertia of the CDW can be neglected). All of the salient features of the dynamics can be captured by treating the short-range interaction with impurities as a δ-function, and keeping as dynamic variables the values of the CDW phase at the impurity sites: $\phi_i = \phi(R_i)$. After some rescaling, equation (3) becomes (here in one dimension)

$$\dot\phi_i = \Delta^2\phi_i + V\sin(\theta_i + \phi_i) + \tfrac{1}{2}E(R_{i+1} - R_{i-1}) \tag{4}$$

where

$$\Delta^2\phi_i = (\phi_{i+1} - \phi_i)/l_i - (\phi_i - \phi_{i-1})/l_{i-1} , \tag{5}$$

and $l_i = R_{i+1} - R_i$, $<l_i> = 1$. For randomly distributed pinning, one has $\theta_i (\equiv QR_i)$ varying randomly from site to site. This is the most important source of disorder in these equations, and one may conveniently set $l_i = 1$ with no qualitative change in the behavior.[4] These and similar equations have been studied in detail by both analytical methods[5] and by numerical simulation [6] [7] [4] [8] [9] These equations are closely related to dynamics of an overdamped Frenkel-Kontorova (FK) model, for which $\theta_i = Qi$.[10]

2.2 Length scales and metastable states

The random pinning produces fluctuations of the phase on a characteristic length scale L_o (the FLR length). Even if the pinning is weak ($V \ll 1$), this length scale is finite, with $L_o \sim V^{-2/3}$ in one dimension. There are many (meta-) stable pinned states; if a local region of size $> L_o$ is displaced by $\sim 2\pi$, the energy change is confined to the boundaries of the region. Thus one estimates that the total number of metastable states is of order $\exp[cL/L_o]$ in a system of size L, with c a constant of order unity.

Because the effective temperature of a pinned CDW is very low (i.e. barriers between states $\gg k_B T$), which metastable state will be occupied is controlled by the sample history. Consider, for example, preparing a state by relaxation from a configuration which is far from static equilibrium. The configuration will evolve by rolling down the energy surface until it finds the first configuration for which $\delta H/\delta\phi = 0$. Such a metastable state will likely be of high energy, and will typically possess very low barriers to nearby states. In fact, numerical calculations[8] have shown that metastable states obtained in this fashion have typically a distribution of barriers extending to arbitrarily low energies; consequently the potential in the vicinity of the minimum is very flat. This leads to an anomalous enhancement of the low frequency linear a.c. susceptibility[11]

$$\varepsilon(\omega) \propto N^{-1}\sum_i \partial\phi_i(\omega)/\partial E(\omega) \sim \varepsilon(0) - (i\omega\tau)^\alpha + \cdots \tag{6}$$

with the exponent α reflecting a power-law distribution of eigenvalues for the modes of oscillation about the metastable state.

Such a state is found[8] to be *marginally* stable , with the nonlinear response divergent as $\omega \to 0$.[8] [7] In particular $<d\phi/dE>_{dc} > \lim_{\omega\to 0}\varepsilon(\omega)$, which is a signature of hysteresis and a "glassy" response. We stress that very few metastable states are marginal - a vast majority have a gap in the excitation spectrum at low energies. Nevertheless, states of marginal stability are *generic* in that they are selected dynamically from typical initial conditions.

281

2.3 Motion in a dc field above threshold

In a dc field exceeding $E_T \sim V^{4/3}$ ($d=1$) a bulk current $j=\langle\dot\phi\rangle$ flows. In contrast to the behavior below threshold, the sliding configuration is *unique* and is also periodic in time : $\phi(t+2\pi/\omega_o)=\phi(t)+2\pi$. The local current oscillations are out of phase between nearby regions so that the average current is uniform in an infinite system: $\langle\dot\phi\rangle=\omega_o$.[6] In a finite system the averaging is not complete and there is a small periodic component to the average current (narrow-band noise, or NBN).

There is no long range order of the phase, with the CDW configuration describing a random walk. However, the velocity correlation function $C_v(j)=\langle(\dot\phi_i(t)-\omega_o)(\dot\phi_{i+j}(t)-\omega_o)\rangle=f(|j|/\xi_v)$ is well behaved, and defines a velocity correlation length ξ_v. The approach to threshold from above can be treated as a dynamic critical phenomenon[7], where $\xi_v \sim (E-E_T)^{-\nu}$ diverges, and the average current vanishes as $\omega_o \sim (E-E_T)^\zeta$. The appearance of non-trivial critical exponents arises as a consequence of the interaction of many long-wavelength modes, and the behavior is inherently that of a many degree of freedom system.

2.4 Motion in pulsed electric fields

We now consider time-dependent driving fields, beginning with an apparently trivial case.[12] We take a sequence of repeated identical rectangular pulses, with an applied field $E(t)=E_1 > E_T$ for a time t_{on}, followed by a period t_{off} where $E(t)=0$.

If $t_{off} \gg 1$, in between each pulse the system will relax completely to an equilibrium static metastable state. Thus the application of each pulse generates a mapping $F:\{\phi_j^A\}\rightarrow\{\phi_j^B\}$, where A and B denote distinct metastable configurations. The equation of motion is deterministic, so the mapping is unique. Because there are so many metastable configurations, one might expect that enormously many pulses would be necessary to have a configuration repeat (once a repetition occurs, the sequence will repeat indefinitely). However, simulations[12] of equation (4) show that following a short transient, the system always locks into a finite, short sequence. The configurations repeat periodically so that after q pulses, the configuration has moved by precisely p wavelengths. Thus the average phase change per pulse is $\Delta\phi=(p/q)\times2\pi$. This phenomenon of mode-locking will be discussed in more detail in the next section.

If one looks at the current response to a single pulse, well defined "ringing" oscillations are seen[13]; for a single long pulse these decay in amplitude to the finite size NBN, although these transient oscillations will be present even in an infinite system. For a periodic application of identical pulses, the phase of the transient oscillations locks to the *end* of the pulse, so that the current is always rising as the pulse ends, *independent* of the pulse length.[14] This "pulse duration memory" is clearly displayed by the experimental data[15] in fig(1a) and is evident, though less apparent, in the results of numerical simulations on small systems (fig1b).[14] Both in experiment and simulation a sequence of a few training pulses is needed to achieve memory. The numerical results shown in the figure are in fact from locked repeating sequences, although this is not strictly necessary to develop the effect.

The transient oscillations occur because the system begins in a pinned configuration ϕ^A where all the degrees of freedom lie at minima of the potential. After the field is turned on, the local velocities are somewhat inhomogeneous (the system is not necessarily close to the equilibrium sliding state) and individual domains generate current oscillations with slightly different periods. As the oscillations dephase, the overall amplitude decays. Note that for a single degree of freedom, a rising cusp in the current at the end of the pulse would imply that the local phase was at a *maximum* of the potential. However for a single degree of freedom this would occur only for special values of the pulse length and height; nevertheless, the existence of such a cusp requires that a disproportionate number of the degrees of freedom do reach the maximum of their local potentials, precisely as the field is turned off. That such a confluence does occur is easiest to demonstrate in the FK model, of particles connected by springs in an incommensurate sinusoidal

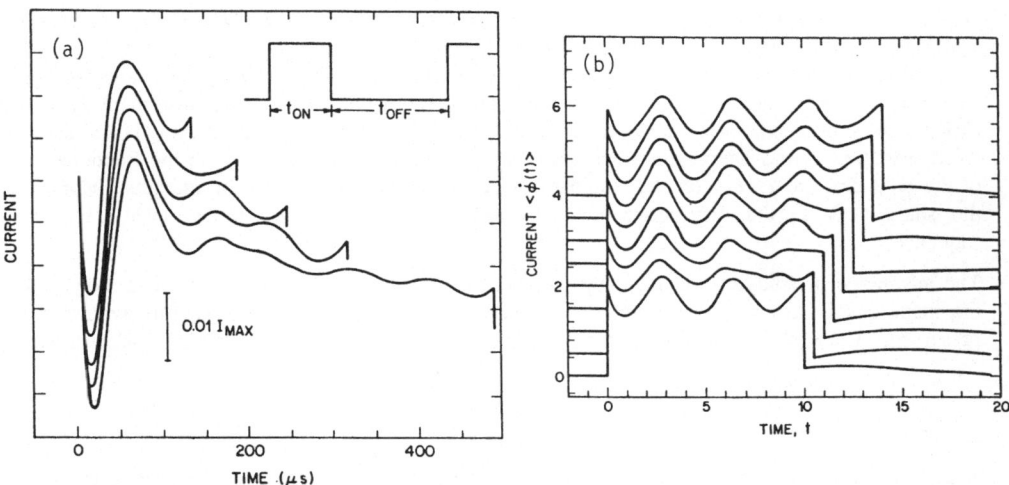

Figure 1. Pulse duration memory, under conditions of pulse driving (inset).(a) Experimental data of Fleming and Schneemeyer[15] on $K_{0.3}MoO_3$. (b) Numerical simulations of equation (4) [14]

Figure 2. Configuration of particles in the FK model after the system has been trained by a sequence of pulses (only a fraction of a long chain is shown).[16]

potential.[16] Figure 2 shows the configuration at the very end of the pulse after the system has been trained; indeed almost all of the particles do lie on the very *top* of the potential well.

The condition that a given state will repeat after a single pulse can be phrased in terms of the width of the initial velocity distribution Δv in a given applied field. If the width $\Delta v \leq 2\pi/t_{on}$ and the average velocity is such that $vt_{on}(mod 2\pi)$ is close to zero, the system will relax to the same initial state, and will thus be harmonically locked. But one can imagine preparing many states which will have a narrow distribution of velocities in a given field; in fact there must be *many* different states which all mode lock under identical driving conditions. Starting from an *arbitrary* initial condition, the system will lock in the first of the many possible states which it finds. For a given value of the pulse height and length we may picture the possible locked states clustered together in a closed region of configuration space. As the configuration evolves slowly toward the cluster, it will be first captured by states which lie on the cluster boundary, and these *generic* configurations will control the observed behavior. But precisely because the generic configurations are on the boundary of the cluster, they will be only *marginally* stable to changes in the driving. (For example in the configuration of figure 2 each particle has to choose whether to relax to the left or the right well - if the pulse were only very slightly different the delicate balance required for repetition would be destroyed) Marginal stability of the locked sequence yields automatically the pulse duration memory with a rising current cusp; the basins of attraction of the local minima of the potential are divided by local maxima.

One can simplify this problem still further by specifying the mapping from one state into another not by a differential equation but by a cellular automaton.[14] One of the simplest cases that can be studied is the two stage map acting on a set of integers[3] $F:\{x_j^n\} \rightarrow \{x_j^{n+1}\}$ with the defining equations

$$y_j^n = k(x_{j+1}^n - 2x_j^n + x_{j-1}^n) + E + d_j , \qquad (7)$$

$$x_j^{n+1} = int(y_j^n + \frac{1}{2}) , \qquad (8)$$

where int(z) denotes the largest integer not greater than z and d_j may be taken to be random variables. The two equations can be considered to model the two stages of coupled motion in the field, followed by relaxation to the bottom of the nearest well (in this case, the nearest integer). This map has been shown[14][3] to exhibit the same phenomenon of pulse duration memory as the FLR model simulations. As the system evolves toward the region of fixed points ("locked states") the distribution of the intermediate variables $\{y_j\}$ develops a sharp peak where the fractional part of the y_j is close to ½, exactly the boundary of the basins of attraction of the integers. Moreover, the sharp peak in the distribution develops well before the system settles into a repeating sequence.

The advantage of studying the mapping equations rather than the full equation of motion is that the maps can be completely characterized, while the essential features of the full dynamics are preserved. The special properties of many-EDF systems are thus not restricted to a single model.

2.5 Mode-locking

In the last section we discussed how under pulsed driving, the average phase change per pulse locks to a value of $(p/q)2\pi$. This mode-locking is conventionally measured under a pure sinusoidal drive.[17] In a field $E(t)=E_o+E_1\cos(\omega t)$, the derivative current voltage characteristics dI/dV develop peaks whenever the CDW washboard frequency is close to a harmonic or subharmonic of the driving frequency ω; under favourable circumstances, true locking (i.e. plateaux in dV/dI) may occur, so that the CDW velocity $\omega_o=(p/q)\omega$ for a range of dc bias.

Calculations[6][12] within the FLR model via a high field perturbation expansion demonstrate the existence of interference features in the linear ac response $\sigma(\omega)|_{E_o}$ when $\omega\approx\omega_o$. Calculations (again within perturbation theory) for the dc conductance in the presence of an ac field display anomalies in dI/dV at both harmonics and subharmonics.[5][6][12][18] With a hydrodynamic form for the relaxation as used by Sneddon, Cross and Fisher (SCF)[5] (i.e. using a convective derivative $D/Dt=\partial/\partial t+v\cdot\nabla$ for the damping term in equation (3)) these anomalies have the form of sharp cusps in dI/dV, and no true mode locking is predicted within perturbation theory. The FLR model neglects the convective derivative and the perturbation theory goes unstable when $\omega=\omega_o$.[18] The subtle difference between the FLR model and the SCF model is not generally important, except perhaps in this specific circumstance; it has been argued that the SCF model is stable against mode-locking to all orders in perturbation theory[12] whereas the FLR model is not. For the SCF model this will likely mean that mode-locking is only pronounced when the total field E(t) is below threshold for enough time in some part of the cycle; even if $E_1\gg E_o$ true mode-locking will not occur when $E_T/\omega E_1$ is small. In the FLR model, a narrow mode-locked region will persist even in the limit of high ac frequency and small ac amplitude. That mode-locking is experimentally enhanced when $\min|E(t)|<E_T$ is indicated by experiments on NbSe$_3$ with square-wave[19] (rather than pure ac) driving. However, in some samples there remains a narrow window of mode-locking even in the high frequency limit,[20] although not all workers agree on this point.[21]

We remarked above that a single degree of freedom model (an overdamped pendulum) displays only harmonic locking. Subharmonic locking does appear as soon as several degrees of freedom are coupled, high-order (q=5) steps existing even for a model of two coupled pendula. Large systems under sinusoidal driving have been studied by Matsukawa and Takayama[22]; by studying sample sizes ranging from N=100 to N=800 they determined that the mode-locking persisted in the infinite-size limit. However the qualitative behavior is at first sight not very different from that exhibited by simple low-EDF models (with the exception that there is no chaos), for example the sine circle map.

The situation under pulsed forcing (as discussed in the last section) is clearer to analyse.[12] Because the system is allowed to relax to static equilibrium at the end of each pulse, multiple metastable states must exist in order for subharmonic locking ($q\neq1$) to occur. In figure 3 is shown

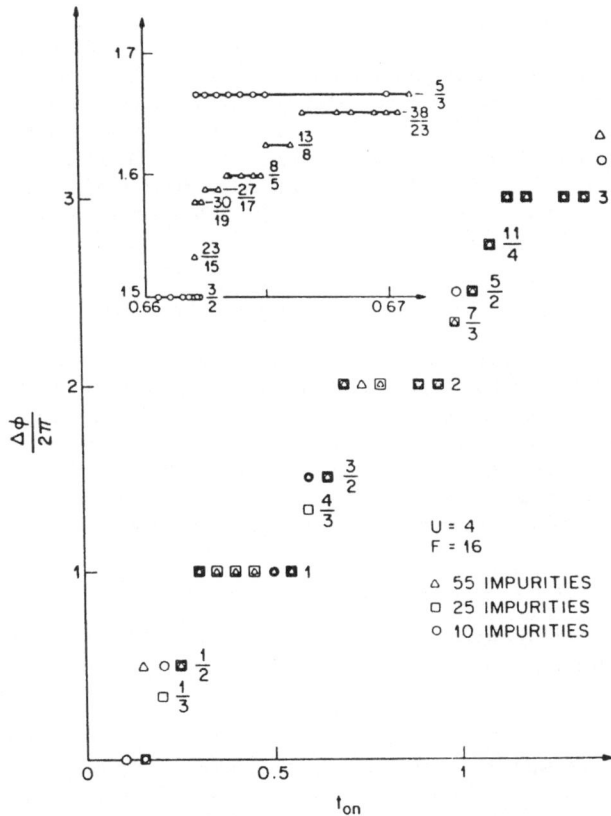

Figure 3. Mode-locking under pulsed driving for fixed pulse height and varying t_{on}, showing the average number of wavelengths $\Delta\phi/2\pi$ moved per pulse.

the results of a sequence of simulations of equation (4), with pulses of different lengths (t_{on}) but identical heights. Plotted is the average phase change per pulse, demonstrating that the system always locks, and that high order subharmonics are easily obtained. The "devil's staircase" is clearly quite irregular, and not easy to classify. Furthermore, as we pointed out above, it is not unique (at least for small samples); there are many *different* configurations which will lock under the *same* driving conditions, and these states may not all lie on the same mode-locked step. One may hope that for N→∞ some statistical regularity will appear; unfortunately this is not easily testable by direct simulation on account of the large amount of computation required.

The cellular automaton (equations 7,8) turns out to yield only harmonic locking; this may be traced to the sharp cutoff in the relaxation step, which is equivalent to replacing a smooth potential by one with cusps at the maxima.[3] Slightly more complex automata can deal with this problem, and do indeed produce subharmonic structure; surprisingly, though, high-order subharmonics are not observed.[3][23] It appears that the maximum denominator q_{max} is related to the complexity of the defining equations.

The experimental situation has been somewhat confused because there are clear differences between the results of different groups. Also, samples which are highly disordered can show qualitatively different behavior (including a period-doubling route to chaos) owing to the influence of "phase-breaking", or amplitude fluctuations induced by strong disorder. One of the surprises was that the observed features were typically much weaker than expected, with only low-order subharmonic features ($q\leq3$) demonstrating complete locking, and large windows of apparently unlocked behavior. It has recently been shown that in at least some part of these unlocked "windows" the motion is actually *intermittent*, with an analysis of the frequency spectrum showing that the system is jumping between different but nearby subharmonics.[21] These fluctuations are induced by broad band noise (BBN) always present in the samples, and "shallow" subharmonic attractors fail to lock. The origin of the BBN (which has a "1/f" character) is obscure. It is clearly produced by the CDW because its amplitude vanishes when the system is locked[21]; one possibility is that the CDW "amplifies" weak sources of noise. Simulations have shown that although the equilibrium sliding state generates no BBN, large amounts of BBN are generated on the *approach* to equilibrium (which may take extremely long times)[4][8].

3. CONCLUSIONS

Charge density waves have proved an extremely rich system for the study of non-linear dynamics. Although the underlying dynamics are purely relaxational, the existence of exponentially-many metastable states and the importance of a large number of degrees of freedom produces complex behavior not present in few-EDF models. The behavior ranges from "glassy" dynamics on long time scales to an apparently coherent response to strong ac driving at moderate frequency. While any one of the myriad phenomena can be modelled phenomenologically when taken in isolation, explicating the totality of the data within a single framework is a challenging problem.

ACKNOWLEDGEMENTS

The authors wish to thank S.Bhattacharya and R.M.Fleming for many useful discussions, and also for communicating the results of experiments prior to publication.

REFERENCES

1. For some current work, see *Charge Density Waves in Solids* Lecture Notes in Physics, vol 217, ed. G.Hutiray and J.Solyom, (Springer, Berlin, 1985); *Proceedings of the Yamada Conference XV* , Physica **143B**, (1986).
2. H.Fukuyama and P.A.Lee, Phys.Rev.B **17**, 535 (1977); P.A.Lee and T.M.Rice, Phys.Rev.B **19**, 3970 (1979).
3. S.N.Coppersmith, Phys.Rev.A **36**, 3375 (1987).
4. For more details of numerical methods, see e.g. P.B.Littlewood, in *Charge Density Waves in Solids* ed. L.P.Gorkov and G.Gruner (Elsevier, Amsterdam) to be published
5. L.Sneddon, M.C.Cross and D.S.Fisher, Phys.Rev.Lett. **49**, 292 (1982)
6. L.Sneddon, Phys.Rev.B **29**, 719,725 (1984)
7. D.S.Fisher, Phys.Rev.B **31**, 1396 (1985).
8. P.B.Littlewood, Phys.Rev.B **33**, 6694 (1986).
9. H.Matsukawa and H.Takayama, Physica **143B**, 80 (1986); H.Matsukawa, to be published.
10. S.N.Coppersmith and D.S.Fisher, Phys.Rev.B **28**, 2566 (1983); and to be published. S.N.Coppersmith, Phys.Rev.B **30**, 410 (1984). L.Sneddon, Phys.Rev.Lett. **52**, 65 (1984).
11. R.J.Cava, R.M.Fleming, P.B.Littlewood, E.A.Rietman, L.F.Schneemeyer and R.G.Dunn , Phys.Rev.B **30**, 3228 (1984).
12. S.N.Coppersmith and P.B.Littlewood, Phys.Rev.Lett. **57**, 1927 (1986)
13. S.N.Coppersmith and P.B.littlewood, Phys.Rev.B **31**, 4049 (1985).
14. S.N.Coppersmith and P.B.Littlewood, Phys.Rev.B **36**, 311 (1987);
15. R.M.Fleming and L.F.Schneemeyer, Phys.Rev.B **33**, 2930 (1986); see also S.E.Brown, G.Gruner and L.Mihaly, Solid State Commun. **57**, 165 (1986).
16. C.Tang, K.Wiesenfeld, P.Bak, S.N.Coppersmith and P.B.Littlewood, Phys.Rev.Lett. **58**, 1161 (1987).
17. S.E.Brown, G.Mozurkewich and G.Gruner, Phys.Rev.Lett. **52**, 2277 (1984). R.E.Thorne, J.R.Tucker and J.Bardeen, Phys.Rev.Lett. **58**, 828 (1987); S.Bhattacharya, J.P.Stokes, M.J.Higgins and R.A.Klemm, Phys.Rev.Lett. **59**, 1849 (1987)

18. H.Matsukawa, J.Phys.Soc.Jpn. **56**, 1522 (1987).

19. S.E.Brown, G.Gruner and L.Mihaly, Solid State Commun. **57**, 165 (1986).

20. R.E.Thorne et al.[17]

21. S.Bhattacharya, M.J.Higgins, J.P.Stokes and R.A.Klemm, to be published

22. H.Matsukawa and H.Takayama, Japanese J.Appl.Phys. **26**, (Supp. 26-3) 601 (1987)

23. S.N.Coppersmith, Phys.Rev.A **38**, 375 (1988).

Stochastic Aspects
of the Sliding Charge-Density-Wave Current Spectra

T. Miyashita[1] *and H. Takayama*[2]

[1]Casio Computer Co., Ltd, Hachioji Research Center,
2951-5, Ishikawa-Cho, Hachioji-City, Tokyo 192, Japan
[2]Research Institute for Fundamental Physics, Kyoto University, Kyoto 606, Japan

It is shown that the simulation on the charge density wave (CDW) dynamics described by the 1D Fukuyama-Lee-Rice model with heat bath effects reproduces qualitatively many stochastic aspects of the CDW current spectra.

1. INTRODUCTION

A variety of peculiar transport phenomena are observed in quasi one-dimensional (1D) conductors such as $NbSe_3$.[1] They are attributed to the sliding motion of charge density waves (CDWs) in the pinning potential due to various defects in the system. Recent experiments on $NbSe_3$ crystals of high quality have revealed interesting stochastic aspects of the CDW current spectra.[2,3] These experimental results lead us to an idea that for a given CDW mean velocity v_{CDW} there exist many quasi-stationary configurations of the sliding CDW, and that thermally activated process between these configurations is responsible to the observed stochastic phenomena. In this work we argue that this type of the CDW glassy dynamics in the nonlinear conduction regime above the threshold field E_{th} is in fact a genuine property of a deformable CDW with randomly-distributed pinning centers. For this purpose we present and discuss the results of our simulation on the 1D Fukuyama-Lee-Rice (FLR) model at finite temperatures.

2. METHOD OF SIMULATION

Our simulation is based on the following Langevin equation of motion for the CDW phase $\phi_i(t)$ which is defined on a discretized chain

$$\Gamma^2 \dot{\phi}_i = (\phi_{i+1} - 2\phi_i + \phi_{i-1}) + \hat{\epsilon}_i \sum_{k=1}^{N_{imp}} \delta_{ik} \sin(\beta_k + \phi_k) + \hat{\epsilon}_f + R_i \qquad (1)$$

(see Ref. 4 for notations and physical meaning of each term). The last term R_i represents the heat bath effects on ϕ_i, and is assumed to obey the following Gaussian form:

$$< R_i(t) >= 0 \; , \; < R_i(t)R_j(t') > = 2\Gamma^2 \hat{T} \delta_{ij} \delta(t - t') \; , \qquad (2)$$

where $\hat{T}$ is the effective temperature.
Equation (1) and (2) are solved by the molecular dynamics (MD) techniques. The spatially averaged CDW current is evaluates by $J(t) = e \sum_i \dot{\phi}_i / \pi N = eI(t)$, N being the total degrees of freedom. Several samples have been examined under fields of the form $\hat{\epsilon}_f = \hat{\epsilon}_{fdc} + \hat{\epsilon}_{fac} \sin(\omega_{ext} t)$. Below we present typical results obtained in a sample with N=1000, N_{imp}=250 and ϵ_i=0.25.

Springer Series in Synergetics, Vol. 43   **Cooperative Dynamics in Complex Physical Systems**
Editor: H. Takayama   © Springer-Verlag Berlin, Heidelberg 1989

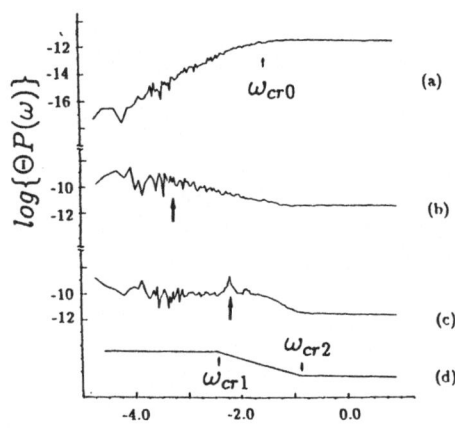

Fig. 1
The current power spectra $P(\omega)$ of the 1D FLR model under $E_{ac}=0$: (a) $E_{dc}=0$, (b) $E_{dc}=1.04E_{th}$, and (c) $E_{dc}=1.37E_{th}$. These $P(\omega)$ are obtained after taking the average in windows of $\log\omega_i \langle \log\omega \langle \log\omega_i + 0.025$. (d) is a schematic behavior of $P_{st}(\omega)$. The crossover frequency ω_{cr0} in (a) corresponds to the pinning frequency under $E_{dc}=0$. The arrow indicates the fundamental NBN frequency $\omega_{WB} \equiv 1/v_{CDW}$.

3. RESULTS

The main results of our simulation are represented by the current spectra $P(\omega)=\Theta \langle |I_\omega|^2 \rangle$ under $E_{ac}=0$, where I_ω is the Fourier component of $I(t)$ and Θ the time duration of observation (typically $2^{20} \sim 10^6$MD steps).

In Fig.1a $P(\omega)$ under $E_{dc}=0$ is shown. It satisfies the fluctuation-dissipation theorem, $P(\omega)=2TRe\sigma(\omega)/N$ where the ac conductivity $\sigma(\omega)$ is independently evaluated by means of the linear excitation mode analysis.

Above E_{th}, whose T-dependence has been discussed in Ref. 5, $P(\omega)$ is considered to consist of two parts: $P(\omega)=P_{dyn}(\omega)+P_{st}(\omega)$. $P_{dyn}(\omega)$ is mainly of a dynamical origin and its counterparts without the broadening of narrow band noise (NBN) spectra is observed also at T=0. The other part $P_{st}(\omega)$ is a purely stochastic origin and its stochastic behavior is shown in Fig.1d. At high frequencies $\omega \rangle\rangle \omega_{cr2}$, $P(\omega) \sim P_{st}(\omega) \infty \omega^0$, and its noise level is almost equal to that under $E_{dc}=0$. This part of $P_{st}(\omega)$ is therefore considered not related to the CDW glassy dynamics of our present interest.

At low frequencies $\omega \langle\langle \omega_{cr1}$ one sees again $P(\omega) \sim P_{st}(\omega) \infty \omega^0$. Its noise level is, however, distinctly larger than that at $\omega \rangle \omega_{cr2}$ so long as E_{dc} is less than a few times of E_{th}. We tentatively interpolate $P_{st}(\omega)$ between ω_{cr1} and ω_{cr2} by the $1/\omega$ noise. The latter behavior is clearly seen in $P(\omega)$ for $E_{dc} \sim E_{th}$ (Fig.1b). The spectrum $P_{st}(\omega)$ at low frequencies $\omega \langle\langle \omega_{cr2}$ is regarded as the broad band noise (BBN) of the present system. This part of power spectrum can be explained by the standard theory of activated random process [6], in which there exists the largest barrier energy proportional to $\ln\omega_{cr1}$. The latter tends to diverge as E_{dc} approaches E_{th}.

We also observe in our simulation that, when the mode locking ocurrs under $E_{ac} \neq 0$, both narrowing of the NBN spectra and depression of the BBN occur simultaneously. This indicates that these two phenomena are of the same origin, i.e., stochastic nature of the sliding CDW current.

4. ACKNOWLEDGEMENTS

This work was financially supported by a Grand-in-Aid for Scientific Research from the Ministry of Education, Science and Culture. One of us (T.M.) wishes to thank the Japan Society for the Promotion of Science for financial support.

5. REFERENCES

1) Proc. Yamada Conf. XV, Lake Kawaguchi, 1986, eds. S. Tanaka and
   K. Uchinokura; Physica 142B (1986).
2) S. Bhattacharya, J. P. Stokes, M. J. Higgins and R. A. Klemm:
   Phys. Rev. Lett. 59 (1987) 1849.
3) G. Lee Link and G. Mozurkewich: preprint.
4) H. Matsukawa and H. Takayama: Solid State Commun. 50 (1984) 283.
5) T. Miyashita and H. Takayama: Jpn. J. Appl. Phys. Suppl. 26 (1987) 603.
6) P. Dutta and P.M. Horn: Rev. Mod. Phys. 53 (1981) 497.

Dynamical Critical Phenomena
in Sliding Charge-Density-Waves

H. Matsukawa

Institute for Solid State Physics, University of Tokyo, 7–22–1 Roppongi,
Minato-ku, Tokyo 106, Japan

The incommensurate charge-density-waves (CDW's) in some quasi-one-dimensional conductors such as $NbSe_3$ are most typical examples of cooperative nonlinear dynamical systems with randomness [1]. Above a certain dc electric threshold field ε_{th} the sliding motion of the CDW's occurs and causes a sharp increment of the dc conductivity with a spontaneous current oscillation, the so called narrow band noise (NBN). Below ε_{th} the CDW's are pinned by impurities. Under ac + dc field the interference between the external ac field and the NBN yields some nonlinear dynamical phenomena, such as mode-lockings (ML's). In the present work the CDW dynamics is investigated numerically based on the FUKUYAMA-LEE-RICE (FLR) model [2], which treats the CDW as a classical deformable object under impurity potentials. We focus our attention on two kinds of phenomena, which are the depinning at ε_{th} and the ML's. Similar phenomena are observed and have been studied well in various systems [3]. The characteristics of the CDW systems mentioned above, however, let us expect that some qualitatively new physics is involved. We discuss that the depinning at ε_{th} and the ML's are accompanied by a new type of dynamical phase transition [5].

The FLR model describes the behavior of the CDW in terms of its phase variable $\Phi(x,t)$ [2]. The overdamped equation of motion of the model for D-dimensional CDW systems is expressed in a normalized form as follows [4]:

$$\dot{\phi} = \nabla^2\phi + \tilde{n}_i^{-1+2/D}\varepsilon_i\sum_j \sin(\beta_j+\phi)\delta(x-X_j) + \tilde{n}_i^{2/D}\varepsilon_f(t). \tag{1}$$

Here the r.h.s consists of the CDW elastic force, the pinning force due to randomly distributed impurities with strength ε_i and concentration n_i, and the external electric field force $\varepsilon_f(t)$. X_j is the site of the j-th impurity, β_j (=QX_j, Q being a CDW wave vector) is regarded as a random variable in mod 2π. Local CDW current density $j(x,t)$ is expressed as $\dot{\phi}(x,t)/\pi$. In the present study (1) is integrated using the Runge-Kutta method [4c].

Each impurity exerts a periodic force on a moving CDW, which yields a local current oscillation. Due to a random distribution of impurities, the phase of the oscillation fluctuates spatially and the coherence length of a CDW current fluctuation, ξ_{dyn}, becomes finite. The amplitude of the NBN in $<j(x,t)>_x$, A_{nbn}, vanishes in infinite volume systems, where < > represents the average with respect to the coordinate noted as subscript. Let us investigate the behavior of ξ_{dyn} near ε_{th}. In a system with linear dimension L, A_{nbn} is estimated as the product of δj_{rms}= $\{<[j(x,t)-<j(x,t)>_{x,t}]^2>_{x,t}\}^{1/2}$ and $(\xi_{dyn}/L)^{D/2}$. We can then evaluate ξ_{dyn} from the data of A_{nbn} and δj_{rms} obtained numerically. The result is shown in Fig.1. ξ_{dyn} diverges toward ε_{th}, which indicates that the depinning process at ε_{th} is a new kind of dynamical critical phenomena, as first proposed by FISHER [6]. The critical exponent of ξ_{dyn} as a function of a dc CDW current is about 0.3.

Under ac+dc field the ML's appear, in which the NBN frequency is drawn into rational value of the external ac frequency in certain finite range of the dc field. Do the ML's also vanish in infinite volume systems? They survive [7], since in ML regimes the external ac field enhances the dynamical coherence of the CDW. That enhancement is clearly seen in the direct investigation of $j(x,t)$ as shown in Fig.2. Out of the ML regime the phase of the local current oscillation fluctuates spatially.

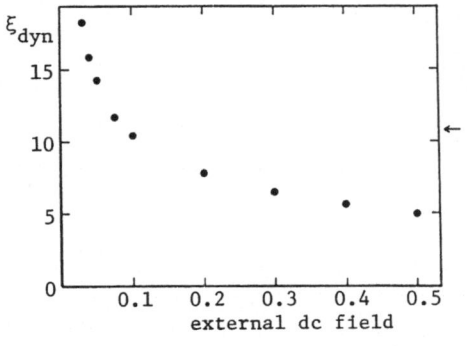

Fig.1. ξ_{dyn} vs external dc field in a 3-dimensional system with $\varepsilon_i=2$, L=40 and $\varepsilon_{th}=0.021$. The arrow indicates the phase coherence length in the pinned state.

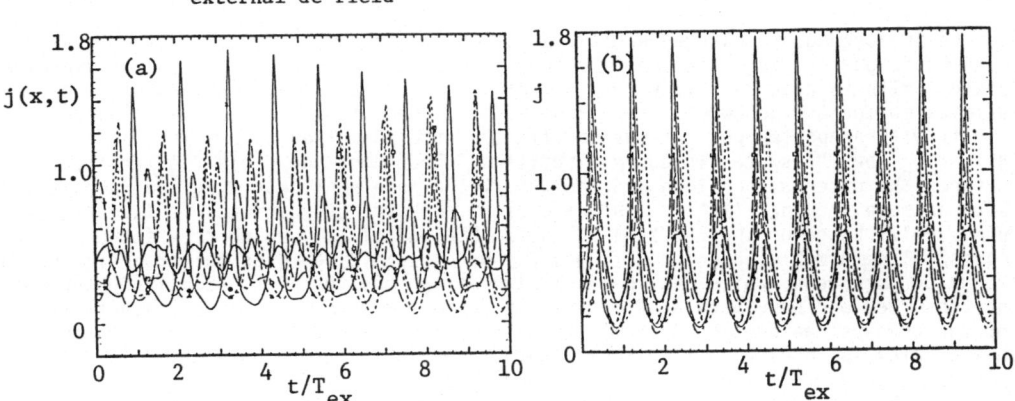

Fig.2. j(x,t) at x=0, 20, 40, 60, and 80 under an ac+dc field in (b) and out (a) of a 1/1 ML regime as a function of t/T_{ex}, where T_{ex} is the period of the external ac field. The system is 1-dimensional with $\varepsilon_i=1$ and L=100.

In the ML regime, however, its phase aligns and has a long-range order [5]. This result is also consistent with the perturbational analysis with respect to the impurity strength, which gives ξ_{dyn} diverging toward the field where the ML's occur [4b]. We thus claim that the ML's in the CDW systems are accompanied by a dynamical phase transition with a long range order induced by external ac field.

In summary we have investigated numerically the depinning at ε_{th} and the ML's in sliding CDW systems based on the FLR model. It has been discussed that both phenomena are accompanied by a new type of dynamical phase transition with the divergence of a certain coherence length.

This work was financially supported by a Grant-in-Aid for Scientific Research from the Ministry of Education, Science and Culture of Japan.

‡Fellowship of the Japan Society for the Promotion of Science for Japanese Junior Scientists

1. a) Charge Density Waves in Solids (Springer-Verlag, Berlin, 1985); b) Proc. of ICSM'88, Synth.Metals (to be published).
2. See for example, P.Bak: In Statics and Dynamics of Nonlinear Systems (Springer-Verlag, Berlin, 1983).
3. H.Fukuyama: J.Phys.Soc.Jpn. 41 (1976) 513; H.Fukuyama and P.A.Lee: Phys.Rev. B17 (1978) 535; P.A.Lee and T.M.Rice: ibid. B19 (1979) 3970.
4. a) H.Matsukawa and H.Takayama: J.Phys.Soc.Jpn. 56 (1987) 1507; b) H.Matsukawa: J.Phys.Soc.Jpn. 56 (1987) 1522, and c) J.Phys.Soc.Jpn. 57 (1988) No.10.
5. H.Matsukawa: In [1b].
6. D.S.Fisher: Phys.Rev.Lett. 50 (1983) 1486 and Phys.Rev.B31 (1985) 1396.
7. H.Matsukawa and H.Takayama: Jpn.J.App.Phys. Suppl.26-3 (1987) 601.

Metastability, Adaptability and Memory in Charge Density Waves

H. Ito

Research Institute for Fundamental Physics, Kyoto University, Kyoto 606, Japan

It is one of the most puzzling problems how a complex system changes itself in order to adapt itself to the external environment. Such adaptive responses have been mainly observed in the biological systems (circadian clock, neural networks, immune system and so on)[1,2]. Recently, however, the adaptation of the charge density waves (CDW) to the pulse fields has been discovered by Ido et al. in quasi-one-dimensional conductors, which is called Ido step memory effect [3]. When the identical pulse fields are repeatedly applied to the sample ($NbSe_3$), the CDW adjusts its sliding motion so that the oscillatory current response to the pulse shows <u>a steady response</u>, which has a common regularity regardless of the pulse width. That is, oscillations always cease at a maximum at the pulse end and the sliding displacement of the CDW within the pulse duration is quantized such as an integral multiple of the CDW wavelength. After the system has shown a steady response, further application of pulses induces no further changes in the response. This fact suggests that the CDW has already remembered the applied pulse width.

Recently we have proposed a possible interpretation of this novel phenomenon. In this short note, we briefly review the basic scenario of our interpretation. More details are presented elsewhere [4]. Selection of static patterns is observed under external perturbations in the fields of the fluid mechanics [5]. It must, however, be emphasized that the dynamical nature of the system is essential in the present phenomenon, and the system <u>selects dynamical motions</u>. A similar phenomenon has been reported by Fleming et al. [6] in $K_{0.30}MoO_3$. Recently, a possible explanation of this experiment has been proposed by Coppersmith and some authors [7-9]. Since Ido step memory effect has many different characters from this experiment [4], we consider that each phenomenon has a different origin. In fact, our explanation is different from the theory of Coppersmith et al..

We use Fukuyama-Lee-Rice model [10], in which the CDW is regarded as a classical deformable object pinned by random impurities. The CDW is composed of many domains, each of which is affected by the independent random impurity pinnings. Because of the periodicity in the pinning potential at each domain, there are quite many metastable states in the system. When the external field is not applied, the CDW is on one of the metastable states. Thus the repeated application of the identical pulse fields induces the successive state transitions between the metastable states. As a result of numerical simulations, we have found that the system can return to the same metastable states only if the sliding motion is stopped when the sliding displacement is an integral multiple of the CDW wavelength. Thus this is a <u>necessary condition</u> for a steady response to the applied pulse field, that is, for a fixed point in the state transition. In the successive transitions, the system is considered to select the states which can be a fixed point under the applied pulses. Once the system gets to a fixed point and shows the steady response to the pulse, the response has a common regularity regardless of the pulse width. We have confirmed such properties by the numerical simulations. The system can be adaptive to a wider range of pulse width, by selecting adequate states within many metastable states.

Next we wonder why the system can always find adequate states for a fixed point among many metastable states. The detailed experimental observations [3,11] suggest the existence of a feedback mechanism which focuses the state transitions on the adequate fixed point. We have introduced a possible scenario for the negative feedback mechanism. Our numerical simulations have shown that when the displacement at the pulse end is not adequate for a steady response, the configuration after the pulse suffers some deformations compared with the starting configuration. In the configuration space around a fixed point, if a special relation is satisfied between the deformations of the configurations and the corresponding changes of the sliding velocities, the successive state transitions converge to the fixed point with the help of the negative feedback (Fig.1). When the system fails to adjust its sliding displacement to the applied pulse width, the deformation of the configuration is accompanied with the state transition and results in a better adjustment to the next response. We have performed some numerical simulations in order to examine the practical possibility of the feedback mechanism. We have really found the states satisfying the suitable special relation, and further confirmed that the successive state transitions are focused on such states when they become fixed points under a certain pulse width. Since the steady responses have a common regularity, the sliding displacements in the steady responses tend to remain constant even if the pulse width is changed. If the configuration space has enough such regions, the successive state transitions can be easily trapped by them to be focused on adequate fixed points. We believe this mechanism is actually taking place in the real sample.

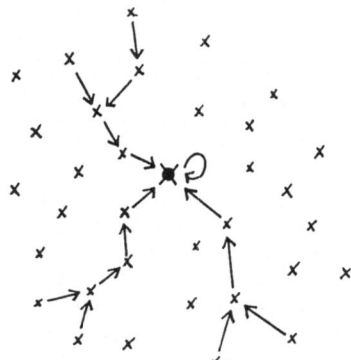

Fig.1: A schematic view of the focusing mechanism. The successive state transitions converge to the fixed point.

This work was partially supported by Japan Society for the Promotion of Science and a Grant-in-Aid for Scientific Research from the Ministry of Education, Science and Culture.

REFERENCES

1   A.T.Winfree, The Geometry of Biological Time (Springer, New York, 1980).
2   Evolution, Games, and Learning: Model for Adaptation in Machines and Nature, ed. by D.Farmer, A.Lapedes, N.Packard and B.Wendroff, Physica 22D (1986).
3   M.Ido, Y.Okajima and M.Oda, J. Phys. Soc. Jpn. 55 (1986) 2106.
4   H.Ito, Ph. D. Thesis, University of Tokyo, Japan, 1987.
    H.Ito, Proceedings of International Conference on Science and Technology of Synthetic Metals, 1988, submitted to Synthetic Metals.
    H.Ito, submitted to J. Phys. Soc. Jpn.
5   Spatio-Temporal Coherence and Chaos in Physical System, ed. by A.Bishop, G.Grüner and B.Nicolaenko, Physica 23D (1986).
6   R.M.Fleming and L.F.Schneemeyer, Phys. Rev. B33 (1986) 2930.
7   S.N.Coppersmith and P.B.Littlewood, Phys. Rev. B36 (1987) 311.
8   C.Tang, K.Wiesenfeld, P.Bak, S.Coppersmith and P.Littlewood, Phys. Rev. Lett. 58 (1987) 1161.
9   P.B.Littlewood, the article in this volume.
10 H.Fukuyama and P.A.Lee, Phys. Rev. B17 (1978) 535;
    P.A.Lee and T.M.Rice, Phys. Rev. B19 (1979) 3970.
11 M.Ido and Y.Okajima, private communication.

Transition to Turbulence via Spatiotemporal Intermittency

H. Chaté and P. Manneville

Institut de Recherche Fondamentale, DPh-G/PSRM, CEN-Saclay,
F-91191 Gif-sur-Yvette, France

The study of low-dimensional dissipative dynamical systems has provided a reasonable understanding of the transition to temporal chaos in strongly confined systems for which the spatial structure can be considered as frozen. The situation is still less advanced for weakly confined systems where chaos has both a spatial and a temporal meaning. In order to approach the specificities of the latter, we have chosen to study first a partial differential equation (PDE) displaying a convective-type nonlinear term, steady cellular solutions as in convection, and a transition to spatiotemporal chaos, namely the damped Kuramoto-Sivashinsky (KS) equation:

$$\partial_t \phi + \eta \phi + \partial_{x^2} \phi + \partial_{x^4} \phi + 2\phi \partial_x \phi = 0.$$ (1)

The pure KS equation /1/ is recovered when $\eta = \eta_{KS} = 0$. A single control parameter then remains: the length L of the interval at which boundary conditions are imposed to the function. In this limit, for L large enough, the solutions are essentially turbulent (phase turbulence). Above $\eta = 1/4$, the trivial solution $\phi \equiv 0$ is stable. As a matter of fact, around this value equation (1) is better written in the form of a modified Swift-Hohenberg (SH) equation, a well known model of convection /2/. Slightly below $\eta = \eta_{SH} = 1/4$, stable steady cellular solutions exist with nonlinearly selected wavelengths of the order of $2\pi\sqrt{2}$ /3/. L being given, the transition to turbulence can be studied by decreasing η from η_{SH} down to η_{KS} /4/. For L small, vis. ~ 10, the transition is rather reminiscent of what occurs in low dimensional systems. In the intermediate region $L \sim 100$, chaos is already spatiotemporal in essence but still sizeable end effects come and play a nontrivial role, thus complicating the picture. On the other hand, at larger L's, say for $L > 800$, the transition to turbulence occurs systematically *via spatiotemporal intermittency*.

Spatiotemporal intermittency (STI) is a fluctuating mixture made of coherent *laminar* patches embedded in a "sea" of incoherent *turbulent* domains. Quantitatively, it is characterized by a statistically stable proportion of laminar and turbulent regions as a function of the control parameter. Below a well-defined value η_{STI}, i.e. on the "phase-turbulence side", this regime remains "for ever" while above η_{STI}, i.e. on the "convection side", it persists only for a finite time and the system eventually reaches a steady asymptotic state /5/. During the transient, laminar domains reach a "macroscopic" size, of the order of the total width of the system and the domains left spatiotemporally intermittent are seen to recede regularly.

Being quite specific of large aspect ratio systems, this scenario can by understood as a contamination process reminiscent of *directed percolation*, as suggested initially by Pomeau /6/. However, a detailed understanding of the transition has yet to be obtained. First hints for a global theory derive from the recognition that it happens in a parameter range for which an inverse Hopf bifurcation takes place, which manifests itself as a finite amplitude periodic oscillation. Thus, two states are in competition, a steady cellular state and a periodic state and, in sufficiently long systems when confinement effects become weak, coherent nearly steady structures can persist for a long time within a dominantly periodic but mostly incoherent region, well separated from it by domain walls. Moreover, since the bifurcation is not continuous, the coherence length of both states, the width of walls between them, and the relaxation time, all remain finite. This suggests considering the whole system as a series of juxtaposed sub-units, interacting with their neighbors at discrete times, i.e. as a chain of coupled maps:

$$X_i^{n+1} = \sum_{j=-p}^{p} W_j f(X_{i+j}^n) \tag{2}$$

where X_i^n is the value at time n of some effective variable X attached to site i, f a map governing the evolution at a given site supposedly isolated, and W_j a coupling weight defined on some neighborhood of each site. As a matter of fact, introducing so-called *coupled map lattices* (CML), i.e. considering arrays of maps coupled by terms to which a spatial meaning is given, is conceptually the simplest way to increase the number of degrees of freedom in a spatiotemporal context /7/.

The transition to turbulence via spatiotemporal intermittency displayed by the PDE has been shown to possess statistical properties reminiscent of *critical properties* found at a second order phase transition, well in line with an interpretation in terms of directed percolation /8/. However, reliable numerical simulation turning out to be very expensive, we have been obliged to rely on the study of well-chosen CML's /9/ in spite of the conjectural features involved in the reduction. The "minimal requirements" for an adequate modeling of spatiotemporal intermittency are as follows:

- first, the local map $f(X)$ must lead to a splitting of the local phase space into two different components, one regular and the other chaotic, the chaotic one being only transient; in our first studies, this was obtained by means of a piecewise-linear map of the form

$$\begin{aligned} f(X) &= rX & \text{when} \quad 0 < X < 1/2 \\ f(X) &= r(1-X) & \text{when} \quad 1/2 < X < 1 \\ f(X) &= k(X - X^*) + X^* & \text{when} \quad X > 1 \end{aligned} \tag{3}$$

with $r > 2$, $|k| \leq 1$, and $X^* = (r+2)/4$.

- second, a spatial coupling chosen to be merely diffusive for simplicity; when this coupling is restricted to nearest-neighbors in one dimension, the complete system can be written as

$$X_i^{n+1} = f(X_i^n) + \frac{\epsilon}{2}\Big(f(X_{i-1}^n) - 2f(X_i^n) + f(X_{i+1}^n)\Big). \tag{4}$$

A straightforward calculation shows that the uniform laminar state remains stable with respect to infinitesimal delocalized perturbations, as expected for a subcritical instability. On the other hand, localized, finite amplitude perturbations can grow at least initially. In this perspective, spatiotemporal intermittency can be understood as a turbulent state resulting from the conversion of a local transient temporal chaos into a global sustained spatiotemporal chaos. Obviously, more complicated possibilities exist (especially that occurring in the PDE since the bifurcated state is probably not well accounted for by the turbulent part of the local maps). However, if some universality were to be found in the transition to turbulence via spatiotemporal intermittency, relevant features should already be displayed by the considered model. Subsequent work has been devoted to a detailed quantitative analysis of the statistical properties of the CML close to the threshold and to the comparison with corresponding results for *probabilistic cellular automata* (PCA) /10/.

Before proceeding with a presentation of our findings, let us recall that cellular automata are discrete-time, discrete-space systems with a finite number of accessible states per site, most often only two, one *active*, the other inactive or *absorbing*. The evolution in time is then defined in terms of transition rules depending on the configuration of the neighborhood of each site, these rules being either deterministic or probabilistic. Directed percolation (DP) is but one concrete realization of a PCA which can model for example the infiltration of a fluid through a porous medium: sites at the ends of pores become wet or dry according to whether they are connected to open or closed pores and to whether some parent sites upstream are wet. The dry state is absorbing since any site connected only to dry sites cannot become wet. Pursuing the analogy in the case of spatiotemporal intermittency we are led, for the local states, to the correspondence

$$\text{turbulent} \leftrightarrow \text{active} \quad \text{and} \quad \text{laminar} \leftrightarrow \text{absorbing}.$$

The probability p of having an open pore being the natural control parameter for directed percolation, there exists a critical value p_c above which sites at infinity are wet with finite probability while remaining dry below. Moreover, close to p_c the fraction of wet sites at infinity grows as $(p - p_c)^{\beta_{DP}}$ where β_{DP} is the corresponding critical exponent. Accordingly, one can define a critical coupling ϵ_c below which spatiotemporal intermittency is only transient. Our first results were obtained with $k = 1$, i.e. a laminar state made of a continuum of fixed points. In that case, the fraction of turbulent sites was observed to vary as $(\epsilon - \epsilon_c)^{\beta_{STI}}$. Our best estimates for exponent β_{STI} are given for $r = 3$ as a function of space dimension d in the table below /9,11/ together with the corresponding values for directed percolation /8/.

d	β_{STI}	β_{DP}
1	0.25	0.28
2	0.50	0.60

As can be anticipated from the very existence of exponent β_{STI}, model (3) with $k = 1$ and a diffusive coupling (4) displays a continuous transition via spatiotemporal intermittency. However, the consideration of values given in the table suggests that the process does not belong to to the universality class of directed percolation, as initially conjectured by Pomeau /6/. This is confirmed by the study of another critical exponent governing at threshold the distribution $\mathcal{N}$ of the lengths l of absorbing domains: $\mathcal{N}(l) = l^{-\varsigma}$. For directed percolation one gets $\varsigma_{DP} \simeq 1.75$ /8/ while for spatiotemporal intermittency (with $r = 3$) we have obtained $\varsigma_{STI} \simeq 1.99$ /9/. The existence of a universality class specific to spatiotemporal intermittency is even questioned by the variation of ς with r since when $r = 2.1$ we get $\varsigma_{STI} \simeq 1.78$. As a matter of fact, the variation of ς may not be a complete surprise since, when $r = 2.1$, the chance of escaping from a turbulent state toward a laminar state at a given site is smaller than for $r = 3$. In all cases, the turbulent part of the map is mixing but since the time spent in the corresponding region of phase space increases when r decreases, the result of the local chaotic evolution may appear more random, thus better approximating a purely stochastic situation.

In order to check the possible role of the continuum of fixed points when $k = 1$, we have also considered cases with $|k| < 1$, thus replacing the slow diffusive relaxation operating inside laminar domains when $k = 1$ by an exponential relaxation toward a single fixed point. Preliminary results show that the continuous character of the transition is preserved at least down to $k \simeq 0.65$ in one space dimension, but the corresponding values of the critical exponents have not yet been obtained. A major modification to the picture comes for smaller values of k from the possibility of *defects*, i.e. localized stable nonlinear solutions of the evolution equation. At small values of the coupling constant, in addition to the uniform laminar state, a new kind of solution is possible with a finite density of defects which can be reached by starting with disordered initial conditions. In one dimension, the defects are solitary waves propagating to the left or to the right with a well defined velocity (if the laminar regime is defined up to a temporal phase, kink-like defects between laminar domains with different temporal phases are expected /7a/). In two dimensions, only steady point-wise defects have been observed up to now.

While the uniform laminar state is linearly stable for all values of ϵ, solutions with defects can be maintained up to some critical value ϵ_s. This value corresponds to the limit of stability of isolated defects above which they are seen to "explode", thus nucleating the turbulent phase. The nucleation process at the origin of the behavior described above has been thoroughly studied in the case of a two-dimensional lattice with $r = 3$, $k = 0.5$, and a diffusive coupling. Our results /11/ can be phrased as follows: at $\epsilon_s \simeq 0.398$ a germ of minimal size, vis. an isolated defect, is sufficient to induce turbulence; below ϵ_s, the size of required nuclei increases rapidly and there exists a value $\epsilon_c \simeq 0.340 < \epsilon_s$ at which this size diverges, that is to say only an infinite turbulent half-space does not decay. Preparing the system as a juxtaposition of two domains, one laminar and the other turbulent, one observes that the turbulent domain increases for $\epsilon > \epsilon_c$ and recedes for $\epsilon < \epsilon_c$. Moreover, around ϵ_c the fraction of turbulent sites remains finite, in practice of the order of 0.75.

Starting with an initially turbulent state corresponding to some large value of ϵ and decreasing it rapidly, one can determine a second threshold value $\epsilon_u < \epsilon_s$ below which the system falls immediately and unconditionally into a laminar state with defects. Decreasing ϵ more slowly down to values slightly above ϵ_u, one is able to maintain spatiotemporal intermittency with a finite, i.e. not small, fraction of turbulent sites. However, for $\epsilon_u < \epsilon < \epsilon_c$, turbulence is only transient and decays irreversibly as soon as a fluctuation brings the density of turbulent sites sufficiently below the average. As can be understood easily, the size of the fluctuation required to make turbulence decay decreases rapidly as ϵ_u is approached from above.

The discontinuous character of the transition and the possibility of describing hysteresis loops between ϵ_u and ϵ_s are strongly reminiscent of first order phase transitions. The analogy is particularly close for the "turbulent $\rightarrow$ laminar" transition. Indeed, ϵ_c can be understood as the threshold for phase coexistence and ϵ_u as the undercooling limit of the turbulent phase. Moreover, the intrinsic fluctuations generated by the chaotic part of the local maps could be thought to play a role analogous to thermal agitation. Unfortunately, in the absence of external noise, the evolution of the laminar phase is strictly deterministic so that the reverse transition must rely on the presence of defects /12/ involving the nucleation and growth processes described above.

Experimental results summarized above are still far from exhaustive but already call for a theoretical interpretation. A first step towards the understanding of the nature of the transition via spatiotemporal intermittency would be the construction of a mean-field theory. In such theories, the effect of the coupling to neighbors is replaced by an effective force deduced from some averaging process and for usual phase transitions, results are valid above some upper critical space dimensionality. To get an idea of the output of a mean-field theory for the studied CML's, as suggested by Kauffman, one can force the averaging by replacing in (2) the coupling to a fixed set of neighbors at the nodes of a lattice by a coupling to a set of sites chosen at random and changing at each iteration. Preliminary results /13/ indicate a discontinuous transition when the connectivity is low (2 neighbors) for model (3) with $r = 2.1$ or 3, $k = 1$ or $k \neq 1$, and 1000 sites. However, reducing the fluctuation level by increasing the number of neighbors, (which is more typical of a true mean-field approximation) makes the turbulent regime hardly observable; the main reason is that the average behavior of the sites is governed by an single effective map analogous to (3) which always leads to the laminar state after a few iterations. This explains basically why no satisfactory mean-field theory of spatiotemporal intermittency has been successfully built so far.

The quite natural reduction to a "laminar/turbulent" alternative suggests a further all-to-nothing reduction in terms of cellular automata. Keeping in mind Pomeau's conjecture, in order to better appreciate the role of the chaotic but deterministic evolution of turbulent sites, we have begun the study of purely stochastic models in terms of PCA's designed to mimic the qualitative behavior of the CML's as closely as possible. One such model can be defined as a totalistic PCA with a 9-site neighborhood, the central site becoming active with probability p when $3, 4, 5, 6$ sites are active in the neighborhood and absorbing in all other circumstances. A standard mean-field approximation /8/ can then be developed yielding a discontinuous transition. Numerical experiments seem to confirm the prediction except in one space dimension for which the transition is continuous, though with critical exponents slightly different from those of directed percolation /14/.

To conclude, spatiotemporal intermittency seems an important process able to reconcile determinism and stochasticity for large aspect ratio systems where confinement effects are weak, in much the same way as scenarios involving strange attractors have done for confined systems. We take the opportunity of this conference to draw   attention to the wealth of "experimental" results we have accumulated and to stress   the fact that, for the first time in the field of   turbulence, usual concepts of thermodynamic phase transitions are directly relevant and standard statistical mechanics seems to offer quite naturally the most adapted framework of interpretation.

References

1. Y. Kuramoto, Chemical Oscillations, Waves, and Turbulence (Springer, Berlin, 1984).
2. J. Swift, P.C. Hohenberg, Phys.Rev.A15, 319 (1977).
3a. Y. Pomeau, P. Manneville, Physics Lett.75A, 296 (1980).
3b. M.C. Cross, P.G. Daniels, P.C. Hohenberg, E.D. Siggia, J.Fluid Mech.127, 155 (1983).
4. for a preliminary review of numerical results, see P. Manneville, AGARD (NATO) Special Course on Modern Theoretical and Experimental Approaches to Turbulent Flow Structure and its Modelling, report No.755 (Neuilly, 1987).
5. H. Chaté, P. Manneville, Phys.Rev.Lett.58, 112 (1987).
6. Y. Pomeau, Physica 23D, 3 (1986).
7a. K. Kaneko, Prog.Theor.Phys.74, 1033 (1985).
7b. G.-L. Oppo, R. Kapral, Phys.Rev.A33, 4219 (1986).
7c. J.D. Keeler, D.J. Farmer, Physica 23D, 413 (1986).
8. W. Kinzel, in Percolation Structures and Processes, Annals of the Israel Physical Society 5, 425 (1983).
9. H. Chaté, P. Manneville, Physica D, in press.
10. W. Kinzel, Zeitschrift für Physik B 58, 229 (1985).
11. H. Chaté, P. Manneville, Europhysics Lett., in press.
12. H. Chaté, P. Manneville, Proceedings of the 1988 annual meeting of CNLS to appear in Physica D.
13. D. Rochwerger, internal report, Ecole Polytechnique, Palaiseau (1988); H. Chaté, P. Manneville, D. Rochwerger, in preparation.
14. R. Bidaux, H.Chaté, in preparation.

Onset of Collective Rhythms in Large Populations of Coupled Oscillators

Y. Kuramoto and I. Nishikawa

Department of Physics, Kyoto University, Kyoto 606, Japan

The kind of complex systems of our present concern is large populations of coupled limit-cycle oscillators with frequency distribution. Such systems are by no means an invention motivated simply by mathematical curiosity. Quite on the contrary, their study would be of considerable practical value, for the same kind of systems are not rare in the real world as we realize when looking into living organisms.[1] Yet we will presently be interested not so much in biological applications as in unique dynamical features shared commonly by random populations of coupled oscillators in general. Among others, we will focus on 'frequency condensation' by which we mean that the natural frequencies with continuous distribution change with the strength of mutual coupling until eventually a finite fraction of the population comes to share a common frequency. This remarkable behavior could be the origin of collective oscillation i.e. the oscillation of the whole population like a single giant oscillator. The same behavior would make it possible for synchronizing waves to propagate with surprizing robustness over arbitrarily long distances, which could be important in information processing in living systems.

The first mathematical difficulty we are confronted with in this sort of study is the fact that limit-cycle oscillations would not allow us to go far with analytical tools. Since we know that even a coupled pair of nonlinear oscillators can exhibit incredibly complex dynamical behavior,[2] it may seem silly to challenge large populations which could be hopelessly complex. We should know, however, that sufficient weakness in coupling washes out most of their dynamical complexities.[3] For a coupled pair of oscillators, for example, we are left only with two basic modes of motion then, namely, entrained (i.e. periodic) and non-entrained (i.e. quasi-periodic) modes. Although such a property may be trivial for an oscillator pair, its great practical use will become obvious when we deal with large populations.

Fortunately, there exists a very simple oscillator model called the phase model[4] (also referred to as the active rotator model) which retains this essential property. In the phase model we describe the motion of a free oscillator only in terms of phase ϕ (mod. 2π) as

$$\dot{\phi} = \omega \quad . \tag{1}$$

When N oscillators are coupled mutually, we simply include interaction terms depending periodically on the phase difference between the interacting pair :

$$\dot{\phi}_i = \omega_i + \sum_{j=1}^{N} \Gamma_{ij}(\phi_j - \phi_i) \quad , \quad \Gamma(x+2\pi) = \Gamma(x) \quad . \tag{2}$$

Although the form in Eq.(2) may look too arbitrary, it is actually a general consequence of an asymptotic method called the phasedynamics when applied to the ordinary differential equation model like

$$\dot{\vec{X}}_i = \vec{F}_i(\vec{X}_i) + \sum_{j=1}^{N} \vec{G}_{ij}(\vec{X}_i, \vec{X}_j) \tag{3}$$

Springer Series in Synergetics, Vol. 43   **Cooperative Dynamics in Complex Physical Systems**
Editor: H. Takayama   © Springer-Verlag Berlin, Heidelberg 1989

which describes a population of *weakly coupled* and *nearly identical* oscillators.[3] Our study below will be based on Eq.(2). For the purpose of specific argument, we have to assume some explicit form for Γ_{ij}. Our choice will be

$$\Gamma_{ij}(x) = K_{ij}\sin(x+\alpha) \quad , \tag{4}$$

and an additional simplification of ignoring α will often be adopted.

Since Eq.(2) is valid for weakly coupled oscillators, its application to an oscillator pair given by

$$\dot{\phi}_1 = \omega_1 + K\sin(\phi_2-\phi_1+\alpha) \quad ,$$
$$\dot{\phi}_2 = \omega_2 + K\sin(\phi_1-\phi_2+\alpha) \quad , \quad K>0 \quad , \tag{5}$$

should lead to either entrained or non-entrained behavior. This property is easy to prove analytically.[5] We find that the actual frequencies $\tilde{\omega}_i$ ($i=1,2$) i.e. the long-time average of $\dot{\phi}_i$, come closer to each other with increasing K up to some critical value K_c beyond which $\tilde{\omega}_1$ and $\tilde{\omega}_2$ remain identical.

There is nothing more to say about two-oscillator systems, and let us thus proceed to large populations. Figure 1 shows an example of a two-dimensional pattern of mutual entrainment, which was obtained as follows.[6] Consider a square lattice, and place our phase oscillator (with $\alpha=0$) on each lattice point. The natural frequencies ω_i change randomly from site to site. The oscillators couple to their four nearest neighbors with the same strength K. When the phase difference between a nearest-neighbor pair of oscillators was found to be bounded as t goes to infinity, those oscillators are regarded as mutually entrained, and a line joining them is assigned ; otherwise no line. The pattern obtained in this way is composed of clusters of various sizes. The notion of frequency condensation is clear now. Suppose that the largest cluster contains N_s oscillators. Its ratio to the total system size N in the limit of large N defines an order parameter r :

$$r = \lim_{N\to\infty} \frac{N_s}{N} \quad . \tag{6}$$

One may then say that the system is in a frequency-condensed state if r is non-zero. Independently of r, one may define a different kind of order parameter W, which is a complex number with amplitude σ and phase Θ, as

$$W = \sigma e^{i\Theta} = \lim_{N\to\infty} \frac{1}{N} \sum_{j=1}^{N} e^{i\phi_j} \quad . \tag{7}$$

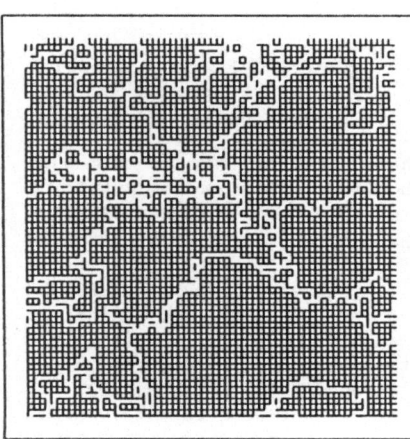

Fig. 1 Two-dimensional pattern
of mutual entrainment.
For explanation, see text.

In order to visualize the meaning of W, imagine a unit circle C in the complex plane, and distribute the N oscillators on C in such a way that their angles coincide with their phases. It is clear that W then gives the simple average of these N complex numbers. If we define the number density $n(\phi)$ of the oscillators of phase ϕ, W equals the first Fourier component of $n(\phi)$, or

$$W = \int_0^{2\pi} n(\phi)e^{i\phi}\,d\phi \quad . \tag{8}$$

W vanishes if the phases are distributed uniformly, while it is generally finite if n is nonuniform. In the latter case, the distribution itself will rotate. In contrast to the order parameter r, the quantity σ measures the amplitude of collective oscillation.

Now that we have seen, though only briefly, that different kinds of long-range order can arise in oscillator communities, we will look more closely into the onset of r and σ on the basis of a particularly simple many-rotator model such as to make detailed analytical study possible. We employ here what we call the mean-field model in which every oscillator couples to every oscillator with equal strength. This may be called a dynamical-system version of the Husimi-Temperly model. The onsets of r and σ will be found simultaneous for this particular case. The model equations are given by

$$\dot{\phi}_i = \omega_i - \frac{K}{N}\sum_{j=1}^{N}\sin(\phi_i - \phi_j) \quad , \tag{9}$$

where ω_i are distributed according to some function $g(\omega)$ which is assumed to be sufficiently smooth and symmetric about ω_0. Although the same model was proposed and studied by one of the present authors more than 10 years ago,[7] some important problems were left unsolved. It is only very recently that they came to be revisited. In what follows, we outline how recent studies led us to a better understanding of the behavior of the model (9).

The basic reason why our system permits analytical treatment is that for a given oscillator the effects from all the others are experienced entirely through W. This is seen by expressing Eq.(9) in the form

$$\dot{\phi}_i = \omega_i - K\sigma(t)\sin(\phi_i - \Theta(t)) \quad . \tag{10}$$

Our goal is to determine $\sigma(t)$ and $\Theta(t)$ in an exact self-consistent manner. As reported in a previous paper,[5] computer simulation of Eq.(9) with N sufficiently large suggests that the order parameter amplitude tends to a constant value (zero or nonzero) and Θ to $\omega_0 t$ as t goes to infinity. With the assumption of the same asymptotic behavior of W, the only quantity to be determined is the constant σ. One may eliminate Θ from Eq.(10) by working with relative phase $\psi_i \equiv \phi_i - \Theta(t)$ as

$$\dot{\psi}_i = \Omega_i - K\sigma\sin\psi_i \quad , \tag{11}$$

where $\Omega_i \equiv \omega_i - \omega_0$ is the relative frequency. We shall hereafter use the term such as "low- (high-) frequency oscillators" which should be understood as referring to those oscillators with small (large) Ω_i. How to find σ self-consistently looks simple : One solves Eq.(11) for each ψ_i, the solutions of course depending on parameter σ. The complete solution set $\{\psi_i\}$ forms the distribution $n(\psi;\sigma)$. Taking its first Fourier component, one gets σ. In this way, σ may be expressed as a function of σ. There seems to be something wrong about this kind of reasoning, however. The question is why $n(\psi;\sigma)$ can be stationary or, in other words, why time-independent σ can be a consistent solution. This question is reasonable because for a given σ-value on the right-hand side of Eq.(11) the solution ψ_i does not always approach a fixed point. Let us look into this point more closely. If we start with constant σ in Eq.(11), the system is subdivided into two parts, i.e. (A) $|\Omega_i/K\sigma| \leq 1$ and (B) $|\Omega_i/K\sigma| > 1$, representing entrained and non-entrained subpopulations, respectively. It is therefore convenient to split various quantities into the corresponding components as

$$N = N_A + N_B \quad , \tag{12}$$

$$n(\psi) = n_A(\psi) + n_B(\psi) \quad , \tag{13}$$

$$W = W_A + W_B$$

$$= \int_0^{2\pi} n_A(\psi) e^{i\psi} d\psi + \int_0^{2\pi} n_B(\psi) e^{i\psi} d\psi \tag{14}$$

etc. There is no consistency problem as to subpopulation A because the corresponding ψ_i's all tend to fixed points, which are given by

$$\psi_i = \sin^{-1} \frac{\Omega_i}{K\sigma} \quad , \tag{15}$$

constituting a stationary distribution $n_A(\psi;\sigma)$. The order parameter component σ_A is thus expressed as some function of σ, denoted by $S(\sigma)$:

$$\sigma_A = S(\sigma) \quad . \tag{16}$$

The explicit form of $S(\sigma)$ will be given shortly. For the oscillators in subpopulation B, the solutions of Eq.(11) cannot be stationary, so that $n_B(\psi)$ is generally time-dependent and hence σ_B is also time-dependent. After all, we arrive at the conclusion that σ is time-dependent from the assumption that σ is time-independent ! Consistency could be recovered, however, if the fluctuation of σ_B is negligible, and this is what actually holds in large-N limit. In order to see why, note first that subpopulation B, if looked upon as forming a single dynamical system, undergoes an ergodic motion on N_B-dimensional torus, because its motion is given by the product of N_B independent circular motions with frequencies $\hat{\Omega}_i (\equiv$ long-time average of $\dot{\psi}_i$). It is clear that the invariant measure ρ on this torus is proportional to the product of $|\dot{\psi}_i|^{-1}$, or

$$\rho(\psi_1, \psi_2, \cdots \psi_{N_B}) \propto \prod_{i=1}^{N_B} |\dot{\psi}_i|^{-1}$$

$$= \prod_{i=1}^{N_B} |\Omega_i - K\sigma\sin\psi_i|^{-1} \quad . \tag{17}$$

The long-time average of any quantity associated with subpopulation B can be calculated from ρ. For instance,

$$\overline{n_B}(\psi) = \frac{1}{\pi} \int_{K\sigma}^{\infty} d\Omega \, g(\omega_0 + \Omega) \sqrt{\Omega^2 - (K\sigma)^2} \, / \, \{\Omega^2 - (K\sigma\sin\psi)^2\} \quad . \tag{18}$$

Since the first Fourier component of $\overline{n_B}(\psi)$ vanishes by symmetry, the average of σ_B vanishes, too. As far as the fluctuation of σ_B is negligible, we therefore have

$$\sigma = S(\sigma) \quad . \tag{19}$$

As shown in a previous paper, the fluctuations of σ and Θ, which come only from B subpopulation, can also be calculated with the use of ρ. The result is simply expressed as

$$|\delta w|^2 = \frac{N_B}{N^2} = \frac{1-r}{N} \quad . \tag{20}$$

We see that the fluctuation of our macrovariable becomes negligible for sufficiently large N as in usual thermodynamic systems. Note that no stochastic assumption has been invoked in deriving this result.

The self-consistent equation (19) which is exact in large-N limit is easy to analyze. Calculation shows that $S(\sigma)$ is expanded for small σ as

$$S(\sigma) = (1+\varepsilon)\sigma - \beta\sigma^3 + O(\sigma^5) \quad , \tag{21}$$

where the first two coefficients are given by

$$\varepsilon = (K-K_c)/K_c \tag{22a}$$

with $K_c=2/\pi g(\omega_0)$, and

$$\beta = -\frac{\pi}{16} K_c^3 g''(\omega_0) \quad . \tag{22b}$$

A pitchfork bifurcation thus occurs at $K=K_c$. If g has a local maximum (minimum) at $\omega=\omega_0$, the bifurcation is supercritical (subcritical).

The results shown above are actually not new but they were essentially known in 1975 though in a much less refined way of presentation.[7] There remained two problems left unsolved, however. The first one is about the stability of the solutions of Eq.(19). It may seem natural to expect that as K increases the zero solution loses stability across K_c and gives way to nonzero solution, where $g''(\omega_0)<0$ is assumed ; still the argument so far tells nothing about stability. In order to study stability, it is most desirable to derive an evolution equation for σ in a closed form. The second problem is about fluctuation. The order parameter fluctuation in Eq.(20) was evaluated by assuming that the order parameter in the 'microscopic' evolution equation (11) never fluctuates. This is clearly a contradiction. Daido[8] first calculated the fluctuation of σ from a computer simulation using 100 rotators. Comparison of the result in Eq.(20) with Daido's shows that disagreement becomes apparent near K_c where the fluctuation exhibits a divergent behavior. We have thus to develop an exact self-consistent theory for fluctuation, too. The two problems stated above are mutually related, but we begin with the first one.

In deriving an evolution equation for σ, we restrict ourselves to the linear dynamics about $\sigma=0$ if such a regime exists at all. It is convenient first to pretend $\sigma(t)$ in Eq.(11) to represent an external force, so that we use notation $h(t)$ for $\sigma(t)$:

$$\dot{\psi}_i = \Omega_i - Kh(t)\sin\psi_i \quad . \tag{23}$$

We shall then study the response of individual ψ_i to a general small-amplitude variation of $h(t)$. From the complete set of responses $\{\psi_i(t)\}$ which will generally depend on the whole history of $h(t)$, one may find the net response $\sigma(t)$ as a *functional* of $h(t)$. For sufficiently small h, $\sigma(t)$ is expected to be a linear functional of $h(t)$, or

$$\sigma(t) = \int_0^\infty M(\tau)h(t-\tau)d\tau \quad , \tag{24}$$

where M is a memory function yet to be specified. We already found the response of σ to static h in the form

$$\sigma = S(h) \simeq S'(0)h \quad . \tag{25}$$

Its comparison with Eq.(24) implies the relation

$$\int_0^\infty M(\tau)d\tau = S'(0) = 1 + \varepsilon \quad . \tag{26}$$

By identifying $h(t-\tau)$ with $\sigma(t-\tau)$, Eq.(24) represents itself a linear evolution equation of σ. The kernel M is easy to calculate by considering the response to a special form of $h(t)$ chosen as

$$h(t) = \begin{cases} h_0 & (t \leq 0) \\ 0 & (t > 0) . \end{cases} \tag{27}$$

If for example $g(\omega)$ is a Lorentzian i.e. $g(\omega)=[\pi\{\omega-\omega_0)^2+1\}]^{-1}$, then $M(t)=e^{-t}$. Note that after switching h off, σ decays like

$$\sigma(t) = \int_t^\infty M(\tau)d\tau \cdot h_0 \equiv L(t)h_0 . \tag{28}$$

A simple way of seeing how the stability changes across K_c and how σ exhibits critical slowing down would be to make a Markoffian approximation on the right-hand side of Eq.(24) and develop $h(t-\tau)$ i.e. $\sigma(t-\tau)$ in powers of τ. The zeroth approximation gives the static limit (Eq.(25)). The first approximation in which $h(t-\tau)\simeq h(t)-\tau\dot{h}(t)$ gives a classical law

$$\dot{\sigma} = \gamma \varepsilon \sigma , \tag{29}$$

where

$$\gamma^{-1} = \int_0^\infty M(\tau)\tau d\tau . \tag{30}$$

We generally expect the existence of γ, and as we saw this is true at least for Lorentzian g.

With a slight generalization of the argument above, one may easily incorporate fluctuation. For simplicity, we only consider K below K_c. Even without variation of h, the order parameter fluctuation, denoted as $f(t)$, arises from the motion on the $N_B(=N)$-dimensional torus :

$$\sigma(t) = f(t) . \tag{31}$$

If $h(t)$ fluctuates in small amplitude, an extra term identical to the right-hand side of Eq.(24) should be included, and we get

$$\sigma(t) = \int_0^\infty M(\tau)\sigma(t-\tau)d\tau + f(t) , \tag{32}$$

where $h(t-\tau)$ in Eq.(24) has been replaced by $\sigma(t-\tau)$. Since the exact statistical properties of $f(t)$ can be calculated from the measure ρ, Eq.(32) is all we need for the study of fluctuation for $K<K_c$. Essentially the same equation was derived by Daido[9] very recently, though from a slightly different point of view. He discussed fluctuation with comparison to his own computer simulation successfully. Our idea sketched above can easily be extended to the supercritical regime where the fluctuation of macroscopic phase Θ is also relevant. This will be reported elsewhere in the near future.

Finally, a few comments will be made on peculiar decaying and ordering mechanisms working at the level of the phase distributions of subpopulations, although nothing singular can be seen from the simple law in Eq.(29). Bearing in mind that the coefficient γ represents physically the inverse time scale of those oscillators which are responsible for the temporal variation of σ, we now consider the decay process of σ when K jumps from slightly above K_c to slightly below K_c. Initially, only a small fraction of the population with low frequencies are phase-locked, giving rise to a small equilibrium value of σ ; unlocked oscillators form an unpolarized phase distribution with vanishing contribution to σ. After relaxation, the system must finally establish the phase distribution in which both n_A and n_B are uniform. One may thus tend to suppose (erroneously) that the initially unlocked part would be irrelevant to the order parameter dynamics all through the relaxation process. If such were true, γ could not be a constant of ordinary magnitude but would be small to the extent that σ is small, because the representative time scale of the locked oscillators is inversely proportional to σ. It was basically this kind of speculation that led us to an incorrect conclusion in the previous work[10] where the absence of a linear decay law of σ in the subcritical regime was claimed. As mentioned earlier, the existence of finite γ is unquestionable ; this we concluded from the consideration of an h-jumped non-interacting system (27). Intuitively, however, the fact that γ remains finite is puzzling even for the case of h-jumped non-interacting systems. This is because we tend to expect that the decay of σ involves only the low-frequency part satisfying $|\Omega_i/Kh|<1$. We thus come to the same contradiction both in cases of interacting and non-interacting systems. A little closer investigation of the h-jumped non-interacting system revealed the following (which holds for our K-jumped interacting system as well). It is surprising to find that even the extremely small-amplitude dynamics of σ is governed essentially by high-frequency oscillators i.e. those with $|\Omega_i/Kh|\gg1$ (or $|\Omega_i/K\sigma|\gg1$). It is true that $n_A(\psi)$ has a very long relaxation time of the order of $|kh|^{-1}$ (or $|K\sigma|^{-1}$). However, the decay of σ proceeds far more quickly by virtue of the fast asymmetric distortion of $n_B(\psi)$ in response to the jump of h (or K), and this gives rise to a negative contribution to σ thus virtually cancelling σ in relatively a short time. This implies that the establishment of stationarity at the level of phase distributions of the subpopulations requires an extraordinarily long time compared to the relaxation time of σ. For ordering processes, the situation is essentially the mirror image of what has been described above. Namely, the virtual order is first created by those high-frequency oscillators which are ultimately not responsible for σ. It is only far later that the true order due to low-frequency oscillators develops. In the past, we seem to have held an erroneous view about the onset of collective oscillation, according to which a fluctuation in σ was imagined to capture spontaneously a few low-frequency oscillators, the fluctuation enhanced in this way attracting more oscillators, and so forth. Actually, however, the order parameter fluctuation would never have time to keep waiting for slow responses of such low-frequency oscillators before decaying itself. Instead, the system makes use of the fast response of high-frequency oscillators to develop a virtual order. The virtual order is then handed down from oscillators to oscillators with increasingly long time scales. Without this kind of mechanism, no phase-transition-like behavior would be possible. Although our argument in the present paper has been based on a very special model, it is our belief that the virtual order (disorder) as described here could universally arise among frequency condensation phenomena due to mutual synchronization.

References

1. A.T. Winfree : The Geometry of Biological Time (Springer, New York, 1980).
2. See e.g. D.G. Aronson, E.J. Doedel and H.G. Othmer : to appear in Physica D (1988).
3. Y. Kuramoto : Chemical Oscillations, Waves, and Turbulence (Springer, Berlin, 1984).
4. The phase model has been employed by many people. S.H. Strogatz and R.E. Mirollo (Physica D31 (1988), 143) provides a fairly complete list of references in this connection.
5. Y. Kuramoto : Prog. Theor. Phys. (Suppl.) 79 (1984), 223.
6. H. Sakaguchi, S. Shinomoto and Y. Kuramoto : Prog. Theor. Phys. 77 (1987), 1005.
7. Y. Kuramoto : In International Symposium on Mathematical Problems in Theoretical Physics, ed. by H. Araki (Springer, New York, 1975), 420.
8. H. Daido : Prog. Theor. Phys. 75 (1986), 1460.
9. H. Daido : Preprint.
10. Y. Kuramoto and I. Nishikawa : J. Stat. Phys. 49 (1987), 569.

Nonlinear Dynamics in Chemical Systems

S.C. Müller and B. Hess

Max-Planck-Institut für Ernährungsphysiologie, Rheinlanddamm 201,
D-4600 Dortmund 1, Fed. Rep. of Germany

1. Temporal and Spatial Patterns in Chemical Reactions

Complexity in space and time is frequently observed in the dynamical evolution
of chemical reactions kept far from thermodynamic equilibrium. First observations were made more than 100 years ago, but systematic investigations on
such phenomena both in theory and experiment have been carried out only during
about the last two decades. The most striking features are chemical oscillations
and the formation of remarkably regular spatial patterns (for recent reviews see
[1-3]). Historically, the first periodic phenomena were found in electrochemical
reactions [4], while liquid phase oscillations, as already reported by Bray in
1921 [5], remained unnoticed by the scientific community for over 40 years. The
discovery of the Belousov-Zhabotinskii (BZ) reaction [6,7] and simultaneous
observations in the area of biochemical oscillations [8] finally led, in the
mid-sixties, to an appreciation of oscillatory phenomena which so far did not
fit into the framework of classical thermodynamics.

Spatio-temporal dynamic behavior was observed very early also in electro-che-
mical systems, for instance in the exotic pulsations of the beating mercury
heart, around 1870 [9]. In the field of spatial organization of chemical species, however, the main early contributions came from other classes of reactions. We may consider Runge as perhaps the first chemist to appreciate the variety and beauty of spatial patterning in chemistry [10]. He produced colorful
"paintings" by impregnating a piece of filter paper with one reactant and then
placing drops of an appropriately chosen chemical counterpart on this paper in a
regular sequence. But still, 40 years passed before in 1896 Liesegang discovered
regular ring- or band-shaped structures of the precipitate of an inorganic salt,
which evolve when two soluble electrolytes react with each other upon
interdiffusion, whereby an insoluble reaction product is formed [11]. Even spiral patterns can form, as shown in Fig. 1A. Several decades of research activity
on this phenomenon followed without achieving a general, satisfactory explanation of the mechanisms involved [11]. In recent years, a reactivation of research in this field can be noticed. Alternative mechanisms have been proposed
[13,14] and new experiments have been reported [16] that emphasize the importance of cooperative phenomena during the nucleation and growth of the new solid
phase. An example is given in Fig. 1B, where no initial concentration gradients
exist, as in the classical Liesegang experiments, but where patterns form
without such gradients.

This system exemplifies phenomena in which diffusion and reaction and their
nonlinear interactions play a crucial role for chemical pattern formation. Other
examples of spatial self-organization are found in the area of enzymic reactions
[17] and especially in the BZ reaction [18]. These reactions not only exhibit
spatially homogeneous oscillations, but show that a state of chemical activity
may propagate through an excitable medium with a well-defined velocity. The active, excited state forms wave fronts of various shapes, for instance circles or
spirals (see Figs. 2 and 3 below). The discovery of this phenomenon triggered a
sequence of ever increasing research activities in the field of chemical dyna-

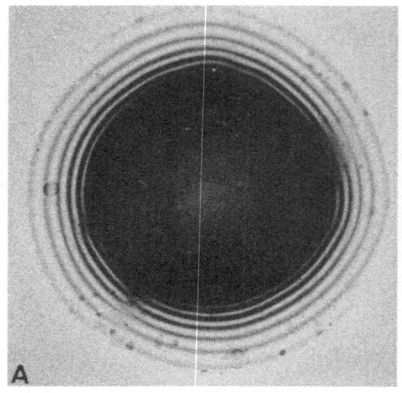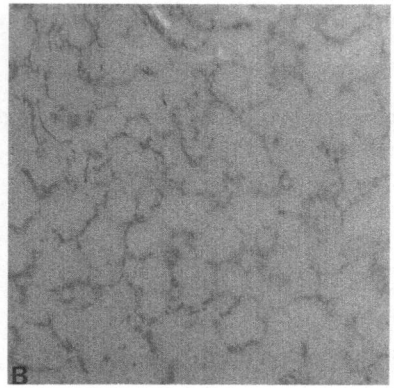

Fig. 1: (A) Precipitation pattern of silver dichromate. The pattern evolves when a drop of concentrated silver nitrate is put on a gelatin layer containing potassium dichromate. The pattern consists of two spiral arms (from [15]). (B) Irregular precipitation pattern of lead iodide in agar gel without imposed gradients of concentration. The pattern forms after placing the solution into a dish at about 90°C and letting the system evolve at room temperature

mics. It also led to a new interest in the "ancient" examples and to the search for new systems showing similar or even quite different patterns of spatially inhomogeneous reactions [2,19,20].

The discovery that chemical oscillations and patterns exist in the liquid phase was made at the right time. With regard to the theoretical background of such behaviour, it was, indeed, the extension of thermodynamics to open and far from equilibrium conditions and the development of tools for the study of nonlinear dynamics which uncovered mechanisms of macroscopic self-organization in time and space and opened experimental pathways into the complexity of these dynamic structures. Important contributions came from Prigogine with his approach to nonequilibrium thermodynamics [21] and from Haken with his concept of synergetics [22]. Also, we have nowadays computational devices at hand which allow modelling and simulations of the highly complicated equations derived from the physical and chemical mechanisms at work. The new concepts and new methodologies in mathematics, physics, chemistry and biology readily bridge the gap between the reduced description of bounded homogeneous systems and a description of open systems displaying a non-intuitive and inexhaustable richness of temporal and spatial patterns in the open world.

In this article, we cannot give an overview of all the interesting chemical systems now under investigation, but we will report on our recent studies on chemical waves in the Belousov-Zhabotinskii reaction and variations of this reaction. Thereby we will emphasize the importance of quantitative space-resolved methodologies, which we introduced recently and which are based on modern developments of computerized video techiques combined with optical precision methods. We will focus on the experimental characterization of chemical spirals and of parameters pertinent to the modelling by appropriate reaction-diffusion equations, analyze in some detail the collision of waves, and briefly mention the influence of hydrodynamic flow on wave propagation.

2. Chemical Waves

Chemical fronts and waves are variations in concentrations of chemical species which travel in space and occur in nonlinear reactive systems far from equilibrium [21]. They are described by solutions of reaction-diffusion equations of the form

$$\frac{\partial \psi}{\partial t} = D \nabla^2 \psi + F(\psi)$$

where ψ is a vector of time and space dependent state variables, such as concentrations of chemical species, D is a matrix of diffusion coefficients, and $F(\psi)$ represents variations in time that arise from the chemical reactions. For a given system both terms can be determined experimentally.

While there are now quite a few examples of chemical systems exhibiting wave propagation [19,23], the BZ reaction has been investigated in the greatest detail with regard to its mechanism. Here, the overall reaction is the oxidative bromination of malonic acid by bromate in acidic solution catalyzed by cerium, ferroin, or other redox compounds. A detailed mechanism in homogeneous solution describing chemical oscillations has been given in [24].

In Fig. 2A we show a photograph of a typical pattern of concentric circular waves, taken about 15 years ago. Such waves are often called "target" patterns. At that time reports were based on photographic observations [25], which have now been replaced by space-resolved photometric methods and digital evaluation of local concentrations. Experimental studies on the detailed properties of the waves, such as dispersion relations [26], characterization of the geometric

Fig. 2: (A) An early photograph of concentric chemical wave patterns in the ferroin-catalyzed Belousov-Zhabotinskii reaction (B. Hess 1972, unpublished work)

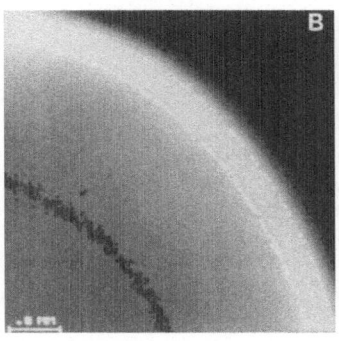
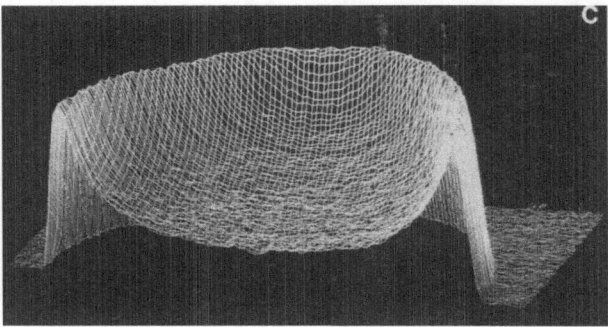

(B) Digital image of a quarter section of a circular wave in the same system obtained with a computerized two-dimensional spectrophotometer (from [30]). (C) Three-dimensional perspective grid image of the ferroin concentration distribution in a half section of the same circular wave (from [31])

shape of fronts [27], relation between curvature and wave velocity [28] have appeared only in the last three years. A major breakthrough came with the application of one- [29] or two-dimensional [30] quantification of light transmission patterns with video cameras connected to large memory computers, which led to a satisfactory description of the spatial distribution of chemical species without disturbing the structures (for instance, by immersing ion-sensitive electrodes).

The outcome of a space-resolved measurement of the concentration of the catalyst ferroin in the BZ reaction, applied to a section of a single circular wave (compare Fig. 2A), is presented in Fig. 2B. In this digital image two narrow concentration levels are marked by assigning the "colors" black and white to the picture elements (pixels) having the corresponding grey levels. In a three-dimensional perspective plot (Fig. 2C) the local concentrations of one half of a circular wave are plotted along a third coordinate perpendicular to the plane of observation in which the pattern evolves. The applied technique, that is the detection of light transmission through the reactive layer at a specific wavelength using a video camera (512 x 512 pixel resolution), proves to be a very helpful tool for the quantification of such patterns in terms of one chemical variable, in this case the concentration of ferroin.

Depending on the initial conditions and the geometry of the container there appear one-, two- or three-dimensional waves [32]. We are here concerned only with patterns evolving in a thin solution layer, that is approximately two-dimensional systems, which we can measure with our apparatus. It turns out that besides the circular geometry also other geometries occur, e.g. spiral-shaped waves [33,34]. These can be created by disrupting a circular front with, for instance, a gentle air blast from a pipette. The disturbance results in open ends of the circular wave fronts which then curl up to form rotating spirals, in which the spiral tip turns inwards while the chemical fronts move outwards. Such spirals are often arranged in symmetric pairs. Upon collision on the symmetry line the waves annihilate each other. A snapshot of one single spiral taken with the video camera of our two-dimensional spectrophotometer is presented in Fig. 3. In this case the catalyst of the BZ reaction was not ferroin, which changes its color in the visible, but cerium, for which an observation in the UV-range (340 nm) was used. The picture contains all the desired information about the concentration distribution of the cerium catalyst.

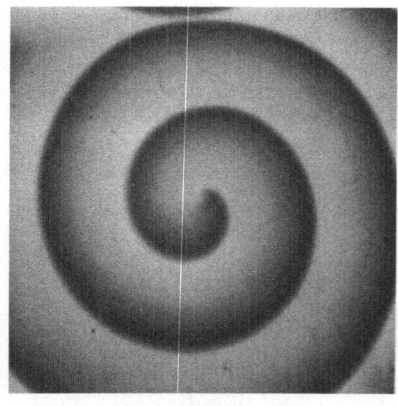

Fig. 3: Spectrophotometric image of a single spiral wave observed by light absorption at 340 nm in the cerium-catalyzed BZ reaction (from [33])

3. Quantitative Characterization of Spiral Waves

We have analyzed in detail the properties of a chemical spiral in the ferroin-catalyzed BZ reaction by evaluating data from two-dimensional spectrophotometric measurements. We summarize in brief the main results obtained for spirals prepared according to a standard recipe [34]. The shape of isoconcentration lines extracted from the video images turns out to be well represented by an Archimedian spiral. Within the experimental error a fit to the measured data by the involute of a circle fits equally well (Fig. 4). This second approach is more appropriate, because it takes into account that there exists a small area around the center of spiral rotation, the core of the spiral, in which the chemical processes are different from those occurring outside this area. Theoretical modelling requires the existence of such an area, as well [36]. So far, its spatial extent can be determined only by experiment.

In fact, by digitally overlaying a sequence of spiral images, taken during several revolutions of the spiral around its center, we obtained the following: there is a small circular area (diameter ≈ 0.7 cm) in which the maximum degree of excitation, which is characteristic for the propagating wave crests, is never reached. In the composite picture of Fig. 5 it is shown how this maximum of excitation decreases towards a singular point that coincides with the center of spiral rotation (diameter < 10 μm). This perspective representation of the upper envelope of excitation clarifies that a chemical spiral possesses a core region resembling very much the shape of a chemical "tornado". At the point of rotation there is a singular site at which the chemical state remains quasi-stationary, while at all points outside this cone-shaped area the chemistry varies in time

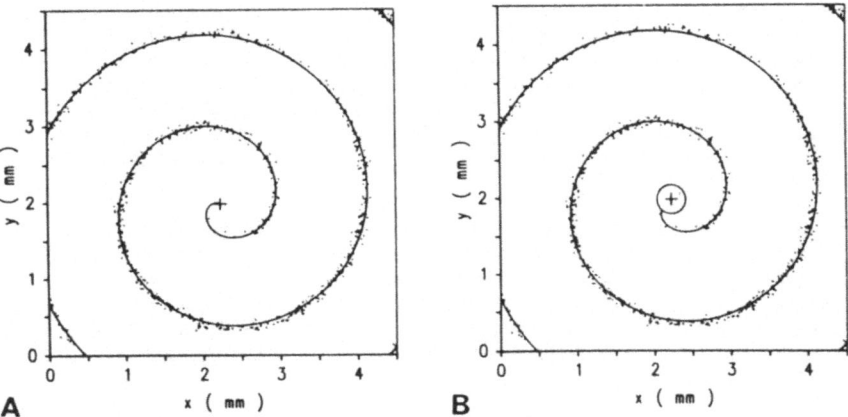

A **B**

Fig. 4: Fits of simple mathematical functions to isoconcentration lines extracted from video images of spiral waves in the ferroin-catalyzed BZ reaction. (A) Archimedian spiral, (B) involute of a circle fitted to the pixels with maximum grey levels (from [27])

Fig. 5: Three-dimensional perspective representation of the core region of a chemical spiral in a standard solution of the BZ reaction. The upper value of excitation is plotted along the third coordinate of a coordinate system, above the plane of the solution layer

between the two extremes of the chemical oscillation induced by wave propagation. This singular point was proven to be exactly at the center of the fits to the isoconcentration lines as shown in Fig. 4. Thus, in chemical spirals we are confronted with the remarkable phenomenon that in a liquid layer there are at least two dynamical states: that of periodic oscillations (in the points outside the core) and that of stationary behavior (in the center of the core). The theoretical descriptions by reaction-diffusion equations yield so far spiral geometry outside the core region [36], but the modelling of the structure of the core itself remains a challenge for the future.

4. Variation of Chemical Parameters

In the BZ reaction we are dealing with a system requiring a multi-dimensional parameter space for its description. Due to the nonlinear character of the interactions, variation of any single parameter may have quite unexpected consequences. We mention here two results:

(1) In the cerium-catalyzed BZ reaction we investigated the influence of various chemical parameters on the chemical wave profiles, in particular by varying the acidity of the reaction, the bromate concentration, and the concentration of the substrate malonic acid. We found that these variations result in remarkable changes of the individual wave profiles regardless of whether the waves are cicular or spiral-shaped. Details are described in [37], and we only show here two representative examples (Fig. 6). While the waves are usually of relaxation type, meaning that a steep front is followed by a slow decrease in terms of the recorded concentration (usually that of the catalyst), we also observed waves in which these profiles are more symmetric. Such studies give insight into the details of the reaction mechanism of the oscillatory reaction cycle [37].

(2) In a systematic study of the concentration dependence of spiral patterns, in this case catalyzed by ferroin, we found that the usually observed Archimedian-type shape of spiral waves is transformed to a more asymmetric shape when the concentration of H^+ and bromate are lower than those of the standard recipe often used in the literature [33,34]. This is in accordance with observations reported in [39,40]. An example of a spectrophotometric measu-

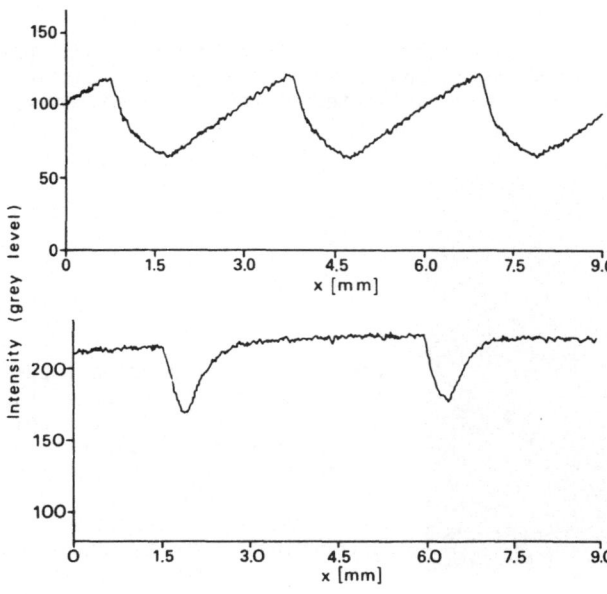

Fig. 6: Two profiles of intensity variation taken along wave patterns evolving in the cerium-catalyzed BZ reaction with different chemical conditions. The concentrations are given in [38]

312

rement of such a structure is given in Fig. 7, obtained by decreasing the acidity by a factor of two. The spiral tip now follows a path which is quite different from that observed in the case of the regular Archimedian spiral. In the "standard" spiral the tip rotates very stably on a circle around the center point (Fig. 8A). But in the case of low H^+-concentration, we can show by digital overlay techniques that the spiral tip follows for a short while an almost straight path before it starts to bend rapidly into a small circle, similar to that of regular circular rotation in the standard case (see Fig. 8B). Thus, if one follows the motion of the spiral tip in time, a "meandering" motion is traced in which almost directed motion alternates with fast turns, such that the trace of the spiral tip results in a curve similar to a cycloid. Our preliminary analysis (Fig. 8B) indicates that, with the given initial concentrations, the spiral tip first moves in a weakly curved fashion into an unexcited area, but then makes a quite rapid turn in order to lead that area to excitation which was previously left unexcited. The experiments demonstrate that dynamics and shape of the waves are governed by a specific part of the reaction mechanism and that a macroscopic pattern might well be reduced to the state of one elementary reaction step of the BZ system. An understanding of the resulting dynamics of spiral motion calls for further work in theory as well as in detailed experiments.

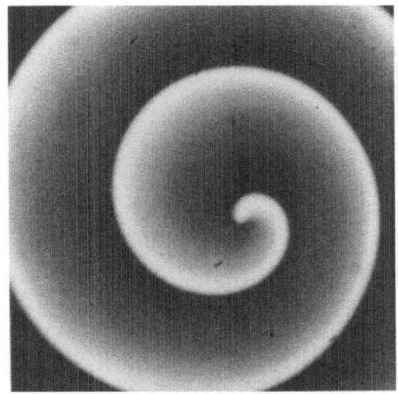

Fig. 7: Typical asymmetric spiral pattern in the ferroin-catalyzed BZ reaction

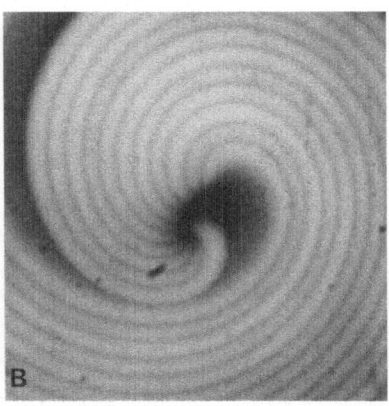

Fig. 8: (A) Overlay of 6 images of a spiral at 3s intervals taken during one rotation around a stable center [34]. The core region (dark spot in the center) is also shown in Fig. 5. (B) Overlay of 8 consecutive snapshots of the spiral of Fig. 7 at 6s intervals showing the motion of the spiral tip

5. Wave Collision

In chemical waves the phenomenon of wave collision is quite different from that commonly known in elastic waves in physics: there are no interference phenomena, but waves running towards each other annihilate when they collide, leaving typical cusp-like structures (Fig. 9).

Models of spiral wave propagation require that besides the dispersion relation [26] the relation between propagation velocity and curvature of wave fronts must play a predominant role [36,42,43]. We found that the formation of the sharp cusps upon wave collision present structures in which this latter relation can be verified experimentally for negative curvatures [28]. A detailed investigation of the geometric shape of isoconcentration lines and a fit of appropriate mathematical curves to the measured points confirmed that the normal velocity of wave front propagation is related to its curvature according to the relation

$$N = c - D \cdot K$$

where N = normal velocity, c = velocity of plane waves, D = diffusion coefficient of the autocatalytic species, and K = front curvature. With the experimental parameters found for the standard solution (c = 100 μm/s, D = 2.0 cm^2/s) nu-

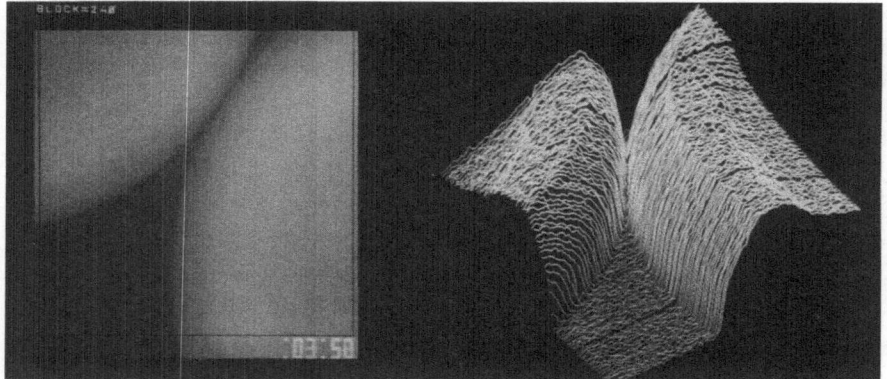

Fig. 9: Collision of two circular waves and a three-dimensional grid plot of the same pattern

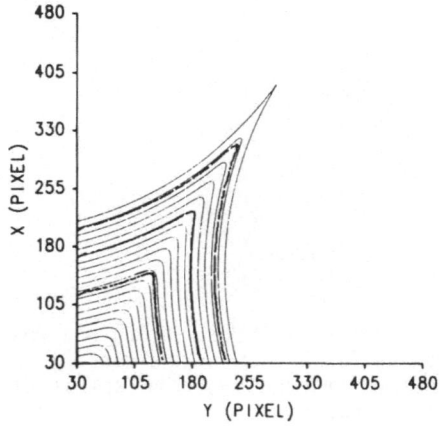

Fig. 10: The temporal evolution of a specific intensity level of the pattern of Fig. 9 (dotted lines) compared to numerical simulations (full curves) based on a model given in [36] (from [28])

merical simulations were carried out with the model proposed by Keener and Tyson [36] and a remarkably good fit between predicted and measured isoconcentration lines could be obtained (Fig. 10).

Thus, the material collected up to now for a quantitative characterization of chemical waves contributes substantially to the verification of theoretical approaches. It should, of course, be extended to questions such as: what is the precise relationship between the chemical parameters and the geometric shape of waves as well as their profiles? How do dispersion and the curvature relation depend on these parameters? What is the precise extent of the singular site at the spiral center? What concentration thresholds have to be crossed so that the spiral location starts to meander?

6. Chemical Waves and Hydrodynamics

The basic transport mechanism necessary to explain chemical wave propagation is the diffusion of the relevant species (especially the autocatalytic variable) in the solution layer. There are, however, other transport mechanisms which may play a significant role, mainly effects from hydrodynamic motion in the fluid layer. Previously observed structures such as the so-called "mosaic" patterns [29,44,45] indicate that under many circumstances such hydrodynamic flows exist. Recently, we have devoted much effort to quantifying such flows and the conditions of their occurrence by two-dimensional velocimetry [46]. Some details of these studies are presented in the contribution by Miike et al. in this volume. It turns out, in fact, that the hydrodynamic stability of the layer is almost never perfectly guaranteed. Convective flow may be induced by temperature or concentration gradients [47] and result in deformation or irregular decomposition of wave fronts, as already reported in [48]. Analysis of transitions from ordered to disordered structures in terms of a single parameter has been performed [49]. But the question of when such flows are of negligible importance and when they create additional structural effects, including the possible generation of spatial chaos, remains a topic of current research not to be addressed further in this article.

7. Conclusions

We have presented several experimental examples of spatial organization in chemistry caused by nonlinear interactions between complex reaction kinetics and physical transport. Spatial patterns are a basic property of biological systems.

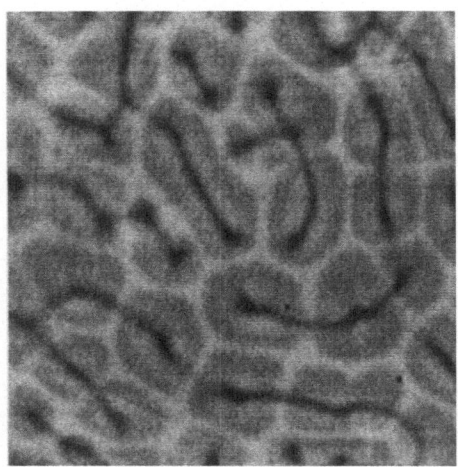

Fig. 11: Pattern in oscillating glycolysis, observed in a thin layer of cytoplasm extracted from yeast cells (from [50])

It is noteworthy that the patterns observed in chemical systems as described above find analogies in biochemical, subcellular, and cellular processes. Classical examples are pattern formation in oscillating glycolysis (Fig.11) [17,50] and in the aggregation of the slime mold Dictyostelium discoideum [51]. Recent studies indicate that macroscopic, spatial organization of this kind is quite common. The analogies between mostly inorganic chemistry under nonequilibrium conditions and cellular systems which evolve always under such conditions are often striking and the research on nonlinear dynamics in chemistry is believed to contribute significantly to our understanding of the mechanisms acting on the more complex levels of biological organization.

This work was in part supported by the Stiftung Volkswagenwerk, Hannover, FRG.

References

1. C. Vidal, A. Pacault (eds.): Non-Equilibrium Dynamics in Chemical Systems, Springer Ser. Syn., Vol. 27 (Springer, Berlin, Heidelberg 1984)
2. R.J. Field, M. Burger (eds.): Oscillations and Traveling Waves in Chemical Systems (Wiley, New York 1985)
3. C. Vidal, P. Hanusse: Int. Rev. Phys. Chem. 5, 1 (1986)
4. M.G. Th. Fechner: Schweiger's J. Chem. Phys. 53, 129 (1928)
5. W.C. Bray: J. Am. Chem. Soc. 43, 1262 (1921)
6. B.P. Belousov: Sb. Ref. Radiats. Med., Medgiz, Moscow, p. 145 (1959)
7. A.M. Zhabotinskii: Biofizika 9, 306 (1964); A.M. Zhabotinskii: Dokl. Akad. Nauk SSSR 157, 392 (1964)
8. B. Hess, A. Boiteux: Annu. Rev. Biochem. 40, 237 (1971)
9. G. Lippmann: Ann. Phys. 2nd Series 149, 544 (1873)
10. F.F. Runge: Der Bildungstrieb der Stoffe veranschaulicht in selbständig gewachsenen Bildern (Oranienburg, Selbstverlag 1855)
11. R.E. Liesegang: Phot. Archiv 21, 321 (1986)
12. E.S. Hedges, J.E. Myers: The Problem of Physico-Chemical Periodicity (Arnold & Co., London 1926)
13. K.H. Stern: Chem. Rev. 54, 79 (1954)
14. R. Lovett, P. Ortoleva, J. Ross: J. Chem. Phys. 69, 947 (1978)
15. S. Kai, S.C. Müller: Sci. Form 1, 9 (1985)
16. S. Kai, S.C. Müller, J. Ross: J. Chem. Phys. 76, 1392 (1982); M.E. LeVan, J. Ross: J. Phys. Chem. 91, 6300 (1987)
17. A. Boiteux, B. Hess: Ber. Bunsenges. Phys. Chem. 84, 392 (1980)
18. H.-G. Busse: J. Phys. Chem. 73, 750 (1962); A.N. Zaikin, A.M. Zhabotinskii: Nature 225, 535 (1970)
19. P. De Kepper, I.R. Epstein, K. Kustin, M. Orban: J. Phys. Chem 86, 170 (1982); M. Orban: J. Am. Chem. Soc. 102, 4311 (1980)
20. D. Avnir, M. Kagan: Nature 307, 717 (1984)
21. P. Glansdorff, I. Prigogine: Thermodynamic Theory of Structure, Stability and Fluctuations (Wiley Interscience, New York 1971)
22. H. Haken: Synergetics, 3rd ed. (Springer, Berlin, Heidelberg 1983)
23. J. Ross, S.C. Müller, C. Vidal: Science 240, 365 (1988)
24. R.J. Field, E. Körös, R.M. Noyes: J. Am. Chem. Soc. 94, 8649 (1972); R.J. Field, H.-D. Försterling: J. Phys. Chem. 90, 5400 (1986)
25. H.-G. Busse, B. Hess: Nature 244, 203 (1973)
26. A. Pagola, C. Vidal, J. Ross: J. Phys. Chem. 92, 163 (1988)
27. S.C. Müller, Th. Plesser, B. Hess: Physica 24D, 71 and 87 (1987)
28. P. Foerster, S.C. Müller, B. Hess: Science 241, 685 (1988)
29. P.M. Wood, J. Ross: J. Chem. Phys. 82, 1924 (1985)
30. S.C. Müller, Th. Plesser, B. Hess: Naturwissenschaften 73, 165 (1986)
31. S.C. Müller, Th. Plesser, B. Hess: Biophys. Chem. 26, 357 (1987)
32. B.J. Welsh, J. Gomatam, A.E. Burgess: Nature 304, 611 (1983); A.T. Winfree, S.H. Strogatz: Physica 13D, 221 (1983)

33. A.T. Winfree: Sci. Am. 248 (no.5), 144 (1983)
34. S.C. Müller, Th. Plesser, B. Hess: Science 230, 661 (1985)
35. Zs. Nagy-Ungvarai, S.C. Müller, Th. Plesser, B. Hess: Naturwissenschaften 75, 87 (1988)
36. J.P. Keener, J.J. Tyson: Physica 21D, 307 (1986)
37. Zs. Nagy-Ungvarai, S.C. Müller, J.J. Tyson, B. Hess: J. Phys. Chem., in press
38. Zs. Nagy-Ungvarai, S.C. Müller: In Propagation in Systems Far from Equilibrium, ed. by J.E. Wesfreid, H.R. Brand, P. Manneville, G. Albinet, N. Boccara, Springer Ser. Syn., Vol. 41 (Springer, Berlin, Heidelberg 1988) p. 100
39. K.I Agladze, V. Panfilov, A.N. Rudenko: Physica 29D, 409 (1988)
40. A.T. Winfree, W. Jahnke, preprint
41. Th. Plesser, S.C. Müller, B. Hess, in preparation
42. Y. Kuramoto: Prog. Theor. Phys. 63, 1885 (1980)
43. V.S. Zykov: Biophysics 25, 329 (1980)
44. A.M. Zhabotinskii, A.N. Zaikin, J. Theor. Biol. 40, 45 (1973)
45. K. Showalter, J. Chem. Phys. 73, 3735 (1980)
46. H. Miike, S.C. Müller, B. Hess: Chem. Phys. Lett. 144, 515 (1988)
47. K.H. Winters, Th. Plesser, K.A. Cliffe: Physica 29D, 387 (1988)
48. K.I. Agladze, V.I. Krinsky, A.M. Pertsov: Nature 308, 834 (1986)
49. M. Markus, S.C. Müller, Th. Plesser, B. Hess: Biol. Cybern. 57, 187 (1987)
50. S.C. Müller, Th. Plesser, B. Hess: In Temporal Order, ed. by L. Rensing, N.I. Jaeger, Springer Ser. Syn., Vol. 29 (Springer, Berlin, Heidelberg 1985) p. 194
51. G. Gerisch: Naturwissenschaften 58, 430 (1971).

The Thermodynamic Behavior
of Elementary Reversible Cellular Automata

S. Takesue

Research Institute for Fundamental Physics, Kyoto University, Kyoto 606, Japan

Here I introduce a family of fully discrete conservative dynamical systems named elementary reversible cellular automata (ERCAs)[1] as the minimal models to study the thermodynamic behavior of large dynamical systems.

Statistical mechanics stands on the assumption that a large dynamical system shows the standard thermodynamic behavior in such respects as the canonical ensemble in an equilibrium state, relaxation to equilibrium obeying the kinetic equation, and transport coefficients given by Kubo's formula. Ergodic theory should link these and underlying dynamics. As the Kolmogorov-Arnol'd-Moser theorem has revealed, however, even ergodicity is not generic for mechanical systems with a finite number of degrees of freedom. On the other hand, it is known that such simple systems as the noninteracting ideal gas, the complete harmonic crystals and one-dimensional hard rod system show Bernoulli property, the highest of the ergodic theoretical hierarchy. To bridge this gap, the study of asymptotic behavior of large systems is needed.

For such studies Hamiltonian dynamics has been used traditionally. However, since the phase space of a Hamiltonian system is too complicated, its use for studying large systems has a number of difficulties. If one focuses on the foundation of statistical mechanics, not the symplectic condition but only the Liouville theorem and energy conservation are necessary. These are satisfied by ERCAs as is seen in the following.

ERCAs are one-dimensional reversible cellular automata in which two Boolean variables on each site evolve according to rules of the form

$$\sigma_i^{t+1} = f(\sigma_{i-1}^t, \sigma_i^t, \sigma_{i+1}^t) \text{ XOR } \hat{\sigma}_i^t$$

$$\hat{\sigma}_i^{t+1} = \sigma_i^t$$

where integers i and t represent the position and time, respectively, f is a Boolean function of three variables, and XOR means the "exclusive OR" operation. All the Booleans take values in the set $\{0, 1\}$, which are also treated as integers. There are $2^{2^3} = 256$ Boolean functions of three variables and as many ERCAs. Each rule is denoted by the number $\sum_{\mu,\nu,\kappa} f(\mu, \nu, \kappa) 2^{4\mu+2\nu+\kappa}$ and an R appended to it. Reflection and Boolean conjugation symmetries classify them into 88 equivalence classes.

Reversibility of the rules is clear from their form. This with the discreteness of states guarantee the Liouville theorem. Some ERCAs conserve an additive quantity of the form

$$\Phi^t = \sum_i F(\sigma_i^t, \sigma_{i+1}^t, \hat{\sigma}_i^t, \hat{\sigma}_{i+1}^t)$$

i.e., $\Phi^{t+1} = \Phi^t$, with an appropriate function F [2]. Such a quantity can be regarded as a kind of energy and one can say that bond $(i, i+1)$ has an energy $F(\sigma_i, \sigma_{i+1}, \hat{\sigma}_i, \hat{\sigma}_{i+1})$. Thus, as the energy conservation and the Liouville theorem are satisfied, the statistical mechanics of the models can be constructed. One thing to be noticed here is that not only the sum but also locally defined quantities themselves might be conserved. If it is the case, such quantities clearly prevent the realization of the statistical mechanics. Rules possessing an additive conserved quantity but no local conservation laws appear in Table I with their additive invariants.

Springer Series in Synergetics, Vol. 43   **Cooperative Dynamics in Complex Physical Systems**
Editor: H. Takayama   © Springer-Verlag Berlin, Heidelberg 1989

Table I. Rules and their additive invariants.

Rules	Additive invariants: $F(\alpha, \beta, \hat{\alpha}, \hat{\beta})$
26R	(a): $(\alpha - \hat{\beta})^2 + (\hat{\alpha} - \beta)^2$
90R	(a) and (d): $\alpha\hat{\beta} - \hat{\alpha}\beta$ [or $(\hat{\alpha} - \beta)^2 - (\alpha - \hat{\beta})^2$]
91R, 123R	(b): $1 + \alpha\hat{\alpha} + \beta\hat{\beta} - \{[1 - 2(1 - \alpha)(1 - \hat{\beta})][1 - 2(1 - \hat{\alpha})(1 - \beta)]\}$ and (d)
77R	(c): $\alpha\hat{\beta}(1 - 2\hat{\alpha} - 2\beta) - \hat{\alpha}\beta(1 - 2\alpha - 2\hat{\beta})$
94R, 95R	(d)

For these models I have carried out the numerical tests for the thermodynamic behavior such as:
i) Canonical distribution of subsystem's energy: A large system can be regarded as the sum of a subsystem and the remaining part as a heat bath. Then, does the distribution of the subsystem's energy, $P(E)$, become the canonical one, that is,

$$P(E) \propto D(E)e^{-\beta E}$$

with the density of state, $D(E)$, and the inverse temperature β which is a function of total energy?
ii) Spatial correlations: Do the ensemble and the time averages really agree with each other?
iii) Relaxation: Can a system starting from an inhomogeneous initial condition eventually come into an equilibrium state?
iv) Thermal conduction: When two heat baths with different temperatures are attached to both ends of the system, can it support the temperature gradient and does the thermal conduction follow the Fourier law? Furthermore, if both the answers are yes, is the thermal conductivity given by the Green-Kubo formula?

As the result, I have obtained the following observations. Concerning equilibrium properties i) and ii), the thermodynamic behavior is realized by all the rules in the thermodynamic limit. This seems due to the same reason as for the ergodicity of the infinite system of the noninteracting ideal gas. In fact, rule 90R can be identified with an ideal gas system. However, some differences are seen among the rules in convergence character and initial condition dependence, reflecting differences in nonequilibrium behavior. Various stages of the thermodynamic behavior are observed with respect to the nonequilibrium nature: As for iii), rules 26R, 94R, and 123R show the relaxation, while others do not. As for iv), rules 26R, 77R, and 94R can support the temperature gradient, but others cannot. Moreover, the Fourier law is followed by rules 26R and 94R, but not by 77R. Thermal conductivity in rule 26R does not contradict the Green-Kubo formula. Details will be published elsewhere [3].

Thus, ERCAs contain a wide range of models from those realizing equilibrium behavior only to rule 26R, which passes all the tests. Accordingly, they can be qualified as the minimal models. Theoretical understanding of the observations is not completed at present. This is just the goal of ergodic theory. The abundance of the observed phenomena promises that to clarify the differences among the rules will make an important contribution to establishing the foundation of statistical mechanics. In addition, this study may bring some useful informations to the computational application of cellular automata such as lattice-gas automata[4] and deterministic Ising dynamics[5].

This work was financially supported by the Grant-in-Aid for Scientific Research from the Ministry for Education, Science and Culture.

References
1. S. Takesue: Phys. Rev. Lett. **57**, 2499 (1987).
2. Y. Pomeau: J. Phys. A**17**, L415 (1984).
3. S. Takesue: to be published.
4. U. Frisch, B. Hasslacher, and Y. Pomeau: Phys. Rev. Lett. **56**, 1505 (1986).
5. M. Creutz: Ann. Phys. (N.Y.) **167**, 62 (1986)

Mixing Property and Lyapunov Spectra of Hamiltonian Dynamical Systems with Many Degrees of Freedom

T. Konishi[1] and K. Kaneko[1;2]

[1]Institute of Physics, College of Arts and Sciences, University of Tokyo,
  Komaba 3-8-1, Meguro-ku, Tokyo 153, Japan
[2]MS B258, CNLS, LANL, Los Alamos, NM 87545, USA

1.INTRODUCTION AND MODEL

In this paper we study a 1-parameter family of Hamiltonian map lattice systems in order to see how a system with conservative deterministic dynamics can obtain thermodynamic behavior. We study a relaxation property to thermal equilibrium, with the use of sub-space box counting, and discuss its relation to other dynamical characteristics. Our model is defined on a 1-dimensional periodic lattice i=1,2, ..., N, subject to the updating rule $(x(i), p(i)) \rightarrow (x'(i), p'(i))$,

$$p'(i) = p(i) + G(x(i + 1) - x(i)) - G(x(i) - x(i - 1)), \bmod M(\text{integer})$$
$$x'(i) = x(i) + p'(i),$$
$$G(x) = (K/2\pi)\sin(2\pi x), K > 0,$$

which satisfies the symplectic condition. Phase space volume and total momentum (sum of p's) are conserved.

2.LYAPUNOV SPECTRUM

The Lyapunov spectrum of our model has the following properties: 1) It has $N - 1$ positive exponents for all $K > 0$ for all initial conditions (except for torus orbits, which have small measure). In other words, the model does not have any additional conserved quantities other than total momentum. 2) Compared with the spectrum of the random updating model, it is concave for $K < 1$. Both spectra agree well for $K \geq 1$. [1]

3.MIXING PROPERTY

As mentioned above, our model has $N - 1$ positive Lyapunov exponents for $K > 0$. Also the total volume of KAM region is observed to be very small. Thus we expect that the system wanders almost uniformly on the plane $\sum p_i$ =const. In other words, the system will achieve thermal equilibrium.

To know the speed of relaxation to thermal equilibrium, we calculated a "mixing rate". A dynamical system is mixing if

$$\lim_{t \to \infty} \mu(T^t(A) \cap B) = \mu(A)\mu(B)$$

for any measurable sets A, B. Here μ is an invariant measure of the dynamics. Instead of carrying out a box-counting algorithm on the whole phase space (which is

Springer Series in Synergetics, Vol. 43   **Cooperative Dynamics in Complex Physical Systems**
Editor: H. Takayama   © Springer-Verlag Berlin, Heidelberg 1989

impossible for large systems), we perform the following "sub-space box counting".
Suppose we have a system with N sites. We take a subsystem with $\dot{N}_{sub}$ sites. We
define the regions $B_i; i = 1, 2, \ldots$ as

$$B_i = \big\{ \, (x(j), p(j)); \quad i \le \sum_{j=1}^{N_{sub}} p(j) < i+1 \big\}.$$

Thus we divide the whole phase space according to the value of momentum in
the subsystem. We measure a global mixing rate γ defined as

$$I_{KL}(t) = I_{KL}(0)\exp(-\gamma t)$$

where $I_{KL}(t)$ is Kullback-Leibler information between equilibrium distribution
$P_0(B_i)$ and sample distribution $P(B_i; t)$;

$$I_{KL}(t) \equiv - \sum_i P(B_i; t)\log\big\{ \, P(B_i; t)/P_0(B_i) \, \big\}.$$

Actual simulation is performed as follows. First we set N_s independent samples
with initial condition in a region $A = \big\{ (x_i, p_i); 0 \le p_i < 1 \big\}$. Updating the N_s
samples simultaneously, we measure the number of samples in each region B_i at
time t, which number is proportional to $P(B_i, t)$. On the other hand, $P_0(B_i)$ is
analytically calculated. (See [2] for details.) Thus we obtain $I_{KL}(t)$. We also
compare the values of $I_{KL}(t)$ with the ones obtained from random updating, similar
to the way we have compared for Lyapunov spectra.

Results of simulation are summarized in Fig.1, where mixing rate γ versus
coupling K is plotted. Squares and crosses correspond to data from deterministic
and random updating, respectively. Random updating yields a mixing rate with
$\gamma \propto K^2$ for all K, whereas the deterministic updating shows a crossover at $K = 1$.
$\gamma \sim K^{3.5}$ for $K \le 1$, and $\gamma \sim K^{2.0}$ for $K \ge 2.0$, the same as the random version.

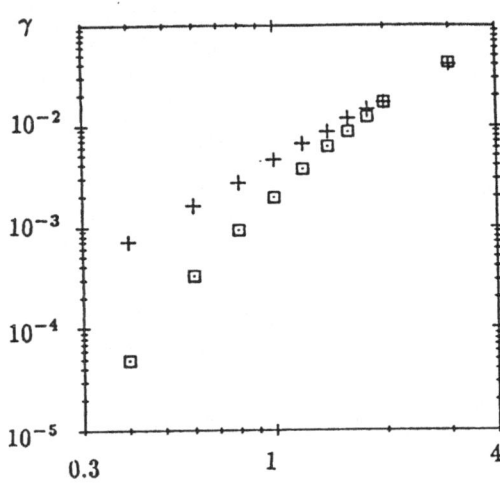

Fig.1 Mixing rate γ vs K. Squares and
crosses represent data from deterministic and random updating models, respectively.

4.RELATION BETWEEN MIXING RATE AND LYAPUNOV SPECTRUM

As we have seen in the previous section, a crossover observed in the change of shape of the Lyapunov spectrum corresponds well to the crossover of mixing rate. This implies that the whole spectrum has more useful information than Kolmogorov-Sinai entropy(H_{ks}). As KS entropy is the rate of loss of initial data, one expects that KS entropy is equal to ralaxation rate. On the other hand, results of simulation show $\gamma = const * H_{ks}^{3.1}$.

The reason of this discrepancy can be understood as follows. KS entropy characterises a rate of stretching processes in the phase space. On the other hand, folding processes are also important for mixing and relaxation process.

REFERENCES

1 . K.Kaneko and T.Konishi, J.Phys.Soc.Japan $\underline{56}$ 2993 (1987).
2 . T.Konishi and K.Kaneko, in preparation.

Defect Turbulence in EHD Convection of Liquid Crystal

S. Nasuno, M. Sano, and Y. Sawada

Research Institute of Electrical Communication, Tohoku University, Sendai 980, Japan

Systems driven far from equilibrium yield fascinating ordered states such as dissipative structure in space and rhythm in time. Such ordered states, however, may lose its stability when changing some external forcing, and the system shows the disordered states such as turbulence or chaos. Here we present experimental studies of an order-disorder transition in electrohydrodynamic convection system with large aspect-ratio $\Gamma=10^3$, using nematic liquid crystal (MBBA).

As the applied AC electric potential across the nematic layer is increased a sequence of convective state appears depending on the frequency [1,2]. At 400Hz the sequence is as follows: 1)Above the threshold voltage Vc=32.0V, there appear parallel convective rolls. 2)This ordered stationary state becomes unstable to weakly disordered state in both space and time. 3)The system yields again an ordered stationary state, i.e. a rectangular structure. 4)This state becomes unstable and a weakly disordered state appears. 5)Finally the system becomes fully turbulent. Here, we restrict ourselves to the transition between 3) and 4).

Figure 1 shows the transition from 3) to 4). By increasing V, there start appearing defects in periodic convective structure. These defects move around irregularly, interacting each other. The number of defects at fixed V is fluctuating in time, due to pair creation and annihilation of defects. Hence the system becomes turbulent in both space and time. To distinguish this turbulent state characterized by the presence of defects from fully turbulent state, it may be called here 'defect turbulence (in grid pattern)' [3,4].

In order to investigate this order-disorder transition in spatially extended system, we measure the spatio-temporal power spectrum P(k,f). An expanded parallel beam from He-Ne laser is incident vertically on the nematic layer and passed through the lens placed just behind the sample. The diameter of the beam is 7mm. By this configuration one can get the spatial Fourier image of the convective pattern on the focal plane of the lens.

By measuring the time evolution of the light intensity at various points on focal plane, one can get easily P(k,f) for various k-value. The results presented here are obtained by the measurement on an axis parallel to a director orientation in the absence of external forcing.

The central results obtained in the present experiments are as follows:
1)In perfect grid pattern regime, the Fourier image is stationary and composed of regularly aligned spots.
2)Near the onset of the defect turbulence regime, spatio-temporal power spectra P(k,f) show clear power law relation f^{-b} only for the spatial Fourier component k_1 corresponding to the spatial period of the grid pattern, as shown in Fig.2. The exponent b at 64.8V is about 1.9. The dependence of b on the external parameters has not yet been studied.
3)Further, the time average of spatial Fourier power spectrum $\langle P(k) \rangle$ appears to have power law relation $\langle P(k) \rangle \sim k^{-a}$ for small k in this regime.
4)In upper region of defect turbulence regime, the ratio of turbulent area in space increases and grid structure remains only as islands. At this stage, the spectra have no power law relation even for k_1.
5)In fully developed turbulence, the clear peaks in the Fourier space disappear.

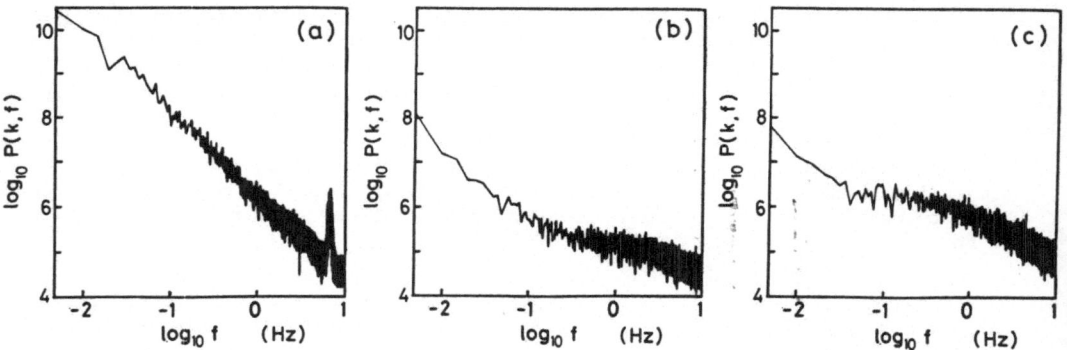

Fig.1 Transition from stationary grid pattern a) to defect turbulence b).

Fig.2 Spatio-temporal power spectra $P(k,f)$ as a function of f for (a)$k=k_1$, (b)$k=k_1/4$, (c)$k=k_1/8$

The physical origin of the result 2) is still unknown. Similar phenomena were reported by KANEKO [3] using a coupled-map lattice model, although it is one dimensional. He showed that in transition regime from spatially periodic state to spatio-temporal turbulence, defects play important roles. In this regime, there appears 'Flicker-like noise' selectively for the spatial Fourier mode whose wave number corresponds to the period of spatial structure.

The result 3) indicates that the system might have no characteristic scale. This must be related to the spatial distribution of defects. These results are interesting in terms of spatio-temporal self-similarity [5].

References

1. S.Kai and K.Hirakawa: Prog. Phys. Supp. 64,   212 (1978)
2. A.Joets, R.Ribotta: J. Physique (Paris) 47, 595 (1986)
3. K.Kaneko: Phys. Lett. A125, 25 (1987)
4. P.Coullet, L.Gil, J.Lega: preprint
5. P.Bak, C.Tang, K.Wiesenfeld: Phys. Rev. Lett. 59, 381 (1987)

Self-Organized Electric Structure
in Uni- and Multicellular Biological Systems

K. Toko, K. Hayashi, T. Fujiyoshi, and K. Yamafuji

Department of Electronics, Faculty of Engineering,
Kyushu University 36, Fukuoka 812, Japan

The electric spatial pattern is often observed in biological systems. The fresh-
water alga Chara, composed of giant internodal cells, develops alternating periodic bands of acidic and alkaline regions along the cell wall under illumination.
Figure 1 shows the formation of 5 alkaline bands along the cell with 70 mm length.
The electric potential near the cell surface and the membrane potential also show
a similar pattern, which implies an electric current flowing from the acidic to
alkaline zone [1-6]. Since the acidic regions elongate, the formation of pattern
is directly related to the growth. The band surrounding the circumferential direction of the cell is formed through a gathering process of many small patches at
the cell surface [4]. Electric oscillations sometimes appearing under weak light
intensity are coherent in the acidic region [3]. This electrochemical spatial
pattern is fairly stable in grown cells, but is unstable in very young cells. Even
if one of the bands formed on the cell is diminished by some electro-mechanical
perturbations, other bands are scarcely changed [7]; the pattern seems a kind of
frozen pattern, which has not been investigated yet in detail, although propa-
gative or oscillatory waves of chemical concentrations have been studied well.

Theoretical investigation [4,8,9] shows that the appearance of patterns can be
regarded as a bifurcation phenomenon brought about by a nonlinear character of H^+
pumps within the cell membrane. The pattern is maintained through the electric
current flowing between acidic and alkaline regions by a supply of chemical substances (e.g., H^+ [10] or ATP) from chloroplasts to H^+ pumps under illumination.
This spatial structure belongs to dissipative structures stabilized under the
above condition far from equilibrium. Computer simulation can reproduce many
typical properties of band pattern in Chara.

A similar pattern can also be found in multicellular systems as roots of bean
[11-13]. The periodic electric pattern is formed with the growth. The electric

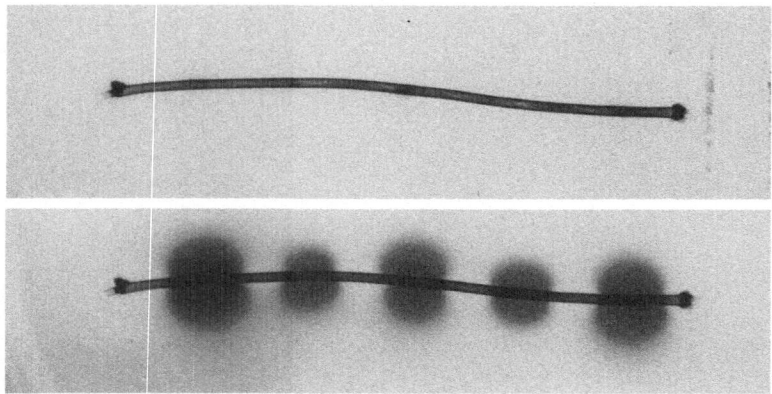

Fig. 1. Periodic alkaline bands formed around the cell of Chara in an aqueous
solution under illumination. The dark areas refer to the alkaline regions visualized by phenol red. Photographs concern one minute after illumination (top)
and 120 minutes (bottom), kindly provided by Dr. K. Ogata

current flows into the root-tip side, accompanied by several local current loops in the mature region. The elongating region exists between 1 and 5 mm from the root tip, and the mature region occupies the remaining part reaching over several centimeters. The enzyme activity and the ATP concentration are distributed spatially in correlation with the electric pattern.

A multi-electrode apparatus [12-14] reveals a self-sustained oscillation of electric potential entirely coherent in the mature region along the root, as shown in Fig. 2. The phase of oscillation in the elongating region differs from the mature region by 180 degrees [12]. A small peak of electric potential is added sequentially to the existing pattern with the elongation (of the speed of about 1 mm/hour) [13,14]. This process is analogous to the precipitation process in chemical systems [15]. The pattern disappears parallel with decline of growth speed when the energy supply is stopped by replacing air by nitrogen gas around the plant [16]. It can therefore be concluded that the electric pattern is a self-organized dissipative structure maintained in an open nonequilibrium system. These kinds of spatio-temporal electric organization in Characean cells and bean roots play important roles on the growth, as found in many biological systems [17].

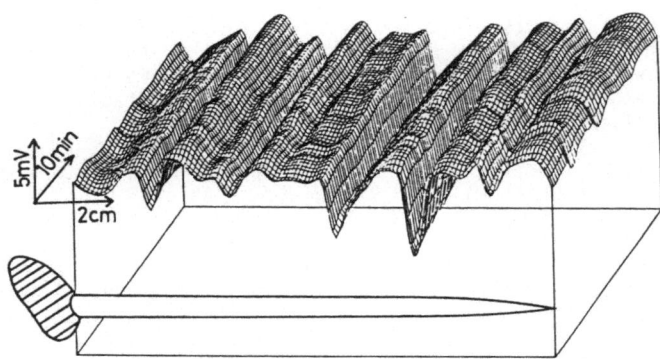

Fig. 2. Periodic electric spatial pattern formed around the root of bean. The root length is 125 mm. Thirty-one electrodes were arranged with 4 mm intervals along and near the root surface. The small-amplitude oscillation with 8 min period is superimposed on the spatial pattern

References

1. N.A. Walker, F.A. Smith: J. Exp. Bot. 28, 1190 (1977)
2. K. Ogata, K. Toko, T. Fujiyoshi, K. Yamafuji: Biophys. Chem. 26, 71 (1987)
3. K. Hayashi, T. Fujiyoshi, K. Toko, K. Yamafuji: J. Phys. Soc. Jpn. 56, 810 (1987).
4. K. Toko, M. Nosaka, T. Fujiyoshi, K. Yamafuji, K. Ogata: Bull. Math. Biol. 50, 255 (1988)
5. K. Toko, K. Hayashi, T. Yoshida, T. Fujiyoshi, K. Yamafuji: Eur. Biophys. J. 16, 11 (1988).
6. H.G.L. Coster, T.C. Chilcott, K. Ogata: In Inorganic Carbon Uptake by Aquatic Photosynthesis Organisms, ed. by W.J. Lucas, J.A. Berry (Waverly Press, Baltimore 1985) p. 255
7. W.J. Lucas, R. Nuccitelli: Planta 150, 120 (1980)
8. K. Toko, H. Chosa, K. Yamafuji: J. Theor. Biol. 114, 127 (1985)
9. K. Toko, T. Fujiyoshi, K. Ogata, H. Chosa, K. Yamafuji: Biophys. Chem. 27, 149 (1987)
10. U.-P. Hansen: Ber. Dtsch Bot. Ges. 92, 105 (1985)
11. S. Iiyama, K. Toko, K. Yamafuji: Biophys. Chem. 21, 285 (1985)
12. K. Toko, K. Hayashi, K. Yamafuji: Trans. IECE Jpn. E69, 485 (1986)
13. K. Toko, S. Iiyama, C. Tanaka, K. Hayashi, Ke. Yamafuji, K. Yamafuji: Biophys. Chem. 27, 39 (1987)
14. K. Toko, K. Yamafuji: Ferroelectrics 77 (1988) in press
15. R. Feeney, S.L. Schmidt, P. Strickholm, J. Chadam, P. Ortoleva: J. Chem. Phys. 78, 1293 (1983)
16. T. Yoshida, K. Hayashi, K. Toko, K. Yamafuji: Ann. Bot. 62 (1988) in press
17. L.F. Jaffe: In Membrane Transduction Mechanisms, ed. by R.A. Cone, J.E. Dowling (Raven Press, New York 1979) p. 199

Interaction of Chemical Waves and Hydrodynamic Flow

H. Miike*, S.C. Müller, and B. Hess

Max-Planck-Institut für Ernährungsphysiologie, Rheinlanddamm 201,
D-4600 Dortmund 1, Fed. Rep. of Germany

Spatial patterns generated by the coupling between chemical reaction and hydrodynamic flow have recently attracted increasing attention [1,2]. Examples are the stationary "mosaic" patterns and the deformation and irregular decomposition of chemical waves observed in an uncovered solution layer of the unstirred Belousov-Zhabotinskii (BZ) reaction [3-6]. Various attempts have been made to explain the origin of these structures by the effects of convective flow caused by evaporative cooling and/or the exothermicity of the reaction. In order to investigate the nonlinear interactions between hydrodynamics and chemical dynamics we carried out direct and quantitative measurements of hydrodynamic flow in an excitable BZ-solution layer using space-resolved microscope video imaging techniques.

An excitable solution of the BZ reaction was obtained by preparing a mixture of 48 mM NaBr, 340 mM $NaBrO_3$, 95 mM $CH_2(COOH)_2$, and 378 mM H_2SO_4. About 5 min after mixing, the catalyst and indicator ferroin (3.5 mM) was added. All solutions were filtered with 0.22 μm Millipore filter. A volume of the mixture was placed in a dust free petri dish of 7.0 cm diamter at 24$\pm$1°C resulting in a layer depth of 0.85$\pm$0.05 mm (except at the dish boundaries). Spiral waves were triggered at a position about 2 cm away from the center of the dish. The dish was carefully cleaned, so that only very few CO_2 bubbles were nucleated and no uncontrolled waves emerged from the dish boundaries. The correlation between chemical pattern dynamics and hydrodynamic flow velocities was investigated by applying 2D-velocimetry and 2D-spectrophotometry based on microscope video imaging techniques. For flow measurements small polystyrene particles serving as scattering centers were mixed into the BZ-solution and illuminated by He-Ne laser light (632.8 nm). Together with a homogeneous transmitted light beam at 490 nm, we could observe hydrodynamic flow and propagation of chemical activity simultaneously [7].

A time trace of flow velocity measured close to the open liquid/gas interface of the BZ-solution is shown in Fig. 1. About 12 min after triggering the spiral waves oscillatory hydrodynamic flow is induced spontaneously. In the early stage of the oscillatory flow (13 to 17 min), the flow period is about 3 times that of the chemical wave trains. The flow amplitude is 50-70 μm, which is comparable to the propagation speed of the waves, in good agreement with recent results [7]. In some cases, however, we observe a much faster oscillatory flow. This is shown in the example of Fig. 1 where after 17 min the flow amplitude increases drastically and the oscillation period starts to coincide with that of the chemical wave trains (about 20 s). The maximum flow amplitude reaches 300 μm/s which surpasses by far the propagation speed of the waves. Thus a pronounced influence on the geometric profile of the waves has to be expected.

In fact, we observed that under such conditions strong geometric distortions occur, as shown for a typical case in Fig. 2. Until 10 min, no remarkable change is observed except for a rather weak static undulation of the propagating wave fronts. After 12 min, however, the oscillatory flow has become large enough to

---

*Permanent address: Yamaguchi University, Ube 755, Japan

Springer Series in Synergetics, Vol. 43   **Cooperative Dynamics in Complex Physical Systems**
Editor: H. Takayama   © Springer-Verlag Berlin, Heidelberg 1989

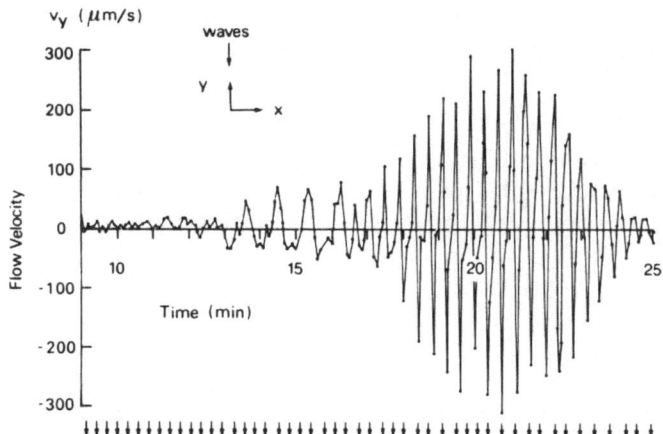

Fig. 1: Temporal evolution of flow velocity. Arrows indicate passage of wave fronts through the detection area

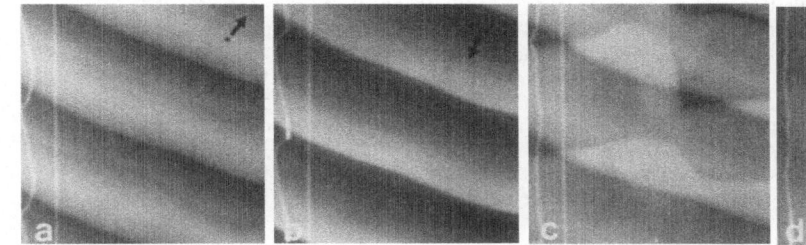

Fig. 2: Evolution of wave front geometry as a function of time

lead to wave front deformations which proceed in an oscillatory manner from broad (a) to sharp shape (b) and vice versa. The motion of a dust particle in the solution confirms the strong correlation between the shape of the wave front and the direction of the flow. When the flow amplitude reaches 200-300 μm/s (after 18 min) a superposition (c) and an irregular decomposition (d) of the chemical waves are observed. Apparently, the decomposition is not caused by preexisting and comparatively slow convective bulk flow but by the pronounced oscillatory flow detected close to the layer surface of the BZ-solution.

Our findings establish the existence of an oscillatory hydrodynamic flow, which is induced spontaneously by wave propagation and is responsible for oscillatory deformation and irregular decomposition of the chemical waves. The oscillation is two [7] or three times or equal to that inherent in chemical wave propagation. These facts suggest that flow is entrained by the periodic passage of chemical wave trains and point to a nonlinear coupling between chemical waves and hydrodynamics.

This work was supported by the Stiftung Volkswagenwerk, Hannover.

1. C. Vidal, P. Hanusse: Int. Rev. Phys. Chem. 5, 1 (1986)
2. P. Borckmans, G. Dewel: In From Chemical to Biological Organization, ed. by M. Markus, S.C. Müller, G. Nicolis, Springer Ser. Syn. (Springer, Berlin, Heidelberg, 1988) p. 114
3. A.M. Zhabotinskii, A.N. Zaikin: J. Theor. Biol. 40, 45 (1973)
4. K. Showalter: J. Chem. Phys. 73, 3735 (1980)
5. K.I. Agladze,V.I. Krinsky, A.M. Pertsov: Nature 308, 834 (1984)
6. S.C. Müller, Th.Plesser, B. Hess: Naturwissenschaften 73, 165 (1986)
7. H. Miike, S.C. Müller, B. Hess: Chem. Phys. Lett. 114, 515 (1988)

Flow Visualization of the Velocity Field of Square Convection Near the Critical Point Using Holographic Interferometry

Y. Harada[1], *A. Tzunoda*[1], *K. Shigemori*[1], *K. Miyasaka*[2], *and M. Ueda*[2]

[1]Department of Applied Physics, Faculty of Engineering,
 Fukui University, 3-9-1 Bunkyo, Fukui 910, Japan
[2]Faculty of Education, Fukui University, 3-9-1 Bunkyo, Fukui 910, Japan

Recently, the appearance of cellular structures and the problems of pattern selection in nonequilibrium systems have received considerable attention /1/. Most of the interest has been focused on the simplest hydrodynamical system, the so-called Rayleigh-Bénard convection system and also subsequent transitions from laminar to turbulent flows especially in the case of lower Prandtl number. For high Prandtl number fluids, convection in fluid layers with strongly temperature dependent viscosity differs from that of lower Prandtl number fluids due to the existence of an additional instability mechanism/2/. An important question is whether patterns emerge in a purely deterministic manner or whether noise plays a central role in pattern selection/3/. In order to solve these problems, new experimental methods for determining the spatial structure of the velocity field should be developed. Recently ,/4/,/5/we have shown that holographic interferometry can be used successfully for the visualization of the spatial structure of the velocity field in Rayleigh-Bénard convection.

In the present paper, we present the first report of the observation of the full velocity fields of steady square-cell convection just above the threshold by means of this technique. In these experiments, we used glycerol as the working fluid. The viscosity of this fluid strongly depends upon the temperature. In order to study the spatial convective structure, we measure the full velocity field at a given instant by using multiple exposure holographic interferometry. A detailed description of the method has been published elsewhere/4/. We have observed subsequent structures of the velocity field in a rectangular box with depth 10 mm, width 40 mm, and length 80 mm . Its side walls are made of transparent polyacryl plate. The aspect ratio of the cell as given by length/depth, is 8.0 which corresponds to an intermediate aspect ratio.

All the reconstructed holographic images have been recorded as a function of the temperature difference. As the temperature difference is increased gradually, a very slow macroscopic flow occurs near the side wall but local convection is not observed, in contrast to the case of optical glass container /5/. When the temperature difference is further increased,we observed abruptly a non-steady roll pattern. This flow remains time dependent and the roll structure is broken into a chaotic state during about 23 hours. Then a steady state convection emerges in the form of square-cell pattern. Upon decreasing the temperature difference, the square-cell pattern change hysterically into a transverse-parallel roll, perpendicular to longitudinal-parallel roll.

Figure 1 presents some photographs of typical reconstructed images from multiple-exposed holograms obtained with the horizontal light sheet. The steady velocity field of the square-cell pattern convection is observed at various heights which are sliced by the sheets of light. The maximum vertical velocity of the flow displays hysteric characteristics against bifurcation parameter accounting in part for the subsequent transition observed. These subsequent transition are related to the imperfect lateral boundary condition. Square-cell pattern can be observed especially as a secondary motion far above onset, in high Prandtl convection. It is also predicted when horizontal plates

Springer Series in Synergetics, Vol. 43   **Cooperative Dynamics in Complex Physical Systems**
Editor: H. Takayama   © Springer-Verlag Berlin, Heidelberg 1989

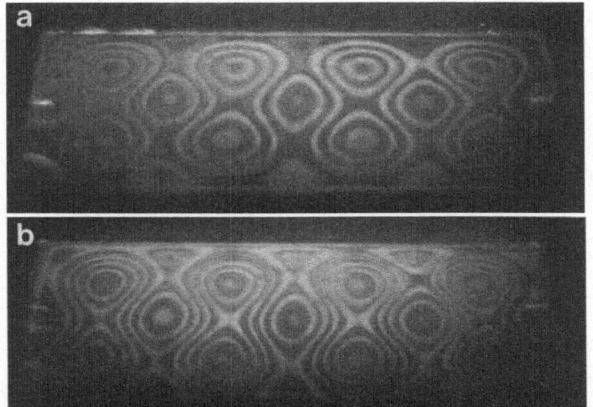

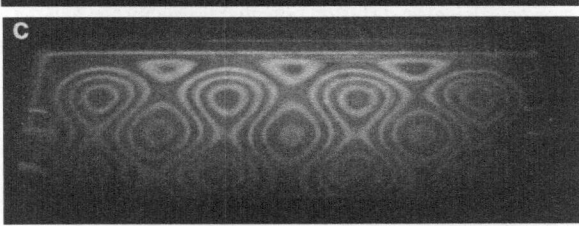

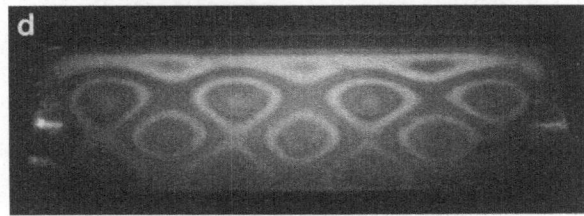

Fig. 1. Reconstructed images (contour map) of the velocity field of square-cell convection at constant Rayleigh number sliced at the following vertical heights (a) 7 mm (b) 5 mm (c) 3 mm (d) 1 mm and with parameters R=1489, P=2589

are poorly conducting /6/. To our knowledge, these are the first experiments which are capable of 3-D velocity field measurements for square-cell convection near the onset of convection. The above experiments thus exemplify our new method for determining the spatial structure of 3-D flows based on multiple exposed holographic interferometry.

REFERENCES

1. In "Spatio-Temporal and Chaos in Physical Systems", ed. by A.R. Bishop, G. Guner, and B. Nicolaenko(North-Holland, Amsterdam, 1986).
2. F.H. Busse and H.Frick: J. Fluid. Mech., 150 451(1985).
3. G.Ahlers, M.C. Cross, P.C. Hohenberg, and S. Safran: J. Fluid. Mech., 110, 297(1981).
4. M. Ueda, K. Kagawa, K.Yamada, C. Yamaguchi, and Y. Harada: Appl. Opt., 18 3269(1982).
5. M. Ueda, M. Hosono, and Y. Harada: Optik, 63 363(1983).
6. F.H.Busse and N.Riahi: J. Fluid. Mech., 96 243(1980).

Pattern Formation by Nonlinear Reactions and Diffusion in Electrochemical Systems

H. Malchow

Humboldt-Universität, Sektion Physik, Bereich 04,
Invalidenstr. 42, DDR-1040 Berlin, GDR [1]

Pattern formation is a typical phenomenon in selforganizing systems, especially chemical and biochemical reaction systems. The ionic character of the reactants is included in the theoretical investigation of the system kinetics here because ions do not react and diffuse independently even in the absence of an external electrical field. Due to unequal diffusion coefficients a diffusion potential is established and the moving ions interact with it. By interference of diffusion and nonlinear reactions stable concentration gradients or traveling ionic waves related to d.c. or a.c. voltages become possible. The concept of ambipolar diffusion can be applied to binary systems for the charge neutrality limit [2]. In the same limit the natural appearance of cross-diffusion coefficients has been shown for systems with more than two kinds of ions [3, 6]. Even negative cross-diffusion increases the possibility of temporal and spatial pattern formation. An ionic wave equation has been derived [4] and applied to the description of self-electrophoresis [5].

The natural effect of concentration dependence of self- and cross-diffusion is included here. The inhomogeneous solution branch first bifurcating from the homogeneous one while varying any critical parameter as e.g. reaction rates, ratio of diffusivities or system size has been given recently as a general expression in the eigenfunctions of the Laplace operator using standard methods of bifurcation theory [6]. The concentration dependence of diffusion influences the amplitudes of the appearing patterns. This effect is called contrast control. Due to the defined dependence of diffusion in electrolytes on concentration [7] the diffusion coefficient has been chosen as a polynomial of third order in concentration which has been used already for the calculation of diffusion coefficients from sorption data [8].

A special three-component model reaction system $\partial X_i/\partial t = f_i - \nabla j_i$; i=1,2,3; with one negative cross-diffusion coefficient and one spatially homogeneous stationary solution $\partial X_{is}/\partial t = 0$; i=1,2,3 ; is treated here. It reads in dimensionless units

$$j_i(r,t) = - D_i(X) \nabla X_i(r,t) + z_i D_i(X) X_i(r,t) E(r,t) \quad ; \quad i = 1,2,3 ; \tag{1}$$

$$f_1 = - X_1 X_2 + \beta \quad ; \quad f_2 = - \alpha X_2 + X_2^2 X_3 - X_1 X_2 + \beta \quad ; \quad f_3 = \alpha X_2 - X_2^2 X_3 , \tag{2}$$

where α and β depend on reaction rates and in- and output. The corresponding charge numbers are $z_1 = -1$, $z_2 = z_3 = 1$. Because of the charge neutrality limit $X_1 = X_2 + X_3$ can be substituted. For the diffusion coefficients we set

$$D_i = D_{i0} \left\{ 1 + S \left(A_i X_i + B_i X_i^2 + G_i X_i^3 \right) \right\} , \tag{3}$$

with $A_i = - X_{is}^2/m_i$; $B_i = (1 + m_i) X_{is}/m_i$; $G_i = -1$; i = 1,2,3 ; D_{io} and m_i are specific constants for each ion. This choice of A_i, B_i and G_i yields $D_i > D_{io}$ for $X_i < X_{is}$ and vice versa for small deviations from X_{is}. S is a technical parameter only switching the concentration dependence on or off. The stationary inhomogeneity for Neumann b.c. after crossing a critical ratio D_3/D_2 reads

$$X_i(r) = X_{is} + \varepsilon \, \phi_{0i} \cos (kr) + \varepsilon^2 \, \phi_{1i} \cos (2kr) ; i = 1,2,3 ; \tag{4}$$

Springer Series in Synergetics, Vol. 43 **Cooperative Dynamics in Complex Physical Systems**
Editor: H. Takayama © Springer-Verlag Berlin, Heidelberg 1989

where ε is a small parameter depending on the distance from the bifurcation point, k the wave number and ϕ_{0i}, ϕ_{1i} contain relations between the chemical and physical parameters. The effect of contrast control by concentration dependence of diffusion is shown in Fig. 1.

The result of integrating the corresponding compartmental system can be seen in Fig. 2. The concentration dependence influences not only the amplitudes but also the symmetry of the Turing structures [9]. The parameters for Figs. 1,2 are $\alpha = 12$; $\beta = 16$; $D_{10} = 0.03$; $D_{20} = 0.002$; $D_{30} = 0.5$; $m_i = 2$.

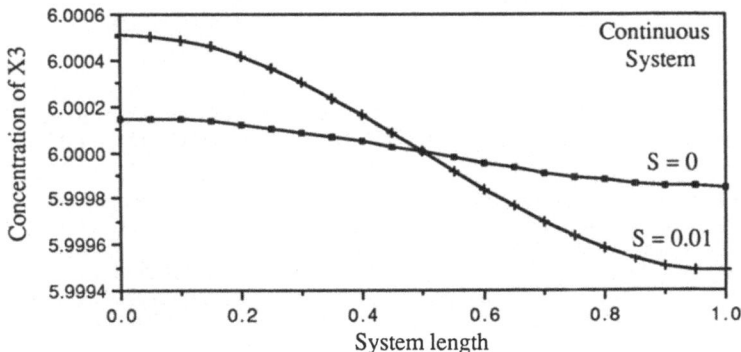

Fig. 1

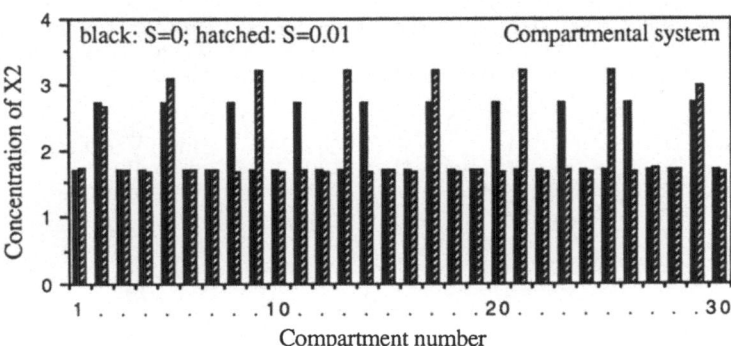

Fig. 2

References

1. Present address: Department of Biophysics, Kyoto University, Kyoto 606, Japan
2. H. Malchow, L. Schimansky-Geier: Noise and diffusion in bistable nonequilibrium systems, Teubner-Texte zur Physik, Vol. 5 (Teubner, Leipzig 1985)
3. J. Jorné: J. theor. Biol. 55, 529-532 (1975)
4. S. Schmidt, P. Ortoleva: J. Chem. Phys. 67, 3771-3776 (1977)
5. R. Larter, P. Ortoleva: J. theor. Biol. 88, 599-630 (1981)
6. H. Malchow: submitted to J. theor. Biol., Dec. 1987, revised March 1988
7. A.L. Horvath: Handbook of aqueous electrolyte solutions, (Horwood Ltd. Publ., Chichester 1986)
8. J.L. Hwang: J. Chem. Phys. 20, 1320-1323 (1952)
9. H. Malchow: Mem. Fac. Sci. Kyoto Univ. (Ser. Biol.), to appear

Dissipative Structure and an External Noise Effect in an Energy Conversion System

S. Kabashima

Department of Energy Sciences, The Graduate School at Nagatsuta,
Tokyo Institute of Technology, 4259 Nagatsuta, Midori-ku, Yokohama 227, Japan

The dissipative structures associated with ionization instabilities, and external noise effects have been examined for plasma placed in an inert-gas-driven MHD generator.

In the disk-type inert-gas-driven MHD generator the electrically conductive gas flows in the radial direction in the applied magnetic field as shown in Fig.1. The Faraday current is induced in the azimuthal direction and it produces Hall voltages between the electrodes which are located in the upstream and downstream regions of the gas flow. The conductive properties of the working gas are obtained by seeding a small amount of alkali metal, which can be easily ionized as compared to the inert gas.

The plasma, which grows under the above MHD conditions, is described by the two-temperature model, where the electron temperature is much higher than the gas temperature.[1] In this model the electrons and other heavy particles such as inert gas and seed atoms are treated by MHD equations with different temperatures. For electrons the following equations with temperature T_e hold:

$$\frac{\partial n_e}{\partial t} + \nabla \cdot (n_e u_e) = \dot{n}_e, \tag{1}$$

$$j + \frac{\beta}{B} j \times B = \sigma \left(E^* + \frac{k}{e n_e} \nabla \cdot p_e \right), \tag{2}$$

where $E = E^* + u \times B$, $\dot{n}_e = \dot{n}_N + \dot{n}_S$,

$$\dot{n}_i = k_{fi} n_e n_i - k_{ri} n_e^2 n_i, \quad (i = N, S)$$

Equations for heavy-particle flow are

$$\frac{\partial \rho}{\partial t} + \nabla \cdot (\rho u) = 0, \tag{4}$$

$$\frac{\partial}{\partial t}(\rho u) + \nabla \cdot (\rho u u) = j \times B - \nabla \cdot p, \tag{5}$$

$$\frac{\partial U_e}{\partial t} + \nabla \cdot (U_e u_e) = \frac{j^2}{\sigma} - A - \nabla \cdot (p_e u_e) \tag{3}$$

$$U_e = \frac{3}{2} n_e k T_e + \varepsilon_N n_N^+ + \varepsilon_S n_S^+,$$

$$A = \frac{3}{2} \delta k n_e (T_e - T_g) \sum_i \frac{m_e}{m_i} \nu_{e-i}.$$

$$\frac{\partial U}{\partial t} + \nabla \cdot ((U + P)u) = J \cdot E, \tag{6}$$

where $U = (C_v T_g + u^2/2)$.

The notation used here is that found in [2,3].

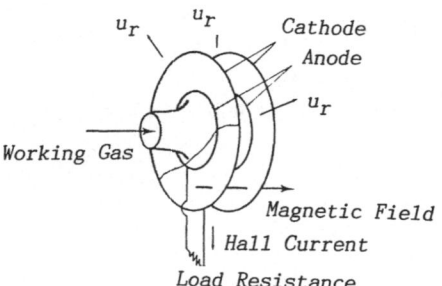

u_r u_r Cathode
Anode
u_r
Working Gas
Magnetic Field
Hall Current
Load Resistance

Fig.1 Configuration of inert gas driven MHD power generator.

Springer Series in Synergetics, Vol. 43   **Cooperative Dynamics in Complex Physical Systems**
Editor: H. Takayama   © Springer-Verlag Berlin, Heidelberg 1989

The above equation systems are known to show ionization and magneto-acoustic instabilities for certain conditions.[3] Two-dimensional simulations were performed based on the above equation systems for conditions shown in Table 1.

Table 1 Conditions and parameters of a model generator

Working gas	He + K	Stagnant gas pressure	3 atm
Stagnant gas temperature	2000K	Generator length	10 cm
Inlet radius	10 cm	Magnetic field	4.0 T
Seed fraction	75 ppm	Load resistance	3.5

An example of plasma structure growth is shown in Fig.2 with the current stream lines. The helical structure appears about 40 s after the initial homogeneous state. It can be found from the tangential component of the current that two states of plasma, one is power generating and the other is power consuming, arise. The structure depends on the load resistance of the generator as shown in Fig.3, which shows that generator matching plays an important role for these plasma structures. In other words, the plasma structure is related to the minimum heating principle of the electric circuit.

The external noise effects are analyzed by a one-dimensional simulation which describes well the power output of the generator[2]. Results are summarized in

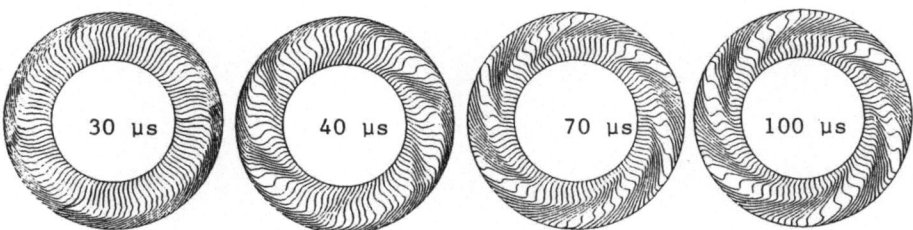

Fig.2 Growth of the plasma structures shown by current stream lines. R_L=1.0 Ω

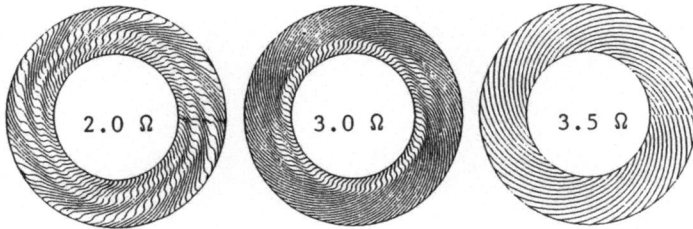

Fig.3 Changes of the plasma structure with the load resistance.

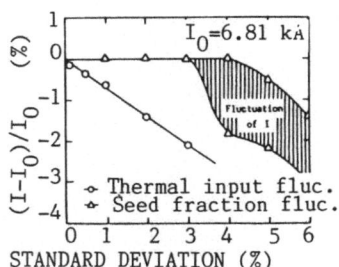

Fig.4 Changes of output current with the external noise.(hatched region denotes fluctuation level)

Fig.4, where we can see that the external noise introduces the fluctuation to the plasma and it reduces the conversion rate of the generator. Disturbances of thermal input always reduce the output power, and large disturbances of seed fraction cause the ionization instabilities and reduce the energy conversion rate, although small disturbances of seed fraction introduce only a small fluctuation to the plasma. These differences are caused by the different roles of thermal input and seed fraction for the energy conversion.

References

1. J.L.Kerrebrock:AIAA J. 2, 1072 (1964)
2. E.Shimizu, Y.Yoshikawa, S.Kabashima: Jpn.J.App.Phys. 26, 492 (1987)
3. Y.Okuno, Y.Endo, S.Kabashima: Trans.I.E.E.Jpn. 108 6 (1988)

Nonlinear Binary Fluid Convection at Positive Separation Ratios

E. Knobloch

Department of Physics, Kyoto University, Kyoto 606, Japan, and
Department of Physics, University of California, Berkeley, CA 94720, USA

When heated from below, binary fluid mixtures exhibit a variety of pattern forming phenomena [1,2]. High Prandtl number mixtures like water/ethanol are described by the non-dimensional equations

$$0 = -\nabla p + R(1+S)\theta \underline{e}_z + SR\phi \underline{e}_z + \nabla^2 \underline{u} \qquad (1a)$$

$$\theta_t + \underline{u}\cdot\nabla\theta = w + \nabla^2\theta \qquad (1b)$$

$$\phi_t + \underline{u}\cdot\nabla\phi = \tau\nabla^2\phi - \nabla^2\theta \quad , \qquad (1c)$$

where $\underline{u}=(u,v,w)$ is the velocity field, p the pressure, and $\phi\equiv\Sigma-\theta$, where θ and Σ are the departures from the linear temperature and concentration profiles due to conduction alone. The dimensionless parameters are the Rayleigh number R, the separation ratio S and the Lewis number τ. Typical boundary conditions of interest are ($D\equiv\partial/\partial z$)

$$a : u = v = w = D\phi = \theta = 0 \qquad \text{on } z = \pm 1 \qquad (2a)$$

$$b : Du = Dv = w = D\phi = \theta = 0 \qquad \text{on } z = \pm 1 \qquad (2b)$$

corresponding to no-slip (a) or stress-free (b), no-mass-flux, fixed temperature boundaries. The no-mass-flux boundary condition $D\phi=0$ forces the wavenumber k of the first steady state instability of the conduction state $\underline{u}=\theta=\phi=0$ to vanish for $S\geq S_\infty(\tau)$ [3]. To take advantage of such long wavelengths we introduce, following [4], a small parameter, ε, $0<\varepsilon<<1$, and scale $(x,y,z) \rightarrow (X/\varepsilon,Y/\varepsilon,z)$, $t\rightarrow T/\varepsilon^4$ and $(u,v,w,\theta,\phi) \rightarrow (\varepsilon U,\varepsilon V,\varepsilon^2 W,\varepsilon^2\Theta,\Phi)$. For supercritical Rayleigh numbers, $R = R_0(1+\mu\varepsilon^2)$, we write $\Phi = \Phi_0(X,Y,z,T) + \varepsilon^2\Phi_2(X,Y,z,T) + \cdots$, etc. One readily finds that $\Phi_0=f(X,Y,T)$ with the following evolution equation for f:

$$f_T = -R_0 S\mu A\nabla^2 f - B\nabla^4 f + C\nabla\cdot|\nabla f|^2\nabla f - E\nabla\cdot f\nabla f \quad . \qquad (3)$$

With the exception of the coefficient E which is due to non-Boussinesq effects not included in (1) the coefficients are listed in Table 1. $S_\infty(\tau)$ is obtained by solving B=0. The point (R_0,S_∞) represents a new type of degeneracy : for $S\geq S_\infty$,

Table 1. The coefficients in equation (3) for the boundary conditions a, b.

	R_0	A	B	C	S_∞
a	$45\dfrac{\tau}{S}$	$\dfrac{1}{45}$	$-\dfrac{131}{231}(1+\dfrac{1}{S})\tau^2+\dfrac{34\tau}{231}$	$\dfrac{10\tau}{7}$	$\dfrac{131\tau}{34-131\tau}$
b	$\dfrac{15}{2}\dfrac{\tau}{S}$	$\dfrac{2}{15}$	$-\dfrac{691}{1386}(1+\dfrac{1}{S})\tau^2+\dfrac{1091\tau}{1386}$	$\dfrac{155\tau}{126}$	$\dfrac{691\tau}{1091-691\tau}$

Springer Series in Synergetics, Vol. 43   **Cooperative Dynamics in Complex Physical Systems**   337
Editor: H. Takayama   © Springer-Verlag Berlin, Heidelberg 1989

k=0, but for $0 < S_\infty - S \ll 1$, $0 < k \ll 1$. The evolution of the instability in this regime is described by an equation of the form

$$f_T = -\mu\nabla^2 f + \nu\nabla^4 f + \nabla^6 f + \nabla\cdot|\nabla f|^2\nabla f - \gamma\nabla\cdot f\nabla f \quad . \tag{4}$$

This equation can be derived by an asymptotic analysis similar to the one leading to (3), provided the non-Boussinesq effects are correspondingly weaker.

Pattern formation described by (4) can be studied for $\mu > -\nu^2/4$, $\nu > 0$ using equivariant bifurcation theory. On the square lattice [5] the linear problem has the solution

$$f_0 = \{z_1 e^{ikX} + z_2 e^{ikY} + c.c. \,|\, (z_1, z_2)\epsilon C^2\} \quad . \tag{5}$$

Nonlinear time-independent solutions of (4) then take the form of Rolls : $(z_1, z_2) = \hat{x}(1,0)$, $\hat{x} > 0$, and Squares : $(z_1, z_2) = \hat{x}(1,1)$, $\hat{x} > 0$. At $\mu = -\nu^2/4$ branches of both Rolls and Squares bifurcate simultaneously from the trivial solution. For $|\gamma| < 7.3485k^3$ both Rolls and Squares bifurcate supercritically, but only the latter are stable; for larger $|\gamma|$ Squares bifurcate subcritically, and neither branch is stable. The point $|\gamma| = 7.3485k^3$ represents a codimension-two singularity. Calculation of the appropriate fifth order term in the amplitude equation shows that for $0 < |\gamma| - 7.3485k^3 \ll 1$, the Squares gain stability at a secondary saddle-node bifurcation leading to a hysteretic transition to a Square pattern. It is reasonable to surmise that this persists for larger non-Boussinesq effects also.

On the hexagonal lattice [6]

$$f_0 = \{z_1 e^{ikX} + z_2 e^{ik(\sqrt{3}Y - X)/2} + z_3 e^{-ik(\sqrt{3}Y + X)/2} + c.c. \,|\, (z_1, z_2, z_3)\epsilon C^3\} \quad . \tag{6}$$

When $\gamma = 0$ a critical cubic term in the amplitude equations for (z_1, z_2, z_3) vanishes [7]. Consequently, as many as six primary solution branches can bifurcate simultaneously at μ_c : Rolls $\hat{x}(1,0,0)$, Hexagons $\hat{x}(1,1,1)$, Patchwork Quilt $\hat{x}(0,1,1)$, Rectangles $\hat{x}(1,1/a,1/a)$, Regular Triangles $\hat{y}(i,i,i)$ and Imaginary Rectangles $\hat{y}(i, i/a, i/a)$. Here $a \neq 0,1,\infty$. All bifurcate supercritically. Consequently their relative stability is determined by certain fifth order terms [7]. The calculation shows that Hexagons are stable, Rectangles do not exist, and the remaining four branches are unstable [7].

When $\gamma \neq 0$ only rolls and hexagons bifurcate at μ_c and both are unstable. One must now distinguish between $H^\pm$ according to sgn $\hat{x}$. Depending on sgn γ, $H^\pm$ gains stability at a secondary saddle-node bifurcation, and $H^\mp$ at a secondary bifurcation to Triangles [7]. Hence at larger amplitude stable $H^\pm$ coexist.

Acknowledgement: The author is grateful to Professor Y. Kuramoto for his kind hospitality in Kyoto, and to the Japan Society for the Promotion of Science for financial support.

Literature

1. V. Steinberg, E. Moses and J. Fineberg, Nuclear Phys. B (Proc. Suppl.) 2, 109 (1987).
2. E. Knobloch, A.E. Deane and J. Toomre: in The Physics of Structure Formation: Theory and Simulation, ed. by W. Güttinger and G. Dangelmayr, Springer Ser. in Syn. (Springer, Berlin, Heidelberg 1987), p.117.
3. E. Knobloch and D.R. Moore : Phys. Rev. A 37, 860 (1988).
4. C.J. Chapman and M.R.E. Proctor : J. Fluid Mech. 101, 759 (1980).
5. M. Silber and E. Knobloch : Phys. Rev. A 38 (1988).
6. M. Golubitsky, J.W. Swift and E. Knobloch : Physica 10D, 249 (1984).
7. E. Knobloch : to be published.

Index of Contributors

Springer Series in Synergetics
Editor: Hermann Haken

Synergetics, an interdisciplinary field of research, is concerned with the cooperation of individual parts of a system that produces macroscopic spatial, temporal or functional structures. It deals with deterministic as well as stochastic processes.